Morphologic Encyclopedia of Palynology

Contribution No. 100.
Program in Geochronology
University of Arizona, Tucson

Morphologic Encyclopedia of Palynology

AN INTERNATIONAL COLLECTION OF DEFINITIONS AND ILLUSTRATIONS OF SPORES AND POLLEN

By

GERHARD O. W. KREMP

PROFESSOR OF GEOCHRONOLOGY, UNIVERSITY OF ARIZONA, TUCSON

THE UNIVERSITY OF ARIZONA PRESS • TUCSON

About the Author . . .
GERHARD O. W. KREMP, founder and co-editor of the Cata-
logue of *Fossil Spores and Pollen*, has been a professor of geochronology at the University of Arizona since 1960. A native of
Germany, he has spent twenty-five years in the field of palynology,
having studied with P. W. Thomson and been a junior fellow
worker under R. Potonié. The author, a naturalized citizen of the
United States, came to the U.S. in 1955 and was associated with
the Department of Geology at Pennsylvania State University
and the U.S. Geological Survey at Denver. Previously he had
been with the German Geological Survey and was a research
associate at the University of Göttingen. A frequent contributor
to geological and paleontological periodicals and yearbooks here
and abroad, Professor Kremp has served on the editorial board of
Grana Palynologica, an international journal of palynology, and
as an associate of the International Committee for the Microflora
of the Paleozoic.

The University of Arizona Press
www.uapress.arizona.edu

ISBN-13: 978-0-8165-0060-4 (cloth)
ISBN-13: 978-0-8165-4023-5 (Century Collection paper)

Library of Congress Cataloging-in-Publication Data are available from the Library of Congress.

Printed in the United States of America
♾ This paper meets the requirements of ANSI/NISO Z39.48-1992 (Permanence of Paper).

To
Robert Potonié,
pioneer in palynology

Foreword

To the young geologists working with Lennart von Post half a century ago, "pollen morphology" meant little more than size and general shape of the pollen grains, plus, perhaps, some additional characters, such as number of furrows or pores. Happily unaware of goniotreme, tetracolporate pollen grains with lalongate ora and suprareticulate tectum with spinuliferous muri they produced good pollen diagrams, certainly worthy forerunners to the very detailed pollen and spore diagrams of today.

Sketches in notebooks from those days often show pollen grains of common forest trees magnified about 250 times. Now, scanning electron microscopy and other advanced techniques have made it possible to visualize pollen grains and spores as they would appear if magnified about 4000×250, that is one million times. In a pollen grain of that size one Angstrom unit corresponds to 0.1 mm, one micron to one metre and 40 microns — the average diameter of a *Tilia* pollen grain — to 40 metres.

If we, in our imagination, could make excursions upon pollen grains or spores enlarged about a million times, it would be possible to study the most delicate details of their fine relief, to gaze at the lustre of the spines — five, six or more metres high — in malvaceous pollen grains, or to glide through holes in the tectum into dark recesses usually crowded with pillars between the upper (outer) and lower (inner) parts of the exine.

A great number of terms are indeed necessary in order to describe in brief, terse language the main morphological features encountered in the lilliput world of the pollen grains. However, as explicitly shown by Dr. Kremp's Encyclopedia, the number of terms is excessive. Judicious weeding by palynologists of unwieldy, intricate, or linguistically erroneous terms is much to be desired.

Dr. Kremp's Encyclopedia is particularly welcome as a substitute for private, perhaps more or less incomplete terminological card indexes and as a well-documented guide to the definition of special terms coined by particular authors in certain years.

Some of these terms are now marooned or rarely used, others are in the mainstream, flourishing. Alas, stability in the use of certain terms has not yet been achieved and can perhaps never be expected: *multa renascentur quae iam cecidere, cadentque quae nunc sunt in honore* . . . .

G. Erdtman
Stockholm

Introduction

This encyclopedia is designed to serve scholars in palynology as a standard text in questions concerning spore morphology and as a source of information for understanding the descriptive literature in palynology. The encyclopedia contains a total of about 1280 terms cited, with approximately 1650 original definitions, 560 additional definition references, and 822 illustrations taken from 70 representative publications. Each entry is augmented with a list of terms of similar meaning with plural forms and etymology. All foreign references have been translated into English.

The work had its beginning when I, as an immigrant from one "terminology circle" to another, compiled for myself a German-English dictionary of technical terms useful in the description of spores and pollen. For each German word I tried to find its English equivalent but soon enough I discovered that this was in many cases impossible. The definitions given for the same term differed from author to author. These differences were in many cases only slight, but in others crucial, making the language barrier only more difficult to overcome.

My difficulty in adjusting to a new terminology may be understood by all palynologists to a varying degree. The field of palynology has developed a descriptive language all of its own. More than 1200 technical words have been counted in palynological literature, of which only 40 percent were in use by botanists in 1928. Since then over 700 additional words have been coined and many older words have been given specialized palynological meanings. Palynologists working in different countries in different languages and at different times can hardly develop a uniform and standardized descriptive terminology.

Confusion has inevitably arisen as a consequence not only of geographic and linguistic difficulties, but also because of the quantity of technical terms which have accrued from the several schools of development.

The number of terms is enormous because palynologists of the world produce many papers each year. From a literature survey which I prepared in 1962 for the International Botanical Congress, I estimated that in 1958 "only" 11 new fossil spores were described each week, these descriptions being scattered in more than 200 different journals from various countries. However, more recent studies indicate that these figures are gross underestimates for the reason that palynologic workers in the west were unaware, even as late as 1958, of the huge amount of Russian publication which has appeared in recent years. Our study shows that about two-thirds of the Russian literature on pre-Pleistocene palynology is not even listed in our otherwise very accurate reference services.

The rate of papers from pre-Pleistocene palynology alone seems to be speeding ahead at the rate of about 330 a year and has reached a total of 4,400 publications, containing a total of about 14,300 new names of fossil pollen and spores described, to the end of the year 1966. (See Text Figs. 1-3).

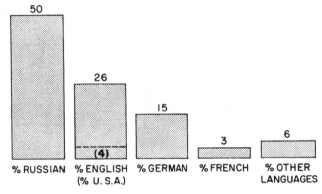

Text Fig. 1. Pre-Pleistocene palynologic publications (1961-1964).

In 1958 the majority of the world's palynologists could be included in one of the five "terminology circles": East Scandinavian, West Scandinavian, German, Russian, and an American mixture of East and West Scandinavian terminology.

A few examples of the confusion in definitions might show the need for a standard definition of each term:

1. *Pore.* Faegri and Iversen (1950) understand a pore as an aperture whose length-width ratio is 2:1 or under. If the greater diameter is more than twice the smaller one, they speak of a colpus. For Thomson and Pflug (1953), an aperture with a length-width ratio of 1:5 is still a pore.

2. *Aperture.* Western palynologists combine under this term pores, colpi, and tetrad-marks. Russian palynologists use the term for what is here commonly called the "pore canal."

3. *Granulate.* A spore described as granulate in Illinois might be defined as "apiculate" in Krefeld.

4. The many diverging terms concerning the sporodermis wall really constitute a chapter in themselves, and the reader is invited to check for himself.

5. Terms not often used like *prevestibulum, pore canal index, torus,* and *discus,* which are found in the various descriptions of fossil spores and pollen, can be understood and evaluated only in accordance with the definition given by the original author.

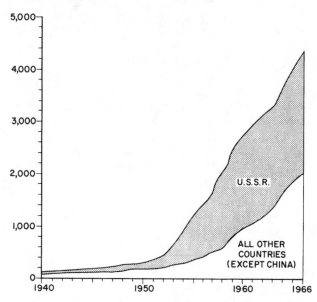

Text Fig. 2. Proliferation of publications in pre-Pleistocene palynology. Cumulative totals.

I am very grateful to Dr. Khlonova for pointing out to me the importance of the pioneer Russian works of B. M. Kozo-Polyanskiy, as well as A. L. Takhtadzhyan and A. A. Yatsenko-Khmelevskiy, both published in 1945. It appears that most of the terms cited in the Encyclopedia under the editorship of I. M. Pokrovskaya, 1950, were already defined in a similar way by those three authors in 1945.

It should be added here that in the Second Edition the citation of the Russian authors and articles was made in accordance with the transliteration of the Russian alphabet as used by the Library of Congress.

A number of errors have been corrected in this Second Edition. However, it has not seemed advisable to follow a suggestion that has been made that "out-dated" definitions be omitted. The work of the pioneers and those who followed them cannot be forgotten or ignored. The special meanings of their terms according to the time they were used must not be omitted lest their work be rendered useless due to semantic difficulties. As long

as the palynologic publications of the forties and fifties of this century are of value to our science the definitions of the terms used at that time are of importance too. (Text Fig. 2).

It is not the object of this encyclopedia to weed out "incorrect definitions" and to tell other people what the author thinks "correct definitions" should be. On the contrary, when the same technical term is used by several authors with shades of different meaning, all interpretations are given and listed in chronological order, i.e., from the earliest use of the term in the literature to the most recent. When definitions are set forth in this way, a more precise, uniform — and simpler — use of terminology for pollen and spore description is facilitated.

It has been suggested also that the development of the literature should be traced into the past beyond the time of the ones we consider to have built the foundation of today's palynology, viz. Potonié, Wodehouse and Erdtman. As valuable as it may be to show the development of a term from the beginning, it would mean that not only would we include the works of Reinsch, 1884, Purkinje, 1830, Fischer, 1890, Frizsche, 1837, and Mohl, 1835, but we would have to go back at least to the

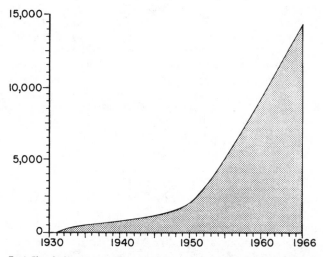

Text Fig. 3. New names for spore discoveries, doubled and redoubled. Cumulative totals.

work of Needham, 1750, and perhaps earlier ones. Obviously this would result in a much larger volume, much of which would be of questionable value.

Jackson's *Glossary of Botanic Terms* constituted the floor of the material chosen for the Encyclopedia. Since he included in his work all the terms he could find, 1928 was the logical starting point.

This encyclopedia was compiled so that a reader may understand better the literature already written and not in order to freeze terminology and prescribe fixed terms never to be changed. Hence the young worker in palynology can read and understand the earliest literature, but need not adopt its terminology. Having before him the similar and related terms and definitions as they have been added, he may make his own choices and judgments.

Our Latinized descriptive terminology has become almost a living international language of its own. It is used by all palynologists and therefore should be kept progressive, flexible, and living. Without advocating that every newcomer start his own descriptive system, it seems justified for authors to change terms as they learn and gain new insights; for them to reject older terms which do not convey desired meaning; and to generate new expressions for new ideas. Thus, in a logical and conservative way, the future may be built upon the past in the matter of describing and defining pollen and spore morphology.

Acknowledgments

This book would not have been possible without the enthusiasm, eager work, devotion, and painstaking patience of many persons who worked on the project in one or another of its phases during the last eight years.

In the first year Dr. Hilde Grebe, Geologisches Landesamt, Krefeld, Germany, helped me in laying the ground work for the project. Dr. Edward L. Stanley, at that time graduate assistant at Pennsylvania State University, devoted countless hours to continuing the project, as did Walter M. Warner, John L. Ablauf, and David M. Peabody in their capacity as graduate assistants at the University of Arizona.

I am grateful to Dr. Ivan Mittin from the U.S. Geological Survey, Denver, and to Dr. Nicholas V. Ponomareff, for providing the needed translations from the Russian literature and to my wife, Eva, for helping me in the many translations from German sources. For their devoted technical assistance my thanks go to Louise Hedricks, Carol L. Jenkins, Ruth Link, Jewel T. Methvin, Eva Suhr, and Judith Wilder.

Dr. G. Erdtman of Stockholm has earned my special thanks for the help he has always provided this project. He has earned my admiration also for his outstanding contributions to palynology, exemplified in this encyclopedia by the citation of so many passages from his work.

Not the least of my acknowledgments is for the valuable advice and actual help I received from Dr. L. E. Kuprianova, Leningrad, Dr. A. F. Khlonova, Novosibirsk, Dr. M. Van Campo, Paris, and from all the other authors who are cited in this encyclopedia, as well as for the many unweighable services of friendship of Dr. William Spackman, Jr., Pennsylvania State University, and of the director of the Geochronology Laboratories, Terah L. Smiley.

The project could not have been undertaken without grants received from Socony-Mobil Oil Co., Inc., National Academy of Sciences (Committee on the Marsh Fund), National Science Foundation (Project No. NSF–G19596), and the University of Arizona Graduate College, and without the mutual support of responsible administrators of the Pennsylvania State University, the U.S. Geological Survey, and the University of Arizona. Indeed, this work was not possible without the full cooperation of many interested and undertstanding persons — to all of them my heartiest thanks.

In the beginning it was planned to publish this Encyclopedia within the framework of the *Catalog of Fossil Spores and Pollen,* for it was that catalog as published by Pennsylvania State University that literally sheltered and supported the first assembling of material. The Encyclopedia might have appeared as one or two volumes of the *Catalog,* but the book grew out of these bounds, and although there is a filial attachment to the *Catalog of Fossil Spores and Pollen,* it was necessary to publish the Encyclopedia separately.

Gerhard O. W. Kremp
Dept. of Geochronology
The University of Arizona

Abbreviations Used in the Text

adj.	adjective
cf.	compare
Gr.	Greek origin
L.	Latin origin
n.	noun
pl.	plural ending
sdb.	similarly defined by
Tr.	translation
[. . .]	inserted by the author
a, b, c,	similar definitions
I, II	quite different definitions

A

ABJECTION, n. pl. -s. (L. abjicio, to throw away). Jackson, 1928: "casting off spores from a sporophore."

ABJUNCTION, n. pl. -s. (L. ab, away from; junctus, joined). Jackson, 1928: "cutting off spores on portions of growing hyphae by septa."

ABPORAL LACUNA, n., pl. -ae. (L. ab, away from; porous, channel; lacuna, hollow).
a. Wodehouse, 1928, p. 933: "those of the circumpolar lacunae, generally three in each hemisphere, which are opposite the germinal pores in lophate grains. Example: *Vernonia jucunda*."
b. Wodehouse, 1935, p. 541: "a lacuna meridionally opposite a germ pore. It may be closed as in the grain of *Scolymus* . . . [Fig. 764], communicate with its adjoining poral lacuna as in that of *Taraxacum* . . . [Fig. 765], or, if the poral lacuna is absent, it may communicate with its meridionally opposite abporal lacuna."

ABSCISSION, n. pl. -s. (L. ab, from; scissus, rent, slit). Jackson, 1928: "detachment of spores from a sporophore by the disappearance of a connecting zone."

ACANTHA, ACANTHON, n., pl. -ae. (Gr. akantha, a thorn). Jackson, 1928: "a spine or prickle."

ACANTHOUS, adj., spinous, spiny, spine-shaped.

ACIDOTUS, adj. (Gr. akidotos, pointed). Jackson, 1928: "when branches or organs end in a spine or hard point."

ACOLPATE, ACOLPATUS, adj. (Gr. a, less, not, without; kolpos, a fold). cf. furrow-less, inaperturate, non-aperturate.
Wodehouse, 1935, p. 541: "without furrows or pores." [*see* COLPATE].
sdb. Takhtadzhyan and Yatsenko-Khmelevskiy, 1945, p. 37; Faegri and Iversen, 1950, p. 160; Pokrovskaya, et al., 1950; Cranwell, 1953, p. 16.

ACUSPORIDE, adj. (L. acus, needle). Krutzsch, 1959, p. 40: *Tr.* The wall is more or less interspersed by needle-stitch-like breaches. In case of concentration an intrabaculate structure may occur. [*see* ORNAMENTATION, Fig. 770].

AECIDIOSPORE, AECIOSPORE, n., pl. -s. (Gr. oikidion, a little house; spora, a seed).
a. Jackson, 1928: "a spore formed in . . . *Aecidium* . . ."
b. Melchior and Werdermann, 1954, p. 15. [*see* SPORE].

AECIDIUM, n., pl. -a. cf. aecium. Jackson, 1928: "a sporocarp consisting of a cup-shaped envelope, its interior surface consisting of a hymenium, from whose basidia the aecidiospores are successively thrown off; the name was propounded by Persoon as a genus of Fungi, but it is now regarded as only a form-genus of Uredineae."

AECIUM, n., pl. -aecia. cf. aecidium.
a. Jackson, 1928: "Arthur's term for *Aecidium*."
b. Webster, 1960: "a cuplike spore fruit produced by certain rusts, in which chains of spores are developed."

AEROSTATIC UMBRELLA, n. pl. -s. cf. marginal frill, velum.

AFZELIUS LAYER, n., pl. -s. (Afzelius, name of a Swedish palynologist). Kupriyanova, 1956: *Tr.* Uppermost, granulated, easily soluble layer of the exine, first described by B. Afzelius. [*see* SPORODERMIS and Fig. 540].

AGAMOSPORE, n., pl. -s. (Gr. a, without; gamos, marriage; spora, a seed). Jackson, 1928: "a spore or gonidium produced asexually."

AGGREGATE, n., pl. -s. (L. aggregatus, assembled). cf. aggregation, tetrad.
a. Jackson, 1928: "collected together, as the flowers of *Cuscuta*."
b. Webster, 1954: n. "formed by collection of particles into a mass or sum." adj. "clustered together in a dense mass or head."

AGGREGATION, n., pl. -s. cf. aggregate, cross tetrad, dyad, dyas, Groups of Sporomorphae, jugate,

1

linear tetrad, massula, monad, pollen fiber, pollen tetrad, pollen tetrahedron, pollina occlusa, pseudomonad, rhombohedral tetrad, rhomboidal tetrad, spore, square, tetrad, tetradium, tetragonal tetrad, tetrahedral tetrad, tetras, tetrasporaceous, tetraspore, tetrasporic, tetrasporine, tetrasporous, triad.

a. Jackson, 1928: "condensation of cell-contents under some stimulus; the coming together of plants into groups (Clements)."

b. Harris, 1955, p. 25: "adherence of spores in fours (tetrads) or other multiples."

AIR SAC, AIR SACK, n., pl. -s. (L. aer, air; saccus, a sack). cf. bladder, cavea, sac, saccus, sack, vesicula, vesicula aerifera, wing.

a. Jackson, 1928: "an empty cavity in the pollen of *Pinus*."

b. Potonié, 1934, p. 9. [*see* SPORODERMIS, air sacs; Fig. 211].

AKINETE, n., pl. -s. (Gr. a, without; kinesis, movement).

a. Jackson, 1928: "in green Algae, single cells whose walls thicken and separate off from the thallus, corresponding to the chlamydospores of Fungi; immobile reproductive cells, formed without true cell-formation, or rejuvenescence."

b. Melchior and Werdermann, 1954, p. 14. [*see* SPORE].

ALETE, ALETUS, adj. cf. contact figure, inaperturate, non-aperturate.

a. Erdtman, 1943, p. 49: "spores without a tetrad scar."

b. Potonié and Kremp, 1955, p. 10: *Tr.* After separating from the tetrad aggregation, the surfaces of the spores may become completely smooth so that no trace of the former contact remains recognizable. Thus, *alete* spores originate, that is spores without contact areas and without dehiscence marks.

sdb. Takhtadzhyan and Yatsenko-Khmelevskiy, 1945, p. 37; Selling, 1947, p. 76; Pokrovskaya, et al., 1950; Cranwell, 1953, p. 16.

ALTITUDE, n., pl. -s. (L. altitudo, height). cf. aperture, colpus, costa colpi, furrow, margo.

I. Jackson, 1928: "used to specify the height above the sea of the vegetation in question."

II. Potonié, 1934, p. 22: *Tr.* In some Tertiary fossils (e.g., *Pollenites pseudocruciatus* . . . [Fig. 333]) which have a great similarity to the previously illustrated subfossil *Fagus* sp. from the Lake Ahlbeck sediments, a more or less distinct line encircles the sulcus at some distance. This borders the sulcus on both sides as an accompanying wall-like protrusion of exine, the *swellings* or *altitudes*. [*see* APERTURE, altitudes].

ALTITUDO, n. (L. height). cf. altitude, dimension, figura, height, morphography, ornamentation, shape classes.

Iversen and Troels-Smith, 1950, p. 37: *Tr.* Height; a dimension perpendicular to the surface of the exine; however, it is used only in connection with sculptures and structures. [*see* DIMENSION].

AMB, n., pl. -s. (L. ambitus, a going around). cf. ambit, ambitus, circumscripto, contour, equatorial contour, equatorial limb, extrema lineamenta, figura, Groups of Sporomorphae, limb, limbus, morphography, optical section, outline, shape classes.

Erdtman, 1952, p. 459: "Amb (. . . Erdtman in Terasmae, 1951 . . . Svensk bot. Tidskr., v. 45; Faegri and Iversen, 1950: equatorial limb; the term 'limb' has also been used by Wodehouse, cf. e.g., 1930, p. 36, and 1935): outline of a spore viewed with one of the poles exactly uppermost, i.e., with the polar axis directed straightly towards the observer. In isopolar, not constricted spores the maximal amb is the same as the equator. In isopolar, equatorially constricted spores the maximum amb does not coincide with the equator."

"Pollen grains with equatorial apertures and ± angular amb are either ANGUL-, PLAN-, or SINU-APERTURATE. In the first case . . . [Figs. 85–87, 91–93, 496–497, 504] the apertures are situated at the angles of the amb (sides of amb convex, straight, or concave); in the second case . . . [Figs. 83, 89, 499, 505] they are situated at the mid-points of the sides (sides of amb ± straight); in the third case equally halfway between the angles (sides of amb ± concave . . . [Figs. 76, 500]). If the amb is lobate, with apertures situated in the ditchlike indentations between the lobes, the grains are FOSSAPERTURATE (L. fossa: ditch . . . [Figs. 82, 88, 501, 507])."

sdb. Pike, 1956, p. 50.

AMBI-APERTURATE, adj. (L. ambo, both). cf. aperture, colpate, furrow, porate, pore, tetrad mark.

Erdtman, 1947, p. 113: ". . . distalo-proximalo-aperturate; with two apertures, one on the distal, the other on the proximal part." [*see* GROUPS OF SPOROMORPHAE].

AMBIT, n. cf. amb.

AMBITUS, n. (L. a going around). cf. amb.

Jackson, 1928: "the outline of a figure, as of a leaf."

ANACATATREME, adj. cf. aperture.

Erdtman and Straka, 1961, p. 66: ". . . 2-treme spores with one aperture with its center at the distal and the other with its center at the proximal pole."

ANACOLPATE, adj. (Gr. ana-, on, up). cf. colpate, furrow, sulcate.

Erdtman, 1958, p. 137. [*see* Fig. 105].

ANAPORATE, adj. cf. aperture, porate, pore, ulcerate, Groups of Sporomorphae.

Erdtman, 1958, p. 137. [see Fig. 105, GROUPS OF SPOROMORPHAE].

ANATOMY, n. (Gr. cutting). cf. morphography, morphology.
a. Jackson, 1928: "in botany, the study of structure."
b. Merriam-Webster, 1960: "the dissecting of a plant or animal in order to determine the position, structure, etc. of its parts, the science of the morphology and structure of plants and animals, a textbook dealing with this science, the structure of an organism or body, or a model of it as dissected, a skeleton, any analysis."

ANATREME, adj. cf. aperture.
Erdtman and Straka, 1961, p. 66: ". . . 1-treme spores with the center of the aperture at the distal pole."

ANDROECIUM, n., pl. -a. (Gr. andros, a man; oikos, a house).
Jackson, 1928: "the male system of a flower, the stamens collectively."

ANEMOGAMAE, n. (Gr. anemos, wind; gamos, marriage).
Jackson, 1928: "wind-fertilized plants; also as ANEMOPHILAE; plants fertilized by the wind (Kirchner)."

ANEMOPHILAE, n. (Gr. phileo, I love).
Jackson, 1928: "wind-fertilized plants."

ANEMOPHILOUS, adj.
Jackson, 1928: "applied to flowers which are wind-fertilized, the pollen being conveyed by the air."

ANGUL-APERTURATE, adj. (L. angulus, corner). cf. colpate, furrow, Groups of Sporomorphae, porate, pore, tetrad mark.
Erdtman, 1952, p. 459: ". . . the apertures are situated at the angles of the amb." [see AMB, and figures cited there].

ANGULAR, adj. (L. angularis, having corners). cf. shape classes.
Jackson, 1928: "used when an organ shows a determinate number of angles . . ."

ANGULAR-ELLIPTICAL, adj.
I. Compare SHAPE CLASSES.
II. Zaklinskaya, 1957: [see Fig. 227].

ANGULAR-OVAL, adj.
I. Compare SHAPE CLASSES.
II. Zaklinskaya, 1957. [see Fig. 227].

ANGULATE, adj. (L. angulatus). cf. shape classes.
Jackson, 1928: "more or less angular."

ANGUSTIMURATE, adj. (L. angustus, narrow; murus, wall). cf. reticulate.
Erdtman, 1952, p. 459: "with narrow muri (mural cross-sections one fifth or less of the diameter of the lumina)."

ANISOPOLAR, adj. (Gr. anisos, unequal).
a. Erdtman, 1947, p. 113: "hemispheres dissimilar." [see GROUPS OF SPOROMORPHAE].
b. Harris, 1955, p. 25: "having the proximal and distal portions of the spore dissimilar."

ANNULAR, adj. (L. annulus, anulus, a ring). cf. halonate, porate, pore.
Jackson, 1928: "used of any organs disposed in a circle."

ANNULATE, adj.
a. Jackson, 1928: "ring-shaped."
b. Norem, 1958, p. 668: "with protruding rim or ridge. Note: Annulate . . . is preferred over zonate . . . The latter term is reserved for apertures that encircle the grain parallel to the equator and separate the surface into zones, or to bands of sculpturing that also parallel the equator . . . [Fig. 159-1]."

ANNULOTRILETE, adj. (L.) cf. annulate.
Norem, 1958, p. 668: "annulate . . . with aperture . . . [Fig. 159-2]."

ANNULUS, ANULUS, n., pl. -i. (L. annulus or anulus, a ring). cf. annulate, annulotrilete, aspis, colpus, edge, costa pori, dissence, endannulus, endanulus, halo, labrum, limes anuli, lip, margo, oculus, operculum, operculum pori, pore lid, pore ring.
I. Jackson, 1928: "in Ferns, the elastic organ which partially invests the theca, and at maturity bursts it;" "in Fungi, a portion of the ruptured marginal veil, forming a frill upon the stipe after the expansion of the pileus;" "in Mosses, the ring of cells between the base of the peristome or orifice of the capsule and the operculum;" "in Diatoms, used by W. Smith for a compressed rim of silex within the frustules of such genera as Rhabdonema Kutz;" "in Equisetaceae, the imperfectly developed foliar sheath below the fruit spike;" "the fleshy rim of the corolla in Asclepiads, as the genus Stapelia."
II. a. Potonié, 1934, p. 19: Tr. The margin of the exoexine immediately surrounding the pore is the pore ring or annulus. [see APERTURE, annulus].
b. Faegri and Iversen, 1950, p. 20: "a pore is frequently surrounded by an annular area (annulus), the exine of which is characterized by differences in the outer layer, the endexine sometimes being thicker, sometimes thinner." [see FURROW, Fig. 357].
c. Thomson and Pflug, 1953, p. 31–41: for detailed discussion see APERTURE.
sdb. Erdtman, 1943, p. 49; Takhtadzhyan and Yatsenko-Khmelevskiy, 1945, p. 37; Kupriyanova, 1948, p. 11; Iversen and Troels-Smith, 1950, p. 33; Pokrovskaya, 1951; Ingwersen, 1954, p. 40;

3

Traverse, 1955, p. 93; Beug, 1961, p. x.

III. Cranwell, 1953, p. 16: "a ring-like (halonate) area around a pore, as in *Gramineae.*"

ANOMOTREME, adj. cf. aperture, Groups of Sporomorphae.
Erdtman and Straka, 1961, p. 65: "with irregular or irregularly placed apertures."

ANOSULCATE, adj. (L. anus, ring, circuit). cf. colpate, furrow.
Erdtman, 1947, p. 105: "as to sulci we still lack knowledge concerning some of the modifications of the monosulcate status. A monosulcate grain has one germinal furrow, borne on the distal side of the grain; a bisulcate grain two sulci (it seems uncertain whether one of the sulci is always borne on the distal, the other on the proximal part of these grains); whilst polysulcate grains, if such exist, should be provided with many sulci. 'Anosulcate' grains . . . should be provided with two confluent sulci."

ANTESPOROPHYLL, n., pl. -ae. (L. ante, before; Gr. spora, a seed; phyllas, a leaf).
Jackson, 1928: "the primitive structure of the spore-bearing organ (Potonié)."

ANTHER, ANTHERA, n., pl. -s, -ae. (Gr. antheros, flowering).
Jackson, 1928: "that portion of a stamen which contains the pollen, usually bilocular, and sessile, or attached to a filament;" "an old term in Fungi, for the Antheridium;" "also used by Linnaeus for the seta and capsule of Mosses, as in *Bryum;*" "used by Parlatore for the loculi in coniferae."

ANTHERIDIUM, n., pl. -a. (Gr. antheros, blooming, dim. suffix - idion).
Jackson, 1928: "the male sexual organs in Cryptogams, the analogue of the anther in Phanerogams;" "in Hymenomycetes, an old term for CYSTIDIUM;" "afterward used for the mother cell of antheridia."

ANTHEROZOID, n., pl. -s. cf. sperm cell.
Jackson, 1928: "Antherozoids . . . male motile cells provided with cilia, produced in antheridia."

ANTHOPHILOUS, adj.
Jackson, 1928: "applied to plants with flower-visiting insects which aid cross-fertilization."

ANULUS, n., pl. -i. (L. a ring). [*see* ANNULUS].

AP-, APO
Jackson, 1928: "prefix of negation (Rothert)."

APERTURA, n., pl. -ae. (L. opening). cf. aperture, colpate, furrow, germinal apparatus, porate, pore, tetrad mark.
Jackson, 1928: "formerly used of the dehiscence of anthers;" "the ostiole of certain Fungi."

APERTURATE, adj. cf. aperturidate, colpate, distalo-proximalo-aperturate, furrow, Groups of Sporomorphae, porate, pore, tetrad mark.
Erdtman, 1947, p. 112: "with apertures." [*see* GROUPS OF SPOROMORPHAE].

APERTURE, n., pl. -s. cf. acolpate, acolpium, acolpatus, altitude, ambiaperturate, anacatatreme, anacolpate, anaporate, anatreme, angulaperturate, annulate, annulus, anomotreme, anosulcate, anulus, apertura, aperturate, aperture, aperturidate, aperturoid, apex colpi, apex marginis, apical, apicula, apocolpium, apoporium, ap-orium, arc, arcuate, arcuate ridge, arcuate rim, arcuatus, arcus, area contagionis, aspidate, aspides, aspidoporate, aspidorate, aspidote, aspis, asulcate, atreme, atrium, bisulcate, bisulcatus, brevicolpate, brevissimicolpate, brevissimirupate, canalus pori, catatreme, centrum colpi, centrum intercolpii, centrum interporii, centrum pori, clinocolpate, colpate, colpato-colporate, colpatus, colpodiporate, colpoid, colpoidorate, colporate, colporatus, colporoidate, colpus, colpus edge, colpus longitudinalis, colpus median, colpus membrane, colpus transversalis, commissure, conclave, constricticolpate, contact area, contact face, contact figure, contact marking, contact point, costa, costa aequatorialis, costa colpi, costa pori, costa transversalis, cover lid, crassimarginate, cribellate, cribellatus, cryptocolpate, cuneus, dehiscence, dehiscence furrow, dehiscence list, dehiscence ridge, demicolporate, demicolpus, diaporus, dicolpate, dicolporate, diorate, diplodemicolpate, diporate, discoideus, disclike, dissence, distalo-aperturate, distalo-proximalo-aperturate, disculate, ditreme, diverse porate, edge of the anulus, edge of the margo, edge of the pore, endannulus, endanulus, endoplica, endoporus, equatorial ruga, exina fissa, exine ruga, exit, exitus, exitus digitatus, exitus opening, exoporus, expansion fold, extraporate, figura contagionis, figura triradiata, fissura, fissura dehiscentis, fissura germinativa triradiata, fissure, fold, foramen, foramina, foraminate, foraminoid, forate, forked Y-mark, formation of lips, forming of lips, fossaperturate, fovea, furrow, furrow-less, furrow membrane, furrow rim, germinicolpate, geniculate, geniculus, germinal, germinal aperture, germinal apparatus, germinal furrow, germinal lid, germinal point, germinal pore, germination, germ pore, halo, halonate, heterocolpate, heteropolar, heterosporous, hexatreme, hilate, hilum, inaperturate, incidence, infundibuliform, intercolpar, intercolparis, intercolpar thickening, intercolpium, interporium, intexine ruga, knee, labrum, laesura, lalongate, leptoma, lid, ligula, ligule, limes annuli, limes colpi, limes marginis, limes pori, linea dehiscens, lip, lolongate, longicolpate, loxocolpate, margo, margo aequatorialis, margo arcuata, margo colpae, medianum colpi longitudinalis, medianum colpi transversalis, medianum intercolpii, medianum interporii, membrana pori, meridional area, meridional furrow, meridional pleat, meridional ruga, mesocolpium, mesoporium, mesorium, monocolpate, monocolpatus, monolept, monolete, monoletoid, monoporate, monopored, monosulcate, monotreme, non-aperturate, obinfundibuliformis, oculus, oligoforate, oncus, opening, open pore, operculum, operculum colpi, operculum pori, orate, oroid, orthocolpate, os, pancolpate, panporate, pantotreme, parasyncolpate, pentacolporate, pentatreme, perforate, pericolpate, periporate, planaperturate, platea, platea luminosa, pleotreme, pleozonotreme, plica, plicate, polar area, pollina trichotoma fissurata, polycolpate, polycolporate, polyforate, polyplicate, polyporate, polyrugate, polysulcate, polytreme, pontoperculate, porate, poratusporate, pore, pore canal, pore canal index, pore membrane, pore-plug, pore ring, poroletoid, pororate, porus, porus annularis, porus collaris, porus simplex, porus vestibuli, postatrium, postcaverna, praevestibulum, prepollen, prevestibulum, primary fovea, primary germinal, primary meridional area, protrudence, protrudent, proximaloaperturate, pseudocolpus, pseudopore, pseudoporus, rimula, ruga, ruga aequatorialis, ruga compressa, rupate, rupus, scar, secondary folds, secondary germinal, secondary meridional area, sillon, simple pore, sinuaperturate, slit, solution, solution area, solution channel, solution notch, solution meridium, solution wedge, spiraperturate, spirate, sulcata, spiroid, stephanaperturate, stephanocolpate, stephanocolporate, stephanoporate, stephanotreme, subapertural thickenings, subisopolar, subzonosulcate, sulcate, sulcatus, sulcoid groove, sulculus, sulcus, sulcus simplex, sutura, suture, syncolpate, syndemicolpate, tarsuspattern, tasicolpate, tenuate, tenuimarginate, tenuitas, tenuity, tetrachotomosulcate, tetracolporate, tetrad mark, tetrad scars, tetratreme, tip of the colpus, transcolpate, transversal furrow, transverse furrow, transverse fold, transverse median of the colpus, trematum, trichotome, trichotomocolpate, trichotomocolpatus, trichotomosulcate, tricolpate, tricolpatus, tricolporate, trilete aperture, trilete mark, trilete marking, triletoid, triletus, triporate, triradial, triradiate crest, triradiate marking, triradiate ridge, trischistoclasic, tritreme, tumescence, ulcerate, ulcus, vertex, vestibulate, vestibulum, vestibulum apparatus, vestibulum pori, Y-mark, Y-split mark, Y-radius, zonaperturate, zoned furrow, zonicolpate, zoniporate, zonisulculate, zonorate, zonosulcate, zonotreme.

I. a. Potonié, 1934, p. 12: *Tr.* Germination of the pollen grain occurs in the following way: the intine penetrates the exine, protrudes, and by growing elongates into a pollen tube.

Y-mark, dehiscence-list and -furrow—In the simplest situation, there is no special place provided on the exine for this process, e.g., in many conifers such as *Taxus, Larix, Sequoia, Taxod-*

ium, Cupressus, and *Juniperus.* All of these have simple, globular pollen with an enveloping, homogeneous, well-developed exine. They agree herein with certain pteridophyte spores. However, in pteridophyte spores the all-enveloping homogeneous exine is in general fairly rare. In contrast to pollen, these spores usually show a tri-radiate mark, the Y-mark, which is formed by the *dehiscence-lists and -furrows* . . . [Fig. 107, 108] and which aids in germination. We find this arrangement in bryophytes too. Comparatively, the uni-radiate dehiscence ridge is a rare type among pteridophytes.

With evolution toward seed plants, we see in the systematically lower representatives, as in cycadophytes, at first only simple, globular pollen grains without recognizable germinal apparatus. A meridional pleat always appears on a mature, shriveled grain, but it disappears upon swelling, and its location can no longer be recognized. Absence of the Y-mark is here to be considered a sign of progress. [Potonié, 1934, p. 13: *Tr.*] In contrast to spores with Y-marks, pollen which lie in groups of four in the mother cell are able to smooth out completely upon separation. The smallest possible, most perfect sphere is better suited to wind dispersal than are many spore types. In contrast to spores, pollen must be lifted higher by the wind to reach the flowers. The aforementioned specialty (namely, the development of air sacs) which appears abundantly for the first time in gymnosperms, seems to facilitate better wind-transport of the pollen grain.

By way of suggestion, this innovation is found already in a Carboniferous spore from the Ruhr, *Sporites pustulatus* (Loose, . . . [Fig. 203]). In this species, three bag-like forms appear between the rays of the Y-mark.

Germinal apparatus—Beginning with angiosperms, *germinal apparatus* or *germinals* appear anew in certain places on the cell membrane. These are *not* homologous to the dehiscence furrows of spores. It will be shown, however, that they are related to dehiscence marks. Thus it is that in pollen, germinal apparatus likewise occur in threes.

It is also noteworthy that with the appearance of germinals the thickness of the exine generally increases.

Tetrad, tetrahedral tetrad—The predominance of just three germinal places is due to the fact that pollen usually lie in groups of four within the pollen mother cell, forming a *tetrad;* they are usually arrayed as a *tetrahedral tetrad* . . . [Fig. 2]. The pollen of Ericaceae, for example, after maturity usually remain closely connected with one another in this arrangement . . . [Fig. 4]. The same is true of many spores . . .

Contact points—Each of the four balls joined in a tetrahedron touches the other three at three points . . . [Fig. 2]. These three *contact points* form an isosceles triangle (that lies in the proximal polar hemisphere). In systematically higher pollen, germinal apparatus begin at these contact points and change position from the proximal polar hemisphere to the equator.

Tetragonal tetrad—In pteridophyte spores, just as in pollen, the tetrad is not in all cases arranged as a tetrahedron. The central points of the four cells may also be arranged in a plane within the mother cell . . . [Fig. 1] so that the cells are enclosed, for example, in a rectangular box. Wodehouse (1929) therefore speaks of a *tetragonal tetrad*. This tetrad is characteristic of monocotyledons. In this situation, each pollen cell touches only two of the grains belonging to the same tetrad. This means that initially only the possibility of development of two germinal apparatus is inherent. Wodehouse and others have, however, shown that in this case more germinal apparatus can be accommodated opposite each contact point, so that pollen with four germinals can be produced. In *Corylus* and other species there are also bi-, tri-, and tetragerminal pollen. This is explained by the fact that both kinds of pollen arrangements occur in the mother cell.

Since in one case, two germinals can be increased to four, so in another instance, three can be increased to six, etc.; thus the multi-germinal pollen types are explained. On the other hand, instead of several, only one germinal apparatus may be formed, as in the monocotyledons. Therefore, it is to be noticed that the [Potonié, 1934, p. 14: *Tr.*] pollen mother cell in many cases bears only one pollen grain, hence becoming a pollen grain itself.

Rhombohedral tetrad—An additional special case of the arrangement of pollen tetrads is that of the rhombohedral tetrad . . . [Fig. 3]. Naegeli designates this as half-tetrahedral. It may be clearer to speak, together with Wodehouse (1929), of a rhomboidal arrangement. The rhomboidal-tetrad makes it understandable how the one and same mother cell might supply pollen with different members of germinals.

Also, there are pollen tetrads which remain attached in one plane after maturity; e.g., *Catalpa bignonioides* displays mature pollen which form tetrahedral tetrads as well as four attached pollen in one plane. In Ericaceae, the latter rarely occurs.

Massulae—Fossil pollen masses have not yet been found like those of Mimosae in which each of the tetrad cells divides once or twice so that there arises a massulae of 8 or 16 pollen which remain together after maturity. This is an exception. The rule is that the tetrad disassociates after maturity.

Pollen fibers—Likewise still unknown as fossils are pollen fibers in which the pollen grains are arranged in a row (*Halophila*).

Although the tetrahedral, as well as tetragonal, tetrad may occur in the same species, the tetrahedral arrangement is the older type. In the Carboniferous, the bilaterally symmetrical, more or less bean-shaped spores (with only a faint, straight dehiscence mark running along the concave side of the "bean" [Fig. 5]) are subordinate to the radiate spores (with 3 dehiscence marks arranged in a Y [Fig. 108]). On the other hand, among the Tertiary pteridophyte spores many bilaterally formed spores are present. If one wishes to designate the proximal pole in bilateral spores as in radiate spores, one must place it in the center of the dehiscence mark.

Pore-plug, lid—Dodge (1924) has shown in the Basidiomycetes *Gymnosporangium* that the pore-plug or lid arises at the place where the spores touch.

We will see that the germinal apparatus is generally characterized as a place where the exine is weakened.

In the pollen mother cell, each of the four daughter cells of the tetrahedral tetrad is connected to the three others by broad channels, so that there are six channels in all (Wodehouse, 1929). All these channels extend from the middle of the facets which result from the mutual pressure of the growing grains. (In the last stage prior to final separation, the cells remain attached to each other by small depressions.)

As a rule, before the division and rounding of daughter cells are completed, in normal grains only a slight or no hint of surface sculpture and no indication of germ pore location are to be observed.

In the mature pollen grain, the weakening of the exine which generally designates the germinal can be accomplished in two ways which in turn allow a number of variations: [Potonié, 1934, p. 15: *Tr.*] (1) The exoexine may be absent at the germinal spot, (2) the whole exine, therefore exo- and intexine, may be strongly reduced (or absent?) so that the content of the pollen cell is isolated from the outside only by the intine which is not preserved in the fossil state (*Pollenites simplex*).

The case mentioned by Dodge probably belongs to type 1. Here the membrane develops a plug which can be discarded, as in *Cucurbita*. We will see that there are homologs of this plug. These are small exoexine-remnants found at the weakened areas of exine which develop upon decomposition of the exoexine; therefore these are type 1 germinal areas. The germinal apparatus of *Cucurbita* consequently can be understood as a special condition of type 1.

Contact area, area contagionis, curvaturae, arc line—Type 1 is quite frequent. Its development is understandable if one starts from certain Carboniferous pteridophyte spores. These spores were, as aforementioned, generally arranged as a tetrahedron. In many cases one can still find them in this attitude in coal. The contact points have enlarged to considerable *contact areas* [Fig. 108]. These *contact areas* are delimited proximally by the rays of the Y-mark; distally by the *arc lines* or "arc-edges." These *curvaturae* are often nicely arched, concave to the proximal pole (one should also compare *Sporites rugosus* R. Potonié, Ibrahim and Loose, 1932, Pl. 20, Fig. 59).

Exospore—It is noteworthy that the contact areas (at least in a great number of Carboniferous spore types) are weakened areas of exospore or of exine. They are the result of pinching-out of the exterior part of the wall of the exospore as in the germinal area of pollen of Type 1. This is made clear by the fact that in Carboniferous pteridophyte spores, the *sculpture* on the surface of the exospore is often distinct, whereas it is very weakened or completely suppressed in the region of the contact area because the latter are pressed tightly together . . . [Fig. 108]. One should also compare Potonié, Ibrahim, and Loose, 1932, Pl. 15, Fig. 14; Pl. 16, Fig. 25; Pl. 18, Fig. 33; Pl. 20, Fig. 57; as well as, above all, Wicher, . . . [Fig. 137–140]. In accordance with this fact, the exoexine is likewise strongly suppressed at the contact areas of Ericaceae tetrads (cf. *Pollenites ericius*).

Primary meridional area, exoexine—We find a similar situation in variations of Type 1 germinal apparatus in those pollen which show three large *primary meridional areas* . . . [Fig. 288]. These meridional areas can be roughly designated as the surfaces of sphere-sectors; however, their points do not reach the poles of the pollen grain, which is spherical in the ideal case. The main agreement with the contact areas of Carboniferous spores is that by omitting the external exine parts (namely, the exoexine) the meridional areas lack surface sculpture (or structure of the exine) which is preserved on the remaining parts of the pollen grain. Their

meridional area tips are distally somewhat removed from the pole, while proximally they are elongated to a corresponding position near the other pole (i.e., beyond the equator). Here it is to be noted that in spores, the curvature (with a point at its center) sometimes tends toward the proximal [should read distal] direction . . . [Fig. 108]. [Potonié, 1934, p. 16: *Tr.*] Furthermore, in contrast to the contact area, the meridional area is reduced, and the Y-mark which proximally limits the contact area is pinched-out. All in all, it is clear that the contact areas of spores are homologous to the meridional areas of pollen.

Subequatorial exitus, porus simplex, germ pore—The relations are also obvious in those cases where the pollen does not produce three large, meridionally-stretched areas, but only three small, generally equatorially-situated *foveae* . . . [Fig. 491], which are circular zones of weakened exine resulting from removal of the exoexine. Here the Carboniferous spore types with smaller, in part almost circular contact areas are to be referred to . . . [Fig. 107]. Kidston had at first described such spores without alluding to the evolutionary importance of their construction. One should for instance compare his *Triletes III* on Pl. III, Fig. 3. The relationship of such contact areas to the foveae of pollen is evident. While the Y-mark is smoothing-out, the contact points move away from one another and now constitute foveae usually equatorially situated . . . [Fig. 389]. That this change of position has really been accomplished is evidenced by those pollen types in which the three germinals have not yet reached the equator, so that they are arranged subequatorially. This condition is realized for example in the Juglandaceae *Carya ovata* (cf. Sears, 1930, Pl. III, Fig. 33). In this respect, there is a Miocene counterpart in *Pollenites simplex* (Potonié, 1931, Fig. 4). However, here the foveae may have changed already into the porus simplex . . . [Fig. 390], viz., not only the exoexine, but the entire exine is pierced. According to H. Fischer (1890) only those germinal apparatus called porus simplex may be precisely designated as germ pore. Fischer suggests namely, ". . . to designate only the actual holes of the 'Aussenhaut' [outer membrane] (i.e., probably the entire exine) as germ pores . . ." Later authors have repeatedly deviated from this suggestion as well as in regard to porus vestibuli, which will be discussed later. The latter does not pierce the exine; nonetheless, Fischer labeled it as a germinal pore too because its structure was not clear at that time.

Secondary folds—Probably favored by the subequatorial position of the pori, *Pollenites simplex* shows three secondary folds . . . [Fig. 390] which have been produced by the exines being pressed together from pole to pole while in the sediment. Such folds, related to the state of preservation, are to be called *secondary folds.* In *P. simplex,* they consist of arcs which begin between the pori at the equator and which are concave toward the pores.

It should be mentioned that Kidston in his *Triletes II* noticed contact areas, each of which showed three or four small pits (Pl. III, Fig. 2C). *Primary germinal, secondary germinal*—In cases where the germinals of pollen have moved not to the equator, but just a little way apart, or not at all, there are secondary germinals developed on the pollen wall on the opposite side from each of the primary germinals, so that we find pollen with 6, 12, etc. germinals (one should also compare the explanation above, concerning types which arise from the dividing mother cell). In the most extreme instance, the pollen grain may be thickly covered with a great number of germinals which have polygonal outlines resulting from being pressed together closely. [Potonié, 1934, p. 17: *Tr.*] This type can be traced directly back to the one marked by three primary fovea which have not separated. The polygonal moulding between the primary fovea can then be compared with the Y-mark of spores.

In pteridophyte spores as well as pollen types, the contact areas can be now bigger, now smaller.

In spores, connection can be maintained after maturity (e.g., in part, the Lycopodiales) or the tetrads can separate from each other. After separation, the surfaces of the spores can smooth out completely so that no trace of the former contact remains; or more or less large, still recognizable contact areas remain; too often only the Y-mark is to be seen.

The situation is quite similar in pollen. There is, however, usually nothing left after separation which can be compared with the Y-mark. Traces of the contact can completely disappear. If, however, they remain as germinals, they can in various ways be secondarily assimilated and supplemented.

Exitus, germinal points, exit places—Those pollen tetrads which are not separated by the process of maturing offer good material for comparison. In this line, we have here thought primarily of Ericaceae. Large contact areas develop from the contact points of immature pollen tetrads. These areas encompass the biggest part of the proximal polar-hemisphere and their

distal limit (i.e., arc-line) approaches very close to the equator. Where the curvaturae come closest to the equator, close to their outside lie the exitus or germinal points . . . [Fig. 4]; which are those particular places of the often rather expanded germinal apparatus where germination will be effected. The term *"exit place"* introduced by H. Fischer is more general. Fischer (1890, p. 16) proposed ". . . to designate the thinned part of the exine as exit place . . ." To this must be added that the protrusion of the pollen tube often takes place at only a certain point within the thinned part of the exine. Furthermore, it is often appropriate to designate more precisely the particular structure of the germinal by the following definitive concepts.

In Ericaceae, if one imagines the four grains being separated from each other, and their pressed areas being equalized, then the grains correspond essentially to most dicotyledonous grains.

Wodehouse (1929) thus concludes: "Obviously in this case the position of the furrows and apertures is determined by the tetrahedral arrangement of the grains in the tetrad group, and is therefore a haptotypic character."

In many instances, the tetrads of Ericaceae pollen divide into twins. In contrast with certain trilete pteridophyte spores, the contact areas immediately disappear when they no longer function. This is a hint that the flattening in Ericaceae is secondary.

In Ericaceae tetrad, a germinal furrow . . . [Fig. 4], the sulcus (see below), runs from each exitus in the direction of the free, therefore distal, pole of each pollen grain. It also represents the opposite direction of that in which the germinal point has traveled from the proximal hemisphere toward the equator zone. [Potonié, 1934, p. 18: *Tr.*] Accordingly, the sulci show up as meridional furrows in singly occurring trisulcate grains. In a significant manner, they do not touch the pole, but only reach toward it, more or less to the *contact point,* i.e., to the place where the exitus could have been developed. Only in a few instances do the sulci seem to touch at the poles (Myrtaceae) and form a pattern similar to the Y-mark in the trilete spores of pteridophytes. In the pteridophytes, the Y-mark also shows furrows which serve in germination and therefore are called *dehiscence furrows.* Dehiscence furrows, however, cannot be considered homologous to sulci according to the discussion undertaken about the origin of Ericaceae. Whereas the sulci of Ericaceae run across the middle of the contact areas, the dehiscence furrows of trilete pteridophyte spores form their meridional boundaries. These meridional boundaries are no longer present in pollen (except for those remaining in the tetrads) because in pollen, after separation of the four grains, the contact areas smooth out or change completely.

Fovea—The simplest case of the strongly localized germinal apparatus is after all, the *fovea.*

The fovea is formed in this way: the exoexine develops a small circular pit whose floor is probably mostly composed of the intexine . . . [Fig. 491]. Thus, for example, the germinal apparatus of *Sagittaria* sp. (from sapropel from sediments of Lake Ahlbeck) consists of many small, shallow, circular depressions regularly distributed over the surface of the spherical pollen cell and which originate by the weakening of the exoexine . . . [Fig. 490]. Sometimes it is not recognizable if the exoexine is really weakened at the foveae-sites (cf. *Pollenites multistigmosus,* Cecilie Mine). The foveae then appear merely as small depressions over the entire exine.

Exit place—In fovea, therefore, is shown an instance of the germinal apparatus which H. Fischer and others called "exit place."

Similar germinal apparatus are found in Chenopodiaceae and Caryophyllaceae.

As an example from the Tertiary, *Pollenites stigmosus* from the Miocene brown coal should be mentioned. In *P. sigmosus,* a few small particles of exoexine are found here and there sticking to the intexineous base of the fovea. Accordingly, one recognizes that here the foveae arise thus; the exine expands and the exoexine, in consequence of dilation of certain especially expansive circular areas, falls apart. The floor of the fovea in *Cucurbita* is not exposed. Here, a delicate circular detachment line forms within which small exoexine fragments remain situated initially as "pore plugs." The exoexine remnants in the fovea of *P. stigmosus,* etc., are therefore homologous to pore plugs.

Germ pore, exitus opening—Directly related to fovea-apparatus, is the development of germinal apparatus, e.g., such as we find in *Juglans* and *Carya. Pollenites simplex* was mentioned as a fossil example. In this fossil type, the exoexine as well as the intexine is pierced so that an effective true *porus,* a *porus simplex,* is present. But it need be said that it is not always easy to ascertain whether we are dealing with a true porus or with a sharply depressed fovea in which the floor consists only of a thin membrane of intexine. [Potonié, 1934, p. 19: *Tr.*] The question is posed as to whether such a thin layer of intexine is generally or always extant, or is removed during fossilization or the maceration of

the fossil material. Then too, in the living grain, the pollen contents beneath the exitus would never be isolated from the outside exclusively by intine. Nevertheless, certain conditions noticed in prepared fossil material satisfy the definition of *porus simplex;* a definition which has been related to the terms *germ pores* and *exitus openings* by previous authors. However, the same authors, in part unconsciously, have from the beginning used these terms in a broader sense. Particularly, they have included porus vestibuli (to be discussed forthwith).

Porus vestibuli, vestibulum—The *porus vestibuli* is formed thus; that the exoexinous margin of the fovea is compressed over its intexinous floor . . . [Fig. 492]. The margin of the depression then looks like a porus (known as porus vestibuli . . . [Fig. 393]), which leads into one of the small, intexine-enclosed ante-chambers, the *vestibulum.*

Vestibulum apparatus—A more thorough examination of germinal apparatus, e.g., *Corylus,* already cited, shows that in this case the germinal apparatus belongs in part to the strongly differentiated type of vestibulum apparatus (e.g., in *Corylus* sp. from sapropel of Lake Ahlbeck sediment). Here the exoexine is clearly separated from the intexine in all parts of the membrane by a distinct separation line. Possibly the intexine detaches itself slightly from the exoexine at the germinal apparatus . . . [Fig. 392], but this is not certain. Thus too a distinct porus vestibuli would grow out of the fovea, pierce the outer exine, and drive into a small, more or less lens-shaped vestibule. In any case, this impression can be conjured up many times. However, in certain optical cross-sections, it can also occur in an apparatus as shown in . . . [Fig. 393]. The ante-chamber, the *vestibulum,* in *Corylus* is in many instances still very small, but it can in other cases be distinctly perceived. An opening which connects the vestibulum and cell-chamber of the pollen grain is not present. However, the intexine is here probably many times thinner than its remaining portions. Hence, the porus leads only from outside into the vestibulum. . . . [Fig. 394] shows the germinal apparatus of *Corylus* sp. in a vertical view of the exine surface. The porus is evident as a small point. At a distance, two parallel circles surround the porus. The diameter of the inner circle is about three times as big as that of the porus. The outer circle is not always clearly defined; it encircles the first at just a little distance. Dokturowsky and Kudrjaschow (1924) give an explanation of vestibulum apparatus by drawings, without text.

H. Fischer particularly defined a break through the whole exine as a germ pore. Accordingly, by this definition, one cannot call the porus vestibuli a germ pore since it pierces only the exoexine. Yet from Fischer's examples, it comes out that he too classifies the pori vestibuli as germ pores.

Annulus, lips—The margin of the exoexine immediately surrounding the pore is the *pore ring* or *annulus*. In *Corylus* sp. the pore ring is often slightly curved outward. In this case one speaks of the *lips* of the porus . . . [Fig. 391]. In *Alnus* this pattern is more pronounced. In the most extreme [Potonié, 1934, p. 20: *Tr.*] cases, beak or chimney is spoken of. Similar structures also appear in types of exitus not classified as pores; they are characterized in the same way. Often, and this also in *Corylus,* the space which serves as vestibulum is acquired only through the formation of lips. Therefore, in such cases the intexine below the vestibulum is not bent in toward the interior of the pollen cell, but is more or less flat or bent weakly outward . . . [Fig. 391, 392, 393].

Rimula—A peculiarity in the differentiation of the germinal apparatus is therein given, that in *Corylus* the pore is sometimes not completely round but is somewhat elongated in the meridional direction. As in meridional folds, compression of the margins of the pores ensues somewhat more in the equatorial direction. The small, incidental elongation of the pore in *Corylus* points to the distinctly stretched *rimula* being developed in other species.

In *Alnus* there is already established in clear form a small, longish slit (a rimula) . . . [Fig. 398]. If one looks perpendicularly at the exine, it is even seen that the meridionally elongated rimula extends a little beyond the inner circle on both sides.

The rimula is to be understood as a special case of porus vestibuli. For that reason, the perimeter of the rimula should likewise be called a pore ring.

Pore ring—The pore ring, hence the strip of exoexine immediately around the germ pore, in *Alnus* is clearly thickened and pushed forward to a beak.

Vertical view, plan view, tangential section of the germinal, arcus—We call the view of germinal or germinal apparatus from a vertical aspect of the exine, its vertical view, its plan view, or its tangential section. In the plan view of the germinal of *Alnus* one can recognize that the thickened edge of the rimula continues at both ends in the form of two diverging arcs . . . [Fig. 399]. In these arcs or arci we are not dealing so much with cell-wall thickenings as with arc-shaped edges which arise through weakened creases in the exine. The arcs swing over to adjacent germinals and are to be recognized in the polar view of the pollen grain as significant features, beneficial to recognition . . . [Fig. 395]. All folds homologous to the arcuate edges of *Alnus* will be called arci, as also the folded arcs of *Pollenites plicatus* . . . [Fig. 396], as well as the almost straight-lined moulding of *P. microexcelsus* . . . [Fig. 397] whose straightness is made possible through the slight protuberance of the equatorial parts of the exine.

The vestibulum and rimula are especially well developed in *Tilia.*

Germinal furrow, sulcus—In the plan view of the germinal, the elongated rimula of *Tilia* extends far over the double circle enclosing the vestibulum. (That is not to say that the double circle need be considered a boundary of the vestibulum as in the above types. In those instances too, focused in a particular way, the lip formation produces such an optic cross-section). In *Tilia,* the rimula is quite clearly an open slit only within the double circle, functioning as an exit from the vestibulum. Outside the double circle, the rimula continues in both directions as only a small, narrow furrow, the *sulcus,* cutting into the exoexine . . . [Fig. 315]. The sulci, pointed at their ends, and pointing toward the poles, terminate quite far from the poles in *Tilia.* However, they are essentially [Potonié, 1934, p. 21: *Tr.*] nothing more than the long furrows of many pollen types differentiated on this basis, in which the furrows approach the poles, and of which there are transitions to those broad meridional areas which appear in fossil pollen grains now evenly expanded, now sharply in-folded. Consequently, there is no boundary between simple sulci and small, in-folded meridional areas.

Folds, rugae—Here it must be inserted that the sulci and the aforementioned, related, plain differentiations of the germinal apparatus will generally be called *folds* or *rugae*. H. Fischer said (1890, p. 16): the exitus places are '. . . simply folds . . .' if '. . . they are elongated and sharply in-folded in the dry grain.' It will be shown in the following what special terms will be needed to designate more precisely the individual peculiarities of rugae.

Compressed fold, ruga compressa—The vestibulum of *Tilia* is strikingly formed. Differentiation of a vestibulum in *Corylus* and still more in *Alnus* resulted essentially from the formation of lips or beaks; in *Tilia,* however, by deep, inward folding of the inner exine and by especially deep impression of a *compressed fold* (ruga compressa). This arrangement is recognizable in the optical cross-section of the germinal apparatus . . . [Fig. 316]. In cases of more distinct development of compressed folds, one recognizes how at first the intexine extends to the edge of the rimula, then bends over as far as possible, lying entirely against the wall (compressed fold) and then turns over to the other edge of the rimula. If in an optical cross-section, one focuses as accurately as possible on the center of the rimula . . . [Fig. 317], the vestibulum seems to be less deep, but broader. This condition also appears in a vertical view of the germinal apparatus . . . The punctate line of . . . [Fig. 317, lower part] corresponds to the optical cross-section through the middle of the rimula just discussed. [Fig. 317, lower part] . . . shows that the double circle bounding the vestibulum produces two small pointed lobes (tips) opposite each other on the equator and which are equatorially directed.

Transverse fold, equatorial ruga—The "tips" in *Tilia* are very indistinct and often hardly recognizable. In this we are dealing with the evolution of an equatorially directed fold of the exine, a *transverse fold*. All differentiations related to it are to be called *equatorial rugae,* which will be discussed later.

An example of a differentiation of germinal apparatus that does not deviate from the principle, *Fagus* is referred to . . . [Fig. 321]. Particularly in the vertical view . . . [Fig. 321, lower right], it is recognizable that the rimula here is extended by distinct sulci far beyond the vestibulum, close to the poles of the pollen grain. These sulci already exhibit more distinct, small in-folds which are changing into meridional areas by swelling of the grains, smoothing of folds, but essentially by expansion of the intexine.

Tenuitas—Very instructive in understanding many of the previously discussed patterns of germinal apparatus is the fossil preservation which has been described as *Pollenites megadolium digitatus* . . . [Fig. 322, 323, 332]. In this case, a long, narrow meridional ruga is present which because of a modest in-fold and its narrowness, is designated as a sulcus. [Potonié, 1934, p. 22: *Tr.*] In the middle of the sulcus, there is a strongly differentiated germinal point or exitus. How this differentiated condition arises is shown in the type figured as *Pollenites sinuatus* . . . [Fig. 330, 334, 335]. Here in the exitus, the meridional ruga is crossed by a very distinctly formed equatorial ruga. In *P. megadolium sinuatus* this equatorial

ruga exists as a slightly inward-bent depression which is crossed by the sulcus and is thus divided into two halves. It seems that in the region of this depression, the exine is thinner; still it could not be determined whether the exine here consists only of intexine or not. If it should be confirmed that we are dealing with a less thick wall division, we would be able to speak of a *tenuitas.*

It should be inserted here that in recent pollen, the intine generally is clearly thickened below the exitus. Here especially it contains much of the strongly expansive pectin which brings about the bursting of the weakened exine. One can also say that the outermost especially pectin-enriched lamella under the germinal is thickened into a pillow.

According to Mangin (1889) the pectin-rich layer of intine beneath the exitus, e.g., in *Cytisus laburnum,* is very distinctly marked. Fritzsche (1832) who did not quite properly recognize the pectin-rich thickening of intine, arrived at the *inter-layer* hypothesis.

Exitus digitatus—In more enlarged specimens, as in fossil *Pollenites megadolium* . . . [Fig. 331], the tenuitas is arched toward the outside, so that the exitus is occupied by a hill over which the sulcus crosses in the form of a furrow. In recent relatives of the pollen under discussion, the hill expands and smooths out considerably by swelling, so that it is no longer divided by the sulcus (for a similar arrangement, see . . . [Fig. 325–328]). Let us now further consider the fossils in which the tenuitas is not protrudent, but is indented. Here the indentation can deepen more and more in such manner that the upper and lower edges of the tenuitas approach each other, more and more covering the hollow forming below them . . . [Fig. 322, 323]. The *exitus digitatus* is developed. The enclosure is accomplished in such a way, that the parts of the exoexine above and below the tenuitas, and right and left of the sulcus, approach one another like four fingers or papillae. The vestibulum which the four fingers enclose can then to a certain degree be compared with the vestibulum of *Tilia* and other species; only, in *Tilia* the equatorial ruga is practically not yet developed. Conditions very similar to those in *Pollenites megadolium* are found in *P. pompeckii* . . . [Fig. 319, 320]; exactly the same conditions are found in many other forms, such as, e.g., *P. exactus,* to be compared with *Castanea* . . . [Fig. 318, 324].

Swellings, altitudes, long folds, meridional rugae, secondary meridional area—In some Tertiary fossils (e.g., *P. pseudocruciatus* . . . [Fig. 333]) which has a great similarity to the previously illustrated subfossil *Fagus* sp. from the Lake Ahlbeck sediment, a more or less distinct line encircles the sulcus at some distance. This borders the sulcus on both sides as an accompanying wall-like protrusion of exine, the *swellings* or *altitudes.* Therefore we have here before us a case which decidedly belongs to *folds* or rugae, and since they are meridionally arranged, [Potonié, 1934, p. 23: *Tr.*] belong to the *meridional rugae* or *long folds.* When the pollen grain swells, the edges of the sulcus separate, the furrow broadens and by expansion of the intexine finally becomes a broad secondary *meridional area* which is more or less pointed at both ends (polewards). Hence the floor of the meridional area is formed by expanded intexine; it is generally completely smooth, while sculpture is situated only on the exoexine.

In the subfossil *Fagus* as in the Tertiary pollen types which are morphologically or systematically related to *Fagus* pollen, the fossils never show meridional areas in an extended state.

Primary meridional area, intexine ruga—However there are also species which in the fossil state regularly exhibit only a broad, long meridional area, and in fact without a differentiated germinal apparatus. Thus we see, e.g., in pollen species morphologically related to the genus *Acer,* merely three simple meridional areas whose points approach the poles and which have a considerable breadth at the equator, i.e., where the differentiation of germinal apparatus would be situated. Sometimes such pollen coats are completely compressed in the sagittal direction, whereby the intexine membrane filling the meridional areas ruptures and is in some instances completely lost. The pollen species of many Liliaceae for example are endowed with only one single, simple meridional area. Therefore it may be expedient to differentiate the expanded meridional areas seen in fossils as *primary* (e.g., Liliaceae) from fossil, unexpanded, *secondary* meridional areas (e.g., Fagaceae). Primary meridional areas are already present as a broader surface in the recent freshly dried pollen grain. They are here folded inward completely, so that particularly this feature merits the name fold or ruga; more precisely, we are dealing with an *intexine ruga.*

The secondary meridional area attains greater breadth by swelling; and this by strong participation of expanding intexine. Here where expansion is not too far advanced, the intexine again contracts by drying or by release of inner pressure.

It is probably a special development which leads to the primary meridional area.

For reasons of evolution, let us first consider the case of the pollen grain with three primary meridional areas. Of all known comparable pollen types, this case shows the most distinct homologies with those Carboniferous trilete spores provided with contact areas. The meridional areas can be readily perceived as contact areas which have secondarily lengthened beyond the equator. The structure- or sculpture-furnished meridional stripes between the meridional areas would then, in their proximal part, be homologous to the Y-mark; they have, to be sure, lost the dehiscence furrow, its function being taken over by the meridional areas. Hence, dehiscence furrows are analogous to meridional areas.

Fissura—We refer to such species as *Taxus, Larix, Sequoia, Taxodium, Cupressus,* and *Juniperous* as pollen without germinal apparatus. When the pollen of these genera are examined more closely, it appears that the exine through stronger swelling of the fresh pollen grain does not burst altogether irregularly. [Potonié, 1934, p. 24: *Tr.*] On the contrary, there arises a quite sharp, straight rent, a *fissura* . . . [Fig. 213]. A corresponding illustration of *Cupressus* was already given by Fritzsche (1832, Pl. 1, Fig. 16). This fissure is very characteristic for certain fossil pollen, we can refer to *Pollenites hiatus* which is probably related to one of these genera. Weakening of the exoexine at a stripe corresponding to the fissure may also have led to the formation of the primary meridional area.

Simple sulcus, sulcus simplex—On the other hand, we also have here the route which could have led to certain pollen types with simple sulcus (Sulcus simplex), not yet discussed. The simple sulcus, i.e., the ordinary furrow, shows no differentiation in its equatorial regions; it lacks any special apparatus, and in some cases could have been a transitional stage leading to the meridional area.

Exine ruga—Also along the line of the simple sulcus, the exine (in the dried living pollen grain as well as in the fossil) can be strongly folded . . . [Fig. 286, 287], and indeed in such manner that the sulci move inward and the equatorial cross-section of the pollen grain will be tribolate. In this case, we have an *exine ruga.* It must be differentiated from intexine ruga, a fold principally of the meridional area . . . [Fig. 329].

Geniculus—In some cases, the sulcus simplex shows a small peculiarity, insofar as it produces equatorially, an in- or out-wardly turned knee, a *geniculus.* Further, the furrow may be equatorially a little broader, i.e., form a kind of rimula. Herewith we are already straying from sulcus simplex and find in this direction the connection to forms with vestibulum apparatus and long sulci.

b. Erdtman, 1946, p. 70: "The main types of apertures in pollen grains may be classed as follows: *sulcus* . . . , *colpa* . . . , *ruga* . . . , and *porus* . . . . The number of apertures are indicated by the prefixes mono-, di-, tri-, tetra-, penta-, hexa-, poly-, etc."

"Typical 'monocotyledonous' pollen grains are provided with a longitudinal furrow borne on the distal part of the grain and crossing [Erdtman, 1946, p. 71] the polar axis at right angles . . . [Fig. 24–26]." "This kind of a furrow is termed a *sulcus*. By shortening and contraction a sulcus may ultimately develop into a porus. Exceptionally a sulcus may present a three- or four-slit opening. Such grains are termed trichotomosulcate . . . and tetrachotomosulcate . . . respectively (cf. Erdtman, 1944)."

"Typical 'dicotyledonous' types of pollen grains are provided with three meridional furrows, meeting, if extended, at the poles and crossing the equator at right angles. Furrows of this kind are termed colpae . . . [Fig. 29]. By shortening and contraction a colpa may ultimately develop into a porus. In tetra- and hexacolpate grains, etc. the colpae do not always cross the equator at right angles, but converge in pairs (cf. Wodehouse, 1935)."

"A polycolpate pollen grain has many furrows which, according to the definition, are equally spaced in the equatorial, more or less extended zone of the grain. If, on the other hand, the grain is provided with many furrows equally distributed over the general surface of the exine the furrows are referred to as *rugae*. In a hexarugate pollen grain the rugae correspond to the six edges of a tetrahedron. By shortening and contraction the rugae may, in the same way as sulci and colpae, exhibit any transition to pori. Some plants produce both tricolpate and hexarugate grains but it is felt, nevertheless, that the term ruga should be retained as a purely descriptive term. . . . [Fig. 33] . . . shows a polyrugate grain."

"The pores (*pori*) are rounded and more or less distinct apertures, as a rule equally distributed on the equator or over the general surface of the grain. '1-po' denotes a grain with one pore. To avoid misunderstandings and the introduction of a new term, '2-po,' '3-po,' '4-po,' etc. should exclusively denote grains with equatorially arranged pores. Any other porate grains should be referred to as 'polyporate' (po-po). This term would thus, according to the definition, include any grains, except monoporate,

with pores not confined to the equator. In practice 'polyporate' would mean the same as 'cribellate,' proposed by Wodehouse (1935) . . . [Fig. 32]. Outline of a pollen grain of *Thelygonum cynocrambe* shows a hexaporate (6-po) grain with six equatorial pores. The hexahedral grain of *Fumaria,* so beautifully illustrated by Fritzsche (1837, Tab. VI, Fig. 11), although provided with six pores, should be termed polyporate (po-po (6))."

"Non-aperturate grains (nap; *pollina non-aperturata*) may [Erdtman, 1946, p. 72] eventually be referred to as asulcate, acolpate, or arugate if derived from sulcate, colpate, or rugate pollen types."

"Spores of mosses and ferns are, as a rule, provided with dehiscence fissures. Monolete spores (1-let) have one fissure severing the straight longitudinal streak, or '*laesura*,' borne on the proximal part of the spore. Trilete spores (3-let) have three fissures longitudinally severing each arm (laesura) of the triradiate streak usually known as the tetrad scar (*cicatrix tetradica*). Alete spores (alet) have no preformed dehiscence fissures."

c. Erdtman, 1947, p. 104: "By an aperture is understood any weak, preformed part of the general surface of the spore which engaged in forming an opening in connexion with the normal exit of intraexinous substance. In tricolporate grains, therefore, not only the pores but the entire colpae should be termed apertures."

d. Erdtman, 1952, p. 459: "any weak, preformed part of the general surface of a spore which may be engaged in forming an opening in connection with the normal exit of intra-exinous substance." [*see also* Fig. 75].

e. Thomson and Pflug, 1953, p. 27: *Tr.* We designate as the "germinal apparatus" the totality of all differentiations of a sporomorph in which we assume, on the basis of extant examples, that they function, or may function, during the development of the pollen tube.

The germinal apparatus usually consists of several similar "germinals" which are situated symmetrically to one another. We want to call the germinaloid differentiations which are located in one place and which are connected, "the germinal." Often a three-part germinal apparatus is present; sometimes the germinal apparatus consists of only one germinal (e.g., among the *Monocolpopoll.*) or the germinal apparatus is completely missing (e.g., in the Inaperturates).

The germinal may be distinguished by the development of solutions and dissences or, more rarely, by special structures or sculptures. The totality of the solutions of one germinal is called the "Exitus." The exitus in each lamella can be differently developed according to form, size and arrangement; the colpus of an exterior lamella may be followed in the interior lamellae by pores, rugae, an atrium or again colpi. Also nonperforated lamellae may be inserted between them.

Especially the vestibula and cavernae in all structure-forms are specific germinaloid dissences. Both can probably be considered in many cases as homologous structures. Also the anulus, endanulus, oculus, and tumescence probably have germinal functions to fulfill. In addition to this function, the extremely expanded anuli of the extratriporates may have the function of a floating apparatus.

The following may be considered "extragerminaloid" differentiations: among solutions, e.g., the solution meridia of the *Extratriporopoll.;* among dissencens, the air sacks and the comb of the *Pityosporit.,* the arci of *Alnus,* the interloculum of the *Extratriporopoll.,* and finally, most of sculpture- and structure-differentiations . . . [Thomson and Pflug, 1953, p. 31: *Tr.*] The often distinctly recognizable anulus forms by lengthwise growth of the rodlets of exolamella d which is enclosed by an outer (exolamella c) and an inner structureless lamella (exolamella e). Another rodlet layer (exolamella b) and a structureless lamella (exolamella a) follow centrifugally. Thus, the anulus is developed from the deeper layer of rodlets of the ektexine. In the anulus, the rodlets are perpendicular to the pore canal whereas the rodlets of exolamella b are radially oriented to the center of the grain and therefore at an oblique angle to the rodlets of the anulus; that is, they both stand perpendicularly between the structureless lamellae . . . [Fig. 544].

Quite analogous to the anulus, the rodlets of exolamella b can also be elongated into a ring-shaped circumpolar zone. This structure is called the "oculus." In contrast to the anulus, the oculus is usually smaller in cross-section, but larger in plan view. It appears in the polar view as a dark circular spot lying on the pore region or, as a dark round disk or ring.

Another layer of rodlets is usually recognizable in the endexine (endolamella b) which is enclosed by two structureless lamellae (endolamellae a and c). If the rodlets are circumpolarly elongated, we speak of an endanulus.

The anular structures can be thickened centripetally or centrifugally. They can show a wedge-shaped, club-shaped, drop-shaped, spher-

ical, or triangular cross-section . . . [Fig. 376–378, 386–388].

The distinct interspace between ektexine and endexine is called interloculum. It is especially large at the equator and widens here toward the pores. A direct connection between the [Thomson and Pflug, 1953, p. 32: *Tr.*] ekt- and endexine is often not at all recognizable and is probably usually localized to small areas. This has the result that in the *Pompeckji*-type, where parts of the endexine are vestigial, the remains of these are easily detached and lie on one another in the pollen lumen.

. . . *Pore canal, vestibulum, atrium*—The anulus encloses the pore canal like a beak. The pore canal may be cylindrical or it may widen conically toward the exterior or the interior. Its length is measured from the outer edge of the ektexine to the inner edge of the endexine. The ratio of "length of the pore canal to pollen diameter" (measured in the axis of the pore canal) is called the "pore canal index." [Thomson and Pflug, 1953, p. 33: *Tr.*] In some groups the pore canal widens in the ektexine to a local "praevestibulum" which forms by splitting of the exolamellae. Between the endexine and ektexine there is either a "vestibulum" or an "atrium" which opens into the interloculum at the side if one is present. The difference between the vestibulum and the atrium which is so important among triangular pollen is defined as follows:

They both have in common the phenomena that the endexine is no longer connected with the ektexine at the pore region and encloses its own porus, the endoporus. Between the endoporus (of the endexine) and the exoporus (of the ektexine) is an intervening space.

We speak of an "atrium" if the endexine is largely dissolved in the pore region so that an endopore forms that is more than three times as large as the exopore.

If the endexine continues independently after splitting and encloses an endoporus which is not more than three times as large as the exoporus, the structure is called a "vestibulum" . . . [Fig. 381].

Also the endexine can be interlamellarly split and then enclose a "postvestibulum" or also form an interlamellar "postatrium" . . . [Fig. 382].

. . . *Solution meridia*—Remarkable is the stripe-like solution process of the endexine, which takes place in a meridional direction and which proceeds to a different extent in the different groups: We differentiate the following states: "solution notch" (=incidence), "solution wedge"

(=cuneus), and "solution channel" (=platea). The overall term is "solution meridium" . . . [Fig. 410–412].

The solution notch (incidence) is the beginning stage of the meridional solution which can be noticed among the Pertrudoidae. [e.g. *Extratriporopollenites pertrudens* Thomson and Pflug (1953)]. Toward the pole of the endoporus, the endexine is absent in a small triangular zone which can be semi-circular in the swollen condition. The length of the notch normally does not exceed the thickness of the endexine. N. Ross (1949) interpreted the solution notch in the *Protrudens*-type as a colpus (*Tricolporites protrudens* Ross). However, this structure has nothing to do with the colpi of the *colporates*. [Thomson and Pflug, 1953, p. 34: *Tr.*] In Hemiperfectoidae [e.g. *Extratriporopollenites hemiperfectus* Thompson and Pflug (1953)], among others, the wedge-shaped solution zone has already moved a third of the way toward the poles (solution wedge); in *Pompeckjoidae* [e.g. *Extratriporopollenites pompeckji* (Potonié, 1931)], the "solution channels" reach the pole.

The endexine has become a very unstable structure . . . [Thomson and Pflug, 1953, p. 36: *Tr.*] If the ektexine in cross-section thickens as a wedge starting from the center of the side, we speak of a "tumescence" [e.g., in . . . *Triatriopollenites rurensis* Thomson and Pflug (1953)].

Some types do not develop an anulus, but evaginate the unthickened ektexine at the pores as a nozzle. This feature is called "labrum."

"Plicae" are doublings of the entire exine that run like two parallel garland-like folds from the pole to the pore region. They are the result of the unequal swelling capacity of the exine. The pore regions expand more strongly than the central part of the side. This type of swelling is systematically important and characteristic for the [*Triatriopollenites*] *Plicatus*-type . . . [Fig. 413–414]. [Thomson and Pflug, 1953, p. 37: *Tr.*] From the plicae, the "endoplicae" can probably be differentiated, despite the fact that the latter are similar in shape and that they probably arise by specific swelling differentials. Here, however, only the endexine is folded so that tube-like detachments occur between the ektexine and the endexine. The endoplicae always open into atria and are attached to these because the atria make the endexine an unstable structure. Endoplicae are characteristic of many *Triatriopollenites*.

In the following, the "arci" of *Alnus* are discussed. These too, are tube-like detachments between end- and ektexine. However, they are probably primary in origin since they open into

vestibuli which do not favor the swelling folds as the atriae do . . . [Fig. 408–409]. [Thomson and Pflug, 1953, p. 38: *Tr.*]

Exitus—. . . Much more important than the outer form is the *germinal structure*. However, here we are at the threshold of our knowledge because the germinal structures of the *tricolporates* are among the most *complicated* exine structures. The convergence of the tricolporate forms is purely superficial. A prime example is the habitual similarity between *Fagus* and *Nyssa*. It can be said that both build basically different germinal types. Certainly we will later subdivide the tricolporates, as has already largely occurred with the triangular [triporate] pollen. [Thomson and Pflug, 1953, p. 39: *Tr.*] The germinals lie in centripetal niches which give the equator a clover-leaf outline. In the exolamella a, therefore in the outermost lamella of the ektexine, only one colpus is ever developed; the author has never found pores in the exolamella a of the Tricolporates, which were very abundant in Tertiary time; the colpus of the exolamella a always crosses the pores.

The pores are found in the deeper lamellae. It is characteristic of many groups that the size and form of the pores are differently shaped in each lamella. Usually two or three pore types lie behind each other; of these, one is round, the other is more equatorially elongated. We differentiate circular, meridionally elongated, equatorially elongated pores, and finally equatorial rugae. The latter are equatorially elongated slits which have the character of colpi . . . [Fig. 368–371].

The great difficulty lies in recognizing the spacial succession of the individual pore types of a germinal. The dimensions of their intervals frequently lie at the limit of resolution of the light-microscope; here only the highest [numerical] apertures can help. Thus, the investigations are in a state of flux. Usually, but not always, the pore-bearing lamellae are free of colpi. On the other hand, the lamella with a colpus can be placed between two pore-bearing lamellae. For reasons of stability one must also assume the presence of intermediate unbroken lamellae; however, these are hard to prove, especially when they are without structure.

The germinal elements are designated, according to their position, as exocolpus a (in exolamella a), exoporous b, endoporus a, etc. In cases that are not exactly known, the centrifugal pore is called the "exopore;" the centripetal pore the "endopore." This is without regard to whether they lie in the ektexine or endexine, but it should be emphasized that the exopore of the

Tricolporates never lies in the exolamella a. The most distinct pore, which is the one that lies in the thickest lamella, is called the "main pore." In many cases, it lies in the separated endexine or is limited to the "a" or "b" lamella of the endexine.

Caverna, cavium—A further differentiation is shown by the meridional splitting of the layers. These are probably homologous to the vestibules, pre- and post-vestibules of the Brevaxones and should be called "cavernae." [Thomson and Pflug, 1953, p. 40: *Tr.*] The cavernae appear in equatorial view as dark stripes that lie symmetrically behind the colpi. Their width is measured at the equator. They can be wider than the pores and then "circumscribe" them (*Nyssa*-type among others) or can also be smaller and thus narrower than the pores (*Fagus*). Often several cavernae lie behind one another: The splitting of the ektexine layers is designated as a "prae-caverna" and that between the ektexine and endexine layers as "caverna" in the strict sense; between the endexine lamellae as a "post caverna." The latter seems to be very common . . . [Fig. 372–375].

The caverna often has the form of two narrow cones which are open toward the equator and whose apices lie at the poles. The cone formation results when the inner lamellae do not part until at the colpus; they then form the outline of a circle . . . [Fig. 375].

f. Krutzsch, 1959, p. 36: *Tr.*

Germinal—An important feature of *Sporites* is their "germinal," their scar, slit-formed and thereby Y-like (trilete), straight-lined (monolete) or also still otherwise shaped (e.g., in transition stages or as in *Divisisporites*), but in part only incomplete breaking of the spore wall. Generally it is confined to one hemisphere. Occasionally so-called Y-double-marks appear, i.e., such which turn out to be irregular to completely regular and mirror-imaged of both halves of a specimen; such shapes of course should no longer be considered among the "typical *Sporites*." Transition stages, cf. Pflug, 1953 . . .

Scar lines—Scar lines can be short or long; their route from pole to equator will herewith be measured and indicated by little r. We differentiate to following lengths:

$$r = \quad \text{to } 1/3$$
$$r = 1/3 \text{ to } 1/2$$
$$r = 1/2 \text{ to } 2/3$$
$$r = 2/3 \text{ to } 3/4$$
$$r = 3/4 \text{ to } 4/5$$
$$r = 4/5 \text{ to } 5/5$$

In r = 5/5 the rays reach the equator or, in cingulate spores, the special equatorial differentiation; in r = 1/3 they are very short. Decisive in trilete scars with variable lengths is then at times the value of the two rays, which are approximately of the same length. If all rays have different lengths, they must be specially recorded.

The rays can be robustly or delicately built out over the whole length, constantly thick or pointed (etc.). They can develop rigidly straight, more or less straight, normally strong or weak, or also abnormally corrugated. At the pole, the apex of the spore and the scar, they often begin with a $\pm$ strong twist, which, in conformity with the manner of deflection of the course of the ray-tips, can be designated as of left- or right-turning (Krutzsch, 1954, p. 297; Geologie, vol. 3, no. 3). The approximate vigor of deflection lends itself to being declared as "angle of rotation" (measured in vertical view).

The rays can also split, thus split-marks arise, by which a split-angle and the length of divided or undivided branches are of importance. Splits appear in trilete as well as monolete spores. Naturally, bursting of the Y-mark has nothing to do with splitting, but there are fossils (especially, e.g., in *Leiotriletes*) in which the bursting occasionally occurs. Farther in the diverse layers of the wall, certain Y-marks have in part a different appearance (similar to certain Carboniferous megaspores; cf. R. Potonié and Kremp, 1954, fig. 63).

g. Beug, 1961, p. X: *Tr.* . . . visible exitus area for the pollen tube.

sdb. Pike, 1956, p. 50.

II. Pokrovskaya, et al., 1950: *Tr.* [Aperture] that is the opening of the pore, the exterior or the interior outline of the pore canal (external and internal openings).

sdb. Takhtadzhyan and Yatsenko-Khmelevskiy, 1945, p. 37.

APERTURIDATE, adj. cf. aperturate.

Erdtman, 1947, p. 112: "with aperturoid areas." [*see* GROUPS OF SPOROMORPHAE].

APERTUROID, n., pl. -s. cf. aperturate, colpate, contact figure, furrow.

Erdtman, 1952, p. 459: "any area, similar to an aperture, on the general surface of a spore."

APEX, n., pl. apices. (L. summit). cf. contact figure, proximal.

I. Jackson, 1928: "an old name for Anther;" "the ostiole of Fungi (Lindley);" "the growing point of a stem or root;" "the tip of an organ."

II. Potonié and Kremp, 1955, p. 10 and Fig. 132, 133, 142: *Tr.* The *proximal hemisphere* can more or less retain the shape of a triangular pyramid; the *distal hemisphere* can more or less stay semi-globular or it can continue to swell . . .

With respect to the arrangement of the mother-cell, each grain has an inner or proximal, and an outer or distal pole, as well as a *proximal* and a *distal* hemisphere. The proximal pole is the *apex;* it lies next to the center of the mother cell. [*see* Y-MARK].

APEX COLPI, n.
Iversen and Troels-Smith, 1950, p. 32: *Tr.* Tip of the colpus. [*see* COLPUS].

APEX FIELD, n., pl. -s. cf. contact figure, torus.
Krutzsch, 1959, p. 38: *Tr.* The area between torus and apex is . . . called the pole or apex field; also it is designated by R. Potonié (1934) as area.

APEX MARGINIS, n.
Iversen and Troels-Smith, 1950, p. 32: *Tr.* Tip of the margo. [*see* COLPUS].

APICAL, adj. (L. apicalis, apex).
Jackson, 1928: "at the point of any structure; Kofoid's term for anterior in Dinoflagellates."

APICAL FIELD, n., pl. -s. cf. apocolpium, polar field.
Beug, 1961: *Tr.* Trigonal to polygonal field of syncolpate pollen grains, depending on the number of colpi, that are situated at the pole in which the colpi terminate.

APICULA, n., pl. -ae. (L. apiculum, a little point).
Jackson, 1928: "a sharp and short, but not stiff point, in which a leaf may end."

APICULATE, adj. (L. apiculatus). cf. shape classes.
I. Jackson, 1928: "furnished with an apicula."
II. a. Faegri and Iversen, 1950, p. 160: "with slightly protuberant poles."
   b. Cranwell, 1953, p. 16: "poles (or ends, e.g., *Freycinetia*) protruding slightly . . . ."

APLANOSPORE, n., pl. -s. (Gr. a, without; planos, wandering).
a. Jackson, 1928: "non-mobile cells which are detached for propagation, formed asexually by true cell-formation and rejuvenescence."
b. Melchior and Werdermann, 1954, p. 15, [*see* SPORE].

APO- (Gr. from, away from, separate).
Merriam-Webster, 1954: "A prefix signifying from, away from, off, as in apogee; or as under, detached, separate, as in apocarpus."

APOCOLPIUM, n., pl. -ia. cf. polar area, polar field.
Erdtman, 1952, p. 459: "area at a pole, delimited toward the equator by the polar limits of the mesocolpia . . . Examples: apocolpia large, . . . [Fig. 457–458], small . . . [Fig. 453–454], rectangular . . . [Fig. 449–451], hexangular . . . [Fig. 459–460, lacking . . . [Fig. 463–464]."
sdb. Pike, 1956, p. 50.

APOLAR, adj. (L. A, privative; Gr. (polar) polos, a pivot). cf. shape classes.

I. Jackson, 1928: "applied by Bertrand and Cornaille, to indeterminate fibrovascular masses without tracheae, in Ferns."
II. a. Erdtman, 1947, p. 112: "without distinct polarity." [*see* GROUPS OF SPOROMORPHAE].
   b. Faegri and Iversen, 1950, p. 160: "(Erdtman, 1947): without distinct polarity, e.g., *Chenopodium.*"
   c. Erdtman, 1952, p. 460: "without distinct polarity (at least as far as the sclerine is concerned). Example: . . . [Fig. 533, 538, 539]." [*see* POLARITY].
   d. Harris, 1955, p. 25: "the proximal and distal surfaces not differentiated."

APOPORIUM, n., pl. -ia. (Gr. apo, from). cf. polar area.
Erdtman, 1952, p. 460: "area at a pole, delimited toward the equator by the polar limits of the mesoporia (q.v.)."

AP-ORIUM, n., pl. -ia. cf. polar area.
Erdtman, 1955, p. 460: "area at a pole, delimited toward the equator by the polar limits of the mesoria (q.v.)."

APOSACCALE AREA, n., pl. -s. cf. aposaccium.
Erdtman, 1957, p. 3: "The surface of the corpus of a pollen grain with n bladders can be divided into the following areas: . . . and finally two aposaccale areas (aposaccia), one at the distal pole, and the other, usually much larger than the former, in the proximal face of the grain with the proximal pole in its centre." [*see* SACCATE].

APOSACCIUM, n., pl. -ia. cf. aposaccale area.
Erdtman, 1957, p. 3: [*see* SACCATE].

APOSPORY, n. (Gr. spora, seed).
Jackson, 1928: "supression of spore-formation, the prothallus developing direct from the asexual generation."

APSILATE, adj. cf. sculptured.
I. a. Erdtman, 1947, p. 112: "not smooth, with sculpturing." [*see* GROUPS OF SPOROMORPHAE].
   sdb. Faegri and Iversen, 1950, p. 160.

ARC, n., pl. -s. (L. arcus, bow, arc). cf. arcus.
Merriam-Webster, 1954: "a bowlike curve or an orbit having such a curvature."

ARC LINE, n., pl. -s. cf. arcuate ridge, contact figure.

ARCUATE, adj. (L. arcuatus).
Jackson, 1928: "bent like a bow, curved."

ARCUATE RIDGE, n., pl. -s. cf. arcuate rim, contact figure, crista arcuata, curvatura, margo arcuata.

ARCUATERIM, n., pl. -s. cf. arcuate ridge.

ARCUATUS, adj. cf. arcuate.

ARCUS, n., pl. -us. (L. bow, arc).
a. Potonié, 1934, p. 20: *Tr.* In plan view of the germinal of *Alnus* one recognizes that the thick-

ened rim of the rimula is continued at both ends into diverging arcs . . . [Fig. 399]. In these arcs or *arci,* we are not dealing so much with cell-wall thickenings as with arc-shaped edges which arise through weakened creases in the exine. The arcs swing over to the adjacent germinals and in the polar view of the pollen grain are to be recognized as significant features, beneficial to recognition . . . [Fig. 395]. All folds homologous to the arcuate edges of *Alnus* will be called arci, as also the folded arc of *Pollenites plicatus* . . . [Fig. 396], as well as the almost straight-lined molding of *P. microexcelsus* . . . [Fig. 397] whose straightness is made possible through the slight protuberance of the equatorial parts of the exine. [*see* APERTURE, Potonié 1934].

sdb. Erdtman, 1943, p. 69; Takhtadzhyan and Yatsenko-Khmelevskiy, 1945, p. 38; Selling, 1943, p. 76; Faegri and Iversen, 1950, p. 160; Pokrovskaya, et al., 1950.

b. Erdtman, 1943, p. 69: "A characteristic feature is the presence of arci. Their real nature is a matter of debate. According to Wodehouse, they are a part of the texture of the grains: i.e., they represent real thickenings which, however, do not protrude above the general surface of the grain. Potonié considers that they belong to the ornamentation of the grain and are to be regarded less as thickenings than as curved margins caused by a slight outward bending of the exine. Potonié, . . . however, when speaking of subfossil *Alnus* pollen, remarks (1934 . . . , p. 59): 'Längs der Arci scheint die Exine schwach verdickt zu sein, so dass sie der Aussteifung des Pollenkorns dienen.' "

"Basing his opinion only on a study of unsectioned pollen grains, the author feels inclined to consider the arci as an ornamental feature, caused by local thickening of the ektexine or, eventually by intercalation of mesexinous strands between the ekt- and endexine. Such strands would correspond to the central layer of the exine stratification of five strata in the pollen grains of *Betula, Corylus,* and *Myrica,* as described by Jentys-Szafer (1928)."

c. Erdtman, 1952, p. 460: "(Potonié, 1934 . . .): band-like, locally thickened parts of the sexine, extending in sweeping curves from aperture to aperture. Examples: . . . [Fig. 402–405]. Cf. also . . . [Fig. 407]."

d. Thomson and Pflug, 1953, p. 21: *Tr.* Primary garland-like tube-shaped partings between ektexine and endexine that connect neighboring vestibula (in *Alnus*) . . . [Fig. 408–409]. [*see* DISSENCE, GERMINAL APPARATUS].

e. Traverse, 1955, p. 95: "Strips of endexinous thickenings of varying length, which are not under rims of furrows or pores."

sdb. Selling, 1947, p. 76.

AREA BASALIS, n., pl. -areae. (L. basalis, area, a space). cf. basal area, figura, shape classes.

AREA CONTAGIONIS, n., pl. -areae. cf. contact figure.

Erdtman, 1943, p. 49–50: "contact area, 'Kontakthof,' 'Pyramidenflaeche;' following the simultaneous divisions of the nucleus of the spore mother cell, the spores are in contact with each other along the contact areas. Each spore has three such areas. The contact areas meet at the proximal pole of the spore. They are laterally separated by the tetrad scar and, as a rule, distally limited by more or less curved lines or fringes (curvaturae)."

sdb. Takhtadzhyan and Yatsenko-Khmelevskiy, 1945, p. 38.

AREA POLARIS, n. cf. polar area.

AREOLA, n., pl. -ae. (L. diminutive of area).

Jackson, 1928: "a space marked out on a surface; a small cell or cavity; a tessellation in the thallus of some Lichens; a lumen in the sporangium of *Achlya* due to the influx of water (Harper)."

AREOLAR, adj. cf. areolate, areolatus.

Jackson, 1928: "marked with areolae, divided into distinct spaces."

AREOLATA, n., pl.

Kupriyanova, 1948, p. 70: *Tr.* If the tubercles have a form of areas (areolata) surrounded by narrow ditches (fossula) a negative reticulate structure is created (areolata, reticulum fossulare). *Dracaena aurea* . . . *Lysichiton kamtschatkense* . . . and *Chamaedorea concolor* . . . as well as some others can serve as examples of the negative reticulate structure. [*see* ORNAMENTATION and Fig. 662].

AREOLATE, adj. (L. areolatus). cf. areolar, areolatus.

a. Jackson, 1928: "marked with areolae, divided into distinct spaces."

b. Erdtman, 1947, p. 106: "areolate (term proposed by Faegri and Iversen (Iversen per lit.)): with small area (areolae) separated by small grooves (fossulae) forming a negative reticulum."

c. Erdtman, 1952, p. 460: "(Erdtman, 1947; term suggested by Iversen): with small sexinous, usually circular or polygonal areas separated by grooves ± forming a 'negative reticulum'." *see* Figs. 775–777, 783–787.

sdb. Kupriyanova, 1948, p. 70; [*see* AREOLATA].
Harris, 1955, p. 19; [*see* ORNAMENTATION].
Faegri and Iversen, 1950, p. 160.
Beug, 1961, p. X.

AREOLATUS, adj. cf. areolar, areolate.

ARISTA, n., pl. -ae. (L. the awn or beard of grain).
cf. cingulum, collar, crassitude, crassitudo, equatorial
collar, flange, frassa, ring, zona, zone.
I. Jackson, 1928: "an awn, the beard of corn."
II. Potonié and Kremp, 1955, p. 15: *Tr.* If a massive ridge is present, which often is wedge-like
in cross section and which distinctly enlarges the
equatorial diameter and circles the entire equator, then we are dealing with a *cingulum;* also
called a ring or arista in a more specific sense.
sdb. Potonié and Kremp, 1956, p. 86; [*see* ZONA].

ARTHROSPORE, n., pl. -s. (Gr. artron, a joint).
a. Jackson, 1928: "one of spores like a chain of
beads, formed by fission."
b. Melchior and Werdermann, 1954, p. 14. [*see*
SPORE].

ASCOSPORE, n., pl. -s. (Gr. askos, a bag).
a. Jackson, 1928: "a spore produced by an ascus,
sometimes termed sporidium, or sporule."
b. Melchior and Werdermann, 1954, p. 16. [*see*
SPORE].

ASCUS, n., pl. -i.
Jackson, 1928: "a large cell, usually the swollen
end of a hyphal branch, in the ascocarp of which
normally eight spores are developed."

ASPIDATE, adj., pl. -us. (L. aspidatus, derivation
from Gr. aspis, shield). cf. aperture, aspidoporate,
aspidorate, aspidote, pollen classes, porate, protrudence.
a. Wodehouse, 1935, p. 541: "bearing aspides."
b. Pike, 1956, p. 50: "Grains in which the apertures (or at least their outermost parts) are
borne on small ± circular areas, protruding as
rounded domes from the general surface of the
grains."
sdb. Erdtman, 1943, p. 50; Takhtadzhyan and Yatsenko-Khmelevskiy, 1945, p. 38; Erdtman, 1947,
p. 113; [*see* ASPIDOPORATE]; Selling, 1947, p.
76; Faegri and Iversen, 1950, p. 160; Pokrovskaya, 1950; Erdtman, 1952, p. 460; [*see* ASPI-
DOTE].

ASPIDES, n., pl. of aspis.

ASPIDOPORATE, adj. cf. aperture, aspidate, aspidorate, aspidote, porate, protrudent.
Erdtman, 1947, p. 112: ". . . with aspidate pores;
the term aspidate (Wodehouse, 1935) is replaced
here by the terms 'aspidorate' and 'aspidoporate.'
The first term designs ora, the second pori with a
more or less well defined shield-shaped area, or
'aspis,' surrounding the ektexinous part of every
germinal aperture (the longest diameter of the ektexinous openings are thus always shorter than that
of the aspides). The application of the terms
postulates that a clear distinction between ektexinous and endexinous layers can be made. In aspides

of the *Betula*-type [Fig. 406, 407] there is no contact between ektexine and endexine; in aspides of
the *Carya*-type [Fig. 400–401] there is apparently
no endexine at all." [*see* GROUPS OF SPOROMOR-
PHAE].

ASPIDORATE, adj. cf. aspidoporate.
Erdtman, 1947, p. 112: "with aspidate ora." [*see*
ASPIDORATE].

ASPIDOTE, adj. cf. aperture, aspidate, aspidoporate,
aspidorate, porate, protrudent.
Erdtman, 1952, p. 460: "(Wodehouse, 1932: ASPI-
DATE) in aspidote grains the apertures (or at least
their outermost parts) are borne on small, ± circular, shield-shaped areas (aspides), protruding as
rounded domes from the general surface of the
grains; . . [Figs. 365–367, 723C]."

ASPIS, n., pl. aspides. (Gr. aspis, shield, a viper,
adder; L. aspidisca, diminutive, small shield; Fr.
aspic, a viper's tongue). cf. annulus, anulus, costa
pori, dissence, endannulus, endanulus, formation of
lips, halo, labrum, lip, margo, oculus, operculum,
operculum pori, pore ring, protrudence.
a. Wodehouse, 1935, p. 541: "a shield-shaped, sub-
exineous thickening surrounding a germ pore."
b. Erdtman, 1943, p. 50: "a shield-shaped area
surrounding a germ pore. Aspidate germ pores
protrude as rounded domes. The protrusions are
due to a thickening of the intine underlying the
region of the pore and sometimes also to a lesser
annular thickening of the exine."
sdb. Takhtadzhyan and Yatsenko-Khmelevskiy, 1945,
p. 38; Selling, 1947, p. 76; Faegri and Iversen,
1950, p. 160; Pokrovskaya, et al., 1950; Erdtman, 1952, p. 460; [*see* ASPIDOTE].

ASULCATE, adj. cf. alete, colpate, furrowless, furrow without a sulcus, inaperturate, non-aperturate.
Kupriyanova, 1948: ". . . such·varied types of single furrowed pollen include pollen grains, furrow-
less or asulcate . . ." [*see* GROUPS OF SPORO-
MORPHAE].

ATREME, adj. cf. aperture, Groups of Sporomorphae.
Erdtman and Straka, 1961, p. 65: ". . . spores:
without aperture(s)."

ATRIUM, n., pl. -ia. (L. a forecourt, hall). cf. pore.
Thomson and Pflug, 1953, p. 33: *Tr.* The difference
between the vestibulum and the atrium which is so
important among triangular pollen is defined as
follows:
They both have in common the phenomena that
the endexine is no longer connected with the ektexine at the pore region and encloses its own
porus, the endoporus. Between the endoporus (of
the endexine) and the exoporus (of the ektexine)
is an intervening space.
We speak of an "atrium" if the endexine is

largely dissolved in the pore region so that an endopore forms that is more than three times as large as the exopore.

If the endexine continues independently after splitting and encloses an endoporus which is not more than three times as large as the exoporus, the structure is called a "vestibulum" . . . [Fig. 379–380].

Also the endexine can be interlamellarly split and then enclose a "postvestibulum" or also form an interlamellary "postatrium" . . . [Fig. 383–385]. [see APERTURE, Thomson and Pflug, 1953].

AURICULA, n., pl. -ae. (L. ear-lap). cf. valva, ear.
I. Jackson, 1928: "a small lobe or ear, an append-age to the leaf, as in Sage, or the Orange; the lobule or minor lobe of the leaf of Hepaticae, often ballon-shaped; formerly and erroneously used for AMPHIGASTRIA; a small lobe or special patch of cells at the basal angle of the leaf in Mosses."
II. Potonié and Kremp, 1955, p. 15: *Tr*. A special development occurs in some cases in the equa-torial region of spores (Zonales) . . . The swelling may be more pronounced at the "cor-ners" of the equatorial section of the spore than on the rest of the equator. Then we speak of *valvae* . . . [Fig. 149]. If the valvae continue to swell into cushions or similar structures which stand out from the outlines, we speak of *little ears, ears,* or *auriculae* . . . [Fig. 151]. [see EQUATORIAL REGION].
   sdb. Potonié and Kremp, 1956, p. 86; [see ZONA].

AURICULATE, adj. (L. auriculatus). cf. auricled, eared.

a. Jackson, 1928: "eared, auricled."
b. Norem, 1958, p. 668: "With earlike appendages . . . [Fig. 159–3]."

AUTOSPORE, n., pl. -s. (Gr. autos, self).
a. Smith, 1950, p. 53: "Aplanospores that have the same distinctive shape as the parent cell are called autospores."
b. Melchior and Werdermann, 1954, p. 16. [see SPORE].

AXIS, n., pl. axes. (L. an axle). cf. axis equatorial, axis poli, dimension, equatorial axis, figura, Groups of Sporomorphae, main axis, polar axis, pole axis, shape classes, symmetry axis.
a. Jackson, 1928: "an imaginary line, around which the organs are developed."
b. Traverse, 1955, p. 91: "The line passing from the center of the distal to the center of the proximal surface."

AXIS AEQUATORIALIS, n. (L.) cf. equatorial axis, symmetry axis.
   Pokrovskaya, et al., 1950: *Tr*. . . . equatorial axis, perpendicular to the polar axis.

AXIS POLI, n. (L.) cf. axis, main axis, polar axis, pole axis.

AZYGOSPORE, n., pl. -s. (Gr. a, not; sygos, a yoke; spora, a seed).
   Jackson, 1928: "the growth of a gamete direct with-out conjugation, a parthenogenetic spore = parth-enospore."

B

BACULARIUM, n., pl. -ia. (L. baculum, a staff). cf. columella, rodlet-layer.

Erdtman, 1952, p. 460: "provisional term used to denote what seems to be groups of ± fused bacula . . . [Fig. 800]."

BACULATE, adj. (L. baculatus). cf. baculatus, duplibaculate, infra-baculate, intra-baculate, microbaculate, multibaculate, oligobaculate, papillate, paxillate, ramibaculate, scabrate, simplibaculate, subpsilate.

Iversen and Troels-Smith, 1950, p. 35: *Tr.* (with) bacula = rodlets: the largest diameter is less than the height; sculpture elements are neither pointed nor wedge-shaped. [*see* ORNAMENTATION, for illustration see Fig. 626].

sdb. Iversen and Troels-Smith, 1950, p. 46; [*see* ORNAMENTATION].

Cranwell, 1953, p. 16.

Thomson and Pflug, 1953, p. 22; [*see* ORNAMENTATION].

Harris, 1955, p. 19; [*see* ORNAMENTATION].

Krutzsch, 1959, p. 39; [*see* ORNAMENTATION].

Beug, 1961, p. X.

Couper and Grebe, 1961, p. 7 and Fig. 680; [*see* ORNAMENTATION].

BACULUM, n., pl. -a. cf. elevation, excrescence, microbaculum, ornamentation, process, projection, protrusion, protuberance, rod, rodlet.

a. Potonié, 1934, p. 9. [*see* SPORODERMIS under "structure, rodlets, etc." *see also* ORNAMENTATION (Potonié, 1934, p. 11); for illustration see Fig. 651].

b. Erdtman, 1952, p. 460: "endosexinous rods supporting any ectosexinous elements . . . cf. . . . [Fig. 628–629]. Also isolated sexinous rods (diameter ± the same at top and base)."

"Potonié uses the term 'Stäbchen' in two senses: -1. as a designation of structural elements 'in Form von vielen dicht neben einander stehenden, radial gerichteten Stäbchen' (syn. columellae, Faegri and Iversen, 1950); -2. as a designation of rod-like sculptural elements (bacula; this term is used in the same special sense by Faegri and Iversen, 1950)."

c. Potonié and Kremp, 1955, p. 14: *Tr.* The rodlets or bacula . . . are cylindrical. The narrowing is only very slight, if present; however, the base can sometimes be a little thicker. In addition the bacula are straight and often end abruptly. [*see* ORNAMENTATION and Fig. 745].

sdb. Iversen and Troels-Smith, 1950, p. 35; [*see* ORNAMENTATION].

Couper and Grebe, 1961, p. 6 and Fig. 680; [*see* ORNAMENTATION].

Beug, 1961, p. 10.

BASAL AREA, n., pl. -s. (L. basis, foundation).

Dijkstra-van V. Trip, 1946, p. 18: *Tr.* Basal area can be defined as the part of the spore surface which is not occupied by the contact figure.

BASIDIOSPORE, n., pl. -s. (L. basidium, a little pedestal; Gr. spora, seed).

a. Jackson, 1928: "a spore produced by a basidium."

b. Melchior and Werdermann, 1950, p. 15. [*see* SPORE].

BASIDIUM, n., pl. -a.

Jackson, 1928: "the spore mother-cells of Hymenomycetous and Gasteromycetous Fungi, having little points from which spores are thrown off; employed by Thaxter for the swollen attachment of the conidium to the conidiophore in *Basidiobolus* Eidam; by older authors employed for the central fertile cells of Uredineae."

BASTIONED, adj. (It. bastione, bulwark).

Potonié, 1934, p. 11: *Tr.* The outline of the grain appears in this case [reticulum] bastionate . . . [Fig. 650], whereby (with not too large a lumen) the bastion prongs appear to be connected by "skins."

sdb. Potonié and Kremp, 1955, p. 15; [*see* ORNAMENTATION, RETICULUM, and Fig. 802].

BILATERAL, adj. (L. bi-, bis-, twice; latus, side). cf. shape classes.

I. Jackson, 1928: "arranged on opposite sides, as the leaves of the yew."

II. a. Erdtman, 1952, p. 460: "with two vertical planes of symmetry; equatorial axes not equilong." [see GROUPS OF SPOROMORPHAE and Fig. 95–98].

b. Pokrovskaya, et al., 1950: *Tr.* Possessing only two mutually perpendicular planes of symmetry (e.g., in *Anona*).

c. Traverse, 1955, p. 95: "Having bilateral, as opposed to radial, symmetry. The bean-shaped endospores of the polypod ferns are an example."

sdb. Erdtman, 1947, p. 113.

Kosanke, 1950, p. 14; [see SPORE and Fig. 110].

Pike, 1956, p. 50.

BISACCATE, adj. cf. disaccate. [see Fig. 206].

a. Jackson, 1928: "having two pouches."

b. Potonié and Kremp, 1954, p. 174: *Tr.* Used for pollen grains with two air sacs belonging to the subturma *Disaccites* Cookson, 1947.

BISULCATE, adj. (L. bisculatus, with two sulci). cf. colpate, furrow.

a. Jackson, 1928: "two-grooved."

b. Erdtman, 1947, p. 105: ". . . with two sulci." [see ANOSULCATE].

c. Kupriyanova, et al., 1948: *Tr.* Disulcate pollen grains *(Pollina bisulcata)* have two symmetrically arranged furrows. [see GROUPS OF SPOROMORPHAE].

BLADDER, n., pl. -s. cf. air sac, air sack, cavea, sac, saccus, sack, vesicula aerifera, wing.

I. Jackson, 1928: "Grew's term for a cell; a hollow membranous appendage on the roots of *Utricularia,* which entraps water insects; similar growths in the frond of some Algae."

II. a. Wodehouse, 1935, p. 541: [see WING].

b. Traverse, 1955, p. 93: "Of a vesiculate pollen grain, the presumably ektexinous protuberances that stand apart from the body."

sdb. Erdtman, 1957, p. 3–4; [see SACCATE].

BODY, n., pl. -ies. cf. central cell, corpus, saccate.

Traverse, 1955, p. 93: "Of a vesiculate pollen grain, such as that of *Pinus,* the main part of the grain, as distinct from the bladders." [see DIMENSION for measurement of grain].

BRANDSPORE, n., pl. -s.

a. Jackson, 1928, "= Uredospore."

b. Melchior and Werdermann, 1954, p. 14. [see SPORE].

BREADTH, n. cf. dimension, figura. [see SACCATE for measurement of disaccate grains].

BREVICOLPATE, adj. (L. brevis, short). cf. colpate, furrow.

Erdtman, 1952, p. 460: "with ± short colpi (length of colpi equal to or shorter than total distance from colpi apices to poles)." [see Fig. 429–430].

sdb. Pike, 1956, p. 50.

BREVISSIMICOLPATE, adj. (L. brevissimi, very short + colpate). cf. aperture, brevissimiporate, colpate, furrow, porate.

Erdtman, 1952, p. 460: "with very short colpi. Colporate grains are brevissimicolpate if the colpi are as long as or shorter than the underlying circular or lalongate os; aspidote grains are brevissimicolpate if the colpi are as long as the aspis diameter or shorter . . . [Fig. 433–435] shows a 'brevissimicolpoidate' grain."

sdb. Pike, 1956, p. 50.

BREVISSIMIRUPATE, adj. cf. aspidote, brevissimicolpate, colpate, furrow, porate, pore.

Erdtman, 1952, p. 461: "see BREVISSIMICOLPATE, RUPI, and . . . [Fig. 365–367]."

BROCHAL, adj. (Gr. broxos, a loop). cf. heterobrochate, homobrochate, mesobrochate, oligobrochate, polybrochate.

Erdtman, 1952, p. 461: "confined to or characteristic of brochi. A brochal aperture is situated in the lumen of a brochus . . . [Fig. 757]."

BROCHUS, n., pl. -i. (Gr. broxos, a noose). cf. lacuna, lumen, mesh.

Erdtman, 1952, p. 461: "the meshes of a reticulum. A brochus consists of a lumen and the adjoining half of the muri which separate that particular lumen from other lumina."

sdb. Beug, 1961, p. X.

BULLA APICALIS, n., pl. bullae. (L. bulla, apex; apicis, summit). cf. contact figure, dehiscence cone, gula, nozzle, top vesical.

Dijkstra-van V. Trip, 1946, p. 19: *Tr.* = top vesical.

C

CANALICULATE, CANALICULATUS, adj. (L. canalis, channel). cf. cicatricose, striate.
   I. Jackson, 1928: "channelled, with a longitudinal groove."
   II. Potonié, 1934, p. 11: *Tr.* If the stripes or lists are *not* arranged like a net, the exine is cicatricose; however, this is only the case if the grooves between the individual muri are just as broad as, or broader than the muri. In the opposite case, the exine is *canaliculate.* [*see* ORNAMENTATION].
   sdb. Thomson and Pflug, 1953, p. 23; [*see* ORNAMENTATION].
      Potonié and Kremp, 1955, p. 15; [*see* ORNAMENTATION].
   III. Faegri and Iversen, 1950, p. 160: ". . . rugulate + striate . . ."

CANALUS PORI, n. cf. pore canal.

CAP, n., pl. -s. (L. cappa). cf. cappa, cappula, discus, disk.
   I. Jackson, 1928: "Grew's term for the husk of a nut; the pileus of Hymenomycetous Fungi; the calyptra of mosses; the short, upper division of the dividing cell in *Oedogonium.*"
   II. a. Wodehouse, 1935, p. 541 and Fig. 206: "Cap or disk, the thickened dorsal surface of the winged grains of the Abietineae and Podocarpineae."
     b. Traverse, 1955, p. 93: "The proximal part of the body of a pine pollen grain, which has thicker exine than the distal part. The author uses the term *cap-ridge* for the thickened edge of this cap."
   sdb. Erdtman, 1957, p. 3; [*see* SACCATE and Fig. 211].
      Pokrovskaja, et al., 1950, Fig. 204.
      Zaklinskaja, 1957; see Figs. 219–221, 228–230.

CAPILLUS, n., pl. -i. (L. a hair). cf. fimbria, hair.
   I. Jackson, 1928: "the width of a hair, taken as 1/12th of a line or about .17 mm."
   II. Potonié and Kremp, 1955, p. 14: *Tr.* The hairs, capilli or fimbriae . . . [Fig. 744] are cylindrical or almost ribbon-like, especially they are not straight but more or less winding or coiled, knobbed, forked, branched or anastomozed . . . [Fig. 157–159] and perhaps somewhat laterally expanded. Even the simple, more or less cylindrical, flexuous hairs are probably almost never completely cylindrical. Their base may be thickened and the narrowing in the main part of the hair, if it occurs, is very small. Sometimes they are rather abruptly pointed. [*see* ORNAMENTATION].
   sdb. Couper and Grebe, 1961, p. 7, and Fig. 683.

CAPPA, n., pl. -ae (L.). cf. cap, discus, disk.

CAPPULA, n., pl. -ae.
   Erdtman, 1957, p. 3: "With respect to the thickness of the exine of the corpus certain [saccate] pollen types . . . exhibit two distinct areas: a proximal, crassi-exinous (referred to as cap, cappa), and a distal, tenuiexinous (referred to as cappula)." [*see* SACCATE].

CAP RIDGE, n., pl. -s. cf. cappa, comb, crest, cristae, crista marginalis, crista proximalis, discus, marginal ridge, proximal crest.
   Traverse, 1955, p. 93: "Thickened edge of the cap." [*see* CAP].

CAPUT, n., pl. capita (L. the head).
   I. Jackson, 1928: "the peridium of some Fungi."
   II. Erdtman, 1952, p. 466: "a pila . . . sculptural elements consisting of a ± swollen apex (caput) and a rod-like neck (collum)." [*see* HEAD, PILUM].

CARINIMURATE, adj. (L. carina, keel; murus, wall). cf. reticulate.
   Erdtman, 1952, p. 461: "with keeled muri . . . [*see* Fig. 757 and 793–794]."

CARPOSPORE, n., pl. -s. (Gr. karpos, fruit).
   Jackson, 1928; "a, a spore; b, a spherical uninuclear spore formed in a sporocarp, arising from the swollen tips of branched filaments resulting from the fertilization of the carpogonium; c, used by Clements for a plant possessing chaffy pappus."
   sdb. Melchior and Werdermann, 1954, p. 14–15; [*see* SPORE].

CATAPORATE, adj. (Gr. kata, down). cf. aperture, Groups of Sporomorphae, hilate, porate, pore.
Erdtman, 1958, p. 137.

CATATREME, adj. cf. aperture.
Erdtman and Straka, 1961, p. 66: "catatreme spores: 1-treme spores with the centre of the aperture at the proximal pole."

CAVATE, adj. (L. cavus, a hollow). cf. pouched, rugulo-saccate, saccate, subsaccate, vesiculate, winged.
a. Faegri and Iversen, 1950, p. 160: "ektexine loosened from endexine, columellae sticking to the under surface of the tectum, e.g., the vesiculi of *Pinus* (Potonié, 1934), the crests of *Trapa*, the verrucae of *Tsuga*."
b. Harris, 1955, p. 25: "The term implies a separation of the outermost layer to form a cavity or cavities and was proposed in respect to the morphology of pollen grains. The present author has extended its use in preference to inventing a new term for a similar condition in spores, though in the latter it may refer to the perispore, a layer not present in pollen grains. This use could be distinguished by employing the prefix perino- if necessary, but the distinction is not made in this manual since only spores are discussed." [*see* PERISPORIUM].

CAVEA, n., pl. -ae. (L. a cavity).
I. Iversen and Troels-Smith, 1950, p. 35: *Tr*. The cavity within the exine which originates by the separation of ektexine from endexine. [*see* STRUCTURE, CAVERNA, DISSENCE].
II. Thomson and Pflug, 1953, p. 21: *Tr*. Cavea or saccus (air sack): Parting in the ektexine with balloon-like expansion of the split exterior lamellae. No important elongation of the rodlet elements occurs. (Among the *Pityosporites*.) [*see* DISSENCE, AIR SAC, AIR SACK, BLADDER, SACCUS, VESICULA].

CAVERNA, n., pl. -ae. (L. a cave). cf. cavea, dissence.
Thomson and Pflug, 1953, p. 21 [*see* Fig. 372–375]: *Tr*. . . . a meridional separation (sometimes between ektexine and endexine, sometimes in the endexine) occurring among nearly all the Tricolpates, Tricolporates, Tetracolporates, and others . . . [Thomson and Pflug, 1953, p. 39–40, *Tr*.] A further differentiation [in the Tricolporates] is shown by the meridional separation of the layers. These are probably homologous to the vestibules, pre- and post-vestibules of the Brevaxones. They should be called "cavernae." The cavernae appear in equatorial view as dark stripes that lie symmetrically behind the colpi. Their width is measured at the equator. They can be wider than the pores and then "circumscribe" them (*Nyssa*-type among others) or can also be smaller, and thus narrower, than the pores (*Fagus*). Often several cavernae lie behind one another: the separation of the ektexine and endexine layers is designated as "caverna" in the strict sense; between the indexine lamellae as a "post caverna." The latter seems to be very common. . . .

The caverna often has the form of two narrow cones which are open toward the equator and whose apices lie at the poles. The cone formation results when the inner lamellae do not part until at the colpus; they then form the outline of a circle in plan view [Fig. 375].

CAVIUM, n., pl. -ia.
Thomson and Pflug, 1953, p. 40: *Tr*. The points of the three cavernae are connected at the tips. Sometimes they open into a large subpolar chamber, the "cavium." [*see* APERTURE].

CELL CONTENTS, n.
a. Jackson, 1928: "of two kinds, living or protoplasmic, and non-living, such as starch, fats, proteids, crystals, cell-sap, and the substances dissolved in it."
b. Cranwell, 1953, p. 16: "occasionally of importance in study of fresh material (e.g., starch grains, of *Agathis*, *Thismia*: hyaline plugs of Juncaceae, etc.); usually destroyed in fossil grains, and in material prepared for their comparative study. Occasionally, the interior may be infested by fungal hyphae (as in *Agathis australis*)."

CENTRAL CELL, n., pl. -s. (L. centrum, the middle; cella, storeroom). cf. body, corpus.
I. Jackson, 1928: "of the archegonium, that in the center from which the oosphere, and ventral canal-cell arise."
II. Harris, 1941, p. 50: "the corpus of a saccate pollen grain." [*see* DIMENSION].

CENTRIFUGAL, adj. (L. fugo, I flee). cf. interior, symmetry.
I. Jackson, 1928: "tending outwards or developing from the centre outwards."
II. Thomson and Pflug, 1953, p. 19: *Tr*. Interior, before.

CENTRIPETAL, adj. (L. petro, I seek). cf. exterior, symmetry.
I. Jackson, 1928: "developing towards the centre from without."
II. Thomson and Pflug, 1953, p. 19: *Tr*. Exterior, behind.

CENTRUM COLPI, n. cf. furrow.
Iversen and Troels-Smith, 1950, p. 32: *Tr*. Center of the colpus median. [*see* COLPUS].

CENTRUM INTERCOLPII, n.
Iversen and Troels-Smith, 1950, p. 34: *Tr*. Point of intersection of the medians of an intercolpium. [*see* INTERCOLPIUM].

CENTRUM INTERPORII, n.
Iversen and Troels-Smith, 1950, p. 34: *Tr.* Point of intersection of the medians of an interporium. [*see* INTERPORIUM].

CENTRUM PORI, n.
Iversen and Troels-Smith, 1950, p. 33: *Tr.* Center of the pore. [*see* PORE].

CHAGRENATE, adj. (F. chagrin, shagreen). cf. laevigate, translucent.
Thomson and Pflug, 1953, p. 22: *Tr.* Homogeneous exine appears not to be composed of particles . . . translucent, individual structure particles are not recognizable. [*see* ORNAMENTATION].
sdb. Krutzsch, 1959, p. 39; [*see* ORNAMENTATION].

CHAGRENATE-CORRUGATE, adj.
For ornamentation of disaccate pollen grains, [*see* SACCATE, Zaklinskaya, 1957, Fig. 223b].

CHLAMYDOSPORE, n., pl. -s. (Gr. chlamys, a cloak).
a. Jackson, 1928: "a spore having a very thick membrane."
b. Melchior and Werdermann, 1954, p. 14; [*see* SPORE].

CICATRICOSE, adj. (L. cicatricosus, scarred). cf. canaliculate, striate.
I. Jackson, 1928: "scarred or scarry."
II. Potonié, 1934, p. 11: *Tr.* If the stripes or lists are not arranged like a net the exine is *cicatricose;* however, this is only the case if the grooves between the individual muri are just as broad as or broader than the muri. In the opposite case, the exine is *canaliculate.* [*see* CANALICULATE].
sdb. Thomson and Pflug, 1953, p. 23; [*see* ORNAMENTATION].
Potonié and Kremp, 1955, p. 15; [*see* ORNAMENTATION].
III. Faegri and Iversen, 1950, p. 160: "rugulate + striate p.p."

CILIUM, n., pl. -a. (L. an eyelash).
Jackson, 1928: "Vibratile whip-like processes of protoplasm by which zoospores and similar bodies move; the hair-like processes in the endostome in Mosses; the marginal hairs of *Luzula.*"

CINGULUM, n., pl. -a. (L. a girdle). cf. arista, collar, crassitude, crassitudo, equatorial collar, flange, frassa, ring, zona, zone.
I. Jackson, 1928: "the neck of a plant, that which is between stem and root, the collum; the connecting zone, girdle, or hoop of Diatom frustules; the girdle in Peridineae which separates the epivalve from the hypovalve (West)."
II. a. Potonié and Kremp, 1955, p. 15: *Tr.* If a massive ridge is present in the equatorial region of spores (Zonales), which often is wedge-like in cross-section, and which distinctly enlarges the equatorial diameter and circles the entire equator, then we have a *cingulum,* also a ring or arista in a more specific sense . . . [Fig. 152–153]. Occasionally the cingulum does not wedge more or less sharply toward the exterior as in *Lycospora* and *Densosporites,* but is in cross-section more or less circular with a narrow neck; according to the definition this is true in *Rotaspora* . . . [Fig. 154]. [*see* EQUATORIAL REGION].
b. Potonié and Kremp, 1956, p. 86: (Crassitudo): *Tr.* The *cingulum* (= arista) is a smaller ± massive equatorial ring or flange with ± wedge-shaped cross-section.
c. Couper, 1958, p. 102: "a flange-like extension of the exine around the equatorial region of the spore."
d. Couper and Grebe, 1961, p. 5: "a surrounding structural feature more or less confined to the equator and in which the exine is relatively thicker than in any other part of the spore . . . [Fig. 184]. Spores with such a feature are described as *cingulate.*"

CIRCULAR, adj. cf. global, orbicular, orbiculate, orbiculatus, shape, shape classes, Fig. 168.

CIRCUMPOLAR LACUNA, n., pl. -ae. (L. circum, round, surrounding). cf. lacuna.
Wodehouse, 1928, p. 933: "The lacunae, generally six in each hemisphere, surrounding the polar lacuna in lophate pollen grains. Example: *Veronia jucunda.*"
sdb. Wodehouse, 1935, p. 541.
Potonié, 1934, p. 11; [*see* ORNAMENTATION].

CIRCUMSCRIPTIO, n. (L. scribo, to write). cf. amb, ambit, ambitus, circumscription, contour, equatorial limb, extrema lineamenta, figura, Groups of Sporomorphae, limb, limbus, morphography, optical section, outline, shape classes.
Pokrovskaya, et al., 1950: *Tr.* Outline. At the polar position the microspores may be round shaped (rotundatus), 2. round-lobate (rotundo-lobatus) . . . etc. . . . .

CIRCUMSCRIPTION, n., pl. -s.
Jackson, 1928: "the outline of any organ; the definition of a form or group of forms, as of species, genera, orders."

CLAVA, n., pl. -ae. (L. a club). cf. club, microclavum, pilum.
a. Iversen and Troels-Smith, 1950, p. 35: *Tr.* clubs: like bacula, but club-shaped. [*see* ORNAMENTATION and Fig. 627].
b. Ingwersen, 1954, p. 40: "club-shaped sculptural elements which in at least one dimension are

>1μ and whose greatest diameter is less than the height."

sdb. Beug, 1961, p. X.

CLAVATE, adj. (L. clavatus, club-shaped). cf. microclavate, pilate.

a. Jackson, 1928: "club-shaped, thickened towards the apex."

b. Faegri and Iversen, 1950, p. 27: ". . . radial projection of sculpturing elements ± isodiametric . . . at least one dimension 1μ or more . . . not pointed . . . height of element greater than greatest diameter of projection . . . upper end of element thicker than base . . . (*Ilex*)." [*see* ORNAMENTATION].

sdb. Kupriyanova, 1948, p. 70; [*see* ORNAMENTATION and Fig. 661].

Iversen and Troels-Smith, 1950, p. 46; [*see* ORNAMENTATION].

Cranwell, 1953, p. 16.

Thomson and Pflug, 1953; [*see* ORNAMENTATION and Fig. 645].

Harris, 1955, p. 19; [*see* ORNAMENTATION].

Traverse, 1955, p. 93.

CLINOCOLPATE, adj. cf. aperture, colpate, furrow, Groups of Sporomorphae.

Erdtman and Straka, 1961, p. 66: "longest axis of the colpi coincident with the middle line of the zone." [*see* GROUPS OF SPOROMORPHAE under Zonatreme].

CLUB, n., pl. -s.

I. Jackson, 1928: "a pluricellular hair, one of the elements of the pulp of the orange or lemon fruit (Crozier)."

II. [*see* CLAVA and PILUM].

COAPERTURATE, adj.

Beug, 1961, p. X. *Tr.* Tetrads in which the apertures of neighboring individual pollen grains join. Therefore, not synonym with the term "syncolpate."

COLLAR, n., pl. -s.

I. Jackson, 1928: "the 'neck' of a plant, the imaginary boundary between the above- and underground portion of the axis; the annulus in Agarics; an encircling outgrowth at the base of the ovule in *Ginkgo* (Potter)."

II. [*see* ZONA].

COLLUM, n., pl. -a. (L. neck).

I. Jackson, 1928: "the collar or neck of a plant, [*see* COLLAR]; the lengthened orifice of the ostiole of Lichens."

II. Erdtman, 1952, p. 466: "Pila: sculptural elements consisting of a swollen apex (caput) and a rod-like neck (collum)." [*see* PILUM].

COLPATE, COLPATUS, adj. (Gr. kolpos, a fold). cf. acolpate, acolpatus, ambi aperturate, anacolpate, angul-aperturate, anosulcate, aperturate, aperturidate, aperturoid, asulcate, bisulcate, brevicolpate, brevis-simicolpate, brevissimirupate, clinocolpate, colpato-colporate, colpatus, colpodiporate, colpoid, colpoidorate, colporate, colporatus, colporoidate, constricticolpate, cryptocolpate, demicolporate, dicolpate, dicolporate, diplodemicolpate, distalo-aperturate, distalo-proximalo-aperturate, disculate, fossaperturate, geminicolpate, heterocolpate, longicolpate, loxocolpate, monocolpate, monocolpatus, monosulcate, orate, oroid, orthocolpate, pancolpate, parasyncolpate, pericolpate, plan-aperturate, plicate, porate, pore, pollina trichotoma fissurata, polyplicate, polyrugate, polysulcate, pontoperculate, proximalo-aperturate, rupate, sinuaperturate, spiraperturate, spirate sulcate, spiroid, stephanaperturate, stephanocolpate, stephanocolporate, subzonosulcate, sulcate, sulcatus, syncolpate, syndemicolpate, tasicolpate, tenuate, tetrachotomosulcate, tetracolporate, tetrad mark, transcolpate, trichotome-fissurate, trichotomocolpate, trichotomocolpatus, trichotomosulcate, trichotomous, tricolpate, trimesocolpate, tricolpatus, tricolporate, trimesocolporate, trischistoclasic, zonaperturate, zonicolpate, zonisulculate, zonosulcate.

I. Wodehouse, 1935, p. 541: "possessing germinal furrows or harmomegathi, generally used with numerical prefixes as mono-, di-, and tri-, signifying the number of furrows."

sdb. Erdtman, 1943, p. 50; Takhtadzhyan and Yatsenko-Khmelevskiy, 1945, p. 39; Pokrovskaya, 1950.

II. a. Selling, 1947, p. 9: "(= with 1 or more colpae, i.e., meridional furrows), either colpate in a restricted sense (each colpa without a pore or a transverse furrow) . . . or colporate (each colpa with 1 or 2 pores or transverse furrows)."

Selling, 1947, p. 76: "(Pollen) with a colpa or colpae devoid of pores or transverse furrows. Prefixes mono-, di-, tri-, tetra-, pantahexa-, octa-, and poly-, denote number of colpae."

b. Faegri and Iversen, 1950, p. 160: "(Wodehouse, 1935) originally comprising *all* pollen types with furrows, after Erdtman (1945 . . .) only when there is no pore in the furrow." [*see* COLPORATE].

c. Erdtman, 1952, p. 461: "(Wodehouse, 1928 . . .) in a wider sense than in the present book: with (colpus or) colpi."

d. Cranwell, 1953, p. 16: "Possessing furrows and/or pores . . . Now kept for grains with meridional furrows only . . ., or for all furrowed grains without reference to orientation . . . Prefixes: mono-, di-, tri-, etc. There seems to be no more satisfactory term than colpate for grains with one aperture, whether the furrow is distal (or rarely, proximal) and sharply defined, or of a transitional furrow-pore type."

COLPATO-COLPORATE, adj. cf. colpate, colporate, demicolporate, furrow, Groups of Sporomorphae, heterocolpate, porate.

Erdtman, 1947, p. 113: "with plain colpae—as in colpate grains—and oriferous colpae alternately." [see GROUPS OF SPOROMORPHAE].

COLPODIPORATE, adj. cf. colpate, colporate, colpus, furrow, Groups of Sporomorphae, porate.

Selling, 1947, p. 76: "[pollen] with colpae, each enclosing two pores or two transverse furrows. Prefix used here: tri, denoting three colpae."

sdb. Faegri and Iversen, 1950, p. 160.

COLPOID, n., pl. -s. cf. colpate, colporate, colpus, furrow, Groups of Sporomorphae, porate.

Erdtman, 1952, p. 461: "apertures ± similar to colpi."

COLPOIDORATE, adj. cf. colpate, colporate, colpus, furrow, geniculus, Groups of Sporomorphae.

Erdtman, 1952, p. 461: "with oriferous colpoids . . . [Fig. 340]."

COLPORATE, COLPORATUS, adj. cf. colpatacolporate, colpate, colpatus, colpodiporate, colpoid, colpoidorate, colporatus, colporoidate, colpus, demicolporate, dicolporate, diorate, diplodemicolpate, diporate, disulcate, diverse porate, extraporate, furrow, Groups of Sporomorphae, heterocolpate, mesocolporate, monocolporate, orate, oroid, oroideferous, pentacolporate, porate, pororate, stephanocolporate, tetracolporate, tri-colpodiporate, tricolporate.

a. Erdtman, 1945, p. 190: "possessing ektexinous germinal furrows (colpae), each provided with a central, endexinous pore or pore-like aperture or structure. Used with numerical prefixes as mono-, di-, tri-, etc., and poly-, signifying the number of furrows."

sdb. Erdtman, 1947, p. 113; [see GROUPS OF SPOROMORPHAE].
Selling, 1947, p. 76.

b. Faegri and Iversen, 1950, p. 22: "When both pores and furrows occur together (*colporate* pollen, Erdtman, 1945) the pores are always located in the furrows, one pore per furrow.* But a number of pollen types is known in which the number of pores is regularly smaller than (most frequently ½ or ¼ of) that of furrows (*heterocolpate*). In *Platycarya* all pores are located outside the furrows (*extraporate*). The homology of these furrows with the regular colpi may be questioned."

* The pollen grains of *Myoporum* have furrows provided with 2 pores or transversal furrows each—Cranwell, 1942; Selling, 1947. Somewhat similar but more irregular features appear in *Ribes grossularia,* where short furrows, provided with one pore each, fuse in pairs. In *Anthyllis vulneraria* each pore seems to be connected with 3 furrows, a smaller lateral one on each side of the main furrow. However, we do not consider the lateral fissures as real furrows.

c. Erdtman, 1952, p. 461: "(Erdtman, 1945 . . . [Svensk Bot. Tidskrift, v. 3, no. 2, p. 187–191])

with oriferous colpi . . . [Figs. 337–339, 345–348]."

sdb. Pike, 1956, p. 50.

COLPOROIDATE, adj. cf. colpate, colporate, colpus, furrow, Groups of Sporomorphae.

Erdtman, 1952, p. 461: "with orodiferous colpi . . . [Fig. 306–311]."

COLPUS, n., pl. -i. [also Colpa, colpae, but the present trend does not follow this usage.] (Gr. kolpos, a fold). cf. dehiscence furrow, fold, furrow, germinal furrow, ruga, sulcus.

I. a. Erdtman, 1943, p. 50: "germinal furrow; a longitudinal groove or opening in the exine of a pollen grain either enclosing a germ pore or serving directly as the place of emission of the pollen tube."

b. Erdtman, 1945, p. 190–191: "As is well known, the furrow of a monocolpate pollen grain is not strictly homologous to the furrows of a tricolpate pollen. In monocolpate grains the furrow is borne on the distal part of the grain whilst in tricolpate grains the furrows are developed as meridional furrows crossing the equator of the grain at right angles. If it be considered that . . . [colpi] should not denote two not strictly homologous features, grains of the typical monocotyledonous monocolpate type may be termed 'monosulcate,' or 'trichotomosulcate,' provided the 'sulcus' presents a three-slit opening.

"Terms such as 'polycolpate' ('nonacolpate,' etc.) may indicate pollen grains of two types, either grains with many equatorial (meridional) furrows, or grains with many furrows equally distributed over the general surface of the exine. Misinterpretations are avoided if the term . . . [colpi] is used to indicate furrows arranged more or less meridionally, and confined to the equatorial, more or less extended zone of the grain. Although originally applied by Potonié (1934) in another sense, the term 'ruga' may be used to denote furrows equally distributed over the surface of the grains. Of the three terms suggested 'ruga' is shortest, . . . ["colpus"] intermediate, and 'sulcus' longest. As to the features, which these terms are supposed to denote, the rugae in a polyrugate grain, as a rule, are shorter than the . . . [colpi] in, e.g., a grain of the common tricolpate type. The . . . [colpi] in their turn, are as a rule comparatively shorter than the sulcus in a typical monocotyledonous monosulcate pollen grain."

c. Erdtman, 1946, p. 71: "Typical 'dicotyledonous' types of pollen grains are provided with three meridional furrows, meeting, if extended, at the poles and crossing the equator at right angles. Furrows of this kind are termed . . . [colpi, Fig. 29]. By shortening and contraction a . . .

[colpus] may ultimately develop into a porus. In tetra- and hexacolpate grains, etc. the . . . [colpi] do not always cross the equator at right angles, but converge in pairs (cf. Wodehouse, 1935)." [see APERTURE].

d. Erdtman, 1947, p. 105: "For the purpose of orientation it has been thought necessary to distinguish between apertures distributed equally over the entire spore surface and apertures confined to a particular spot or zone of the spore. Thus germinal furrows are termed . . . "colpi" if they are located in the equatorial zone of a spore and extend more or less longitudinally."

e. Selling, 1947, p. 76: "a meridional groove or (pre-formed) opening in the exine, with or without one or two pores (or transverse furrows) in each."

f. Faegri and Iversen, 1950, p. 20: "Most pollen grains possess openings in, or thin parts of the exine, through which the pollen tube emerges. Two different types of apertures can be recognized, which are generally called *pores* and *furrows* (*colpi*, Wodehouse, 1935)." [see FURROW].

Faegri and Iversen, 1950, p. 160: "Although the term is rather meaningless in fossil grains, in which the furrows do not bulge out (like the living grains studied by Wodehouse), we have kept it because of the very useful derivations." For illustration see Fig. 349–360.

g. Iversen and Troels-Smith, 1950, p. 31 and Fig. 349, *et seq.*: *Tr.* C = colpus: Area which forms, or contains, the normal exitus place of the pollen tube and whose arbitrary length-width ratio is greater than two. The colpus is designated in three different ways in relation to the surrounding exine:

C (ex = O): by the absence of the exine (for instance by casting off an operculum).

C (mb): by the thinning of the exine.

Iversen and Troels-Smith, 1950, p. 32: *Tr.*

C (mb, ekt = O): ektexine elements are absent.

C (mb, ekt): ektexine elements are present.

C (op): by the demarcation of a part of the normal exine by a furrow or suture.

C, 1 = limes colpi = colpus edge: demarcation line of the colpus; therefore, either the above mentioned furrow or suture or, in the case of a thinned or absent exine, the exterior limit of the lighter area to which it gives rise.

C, ap = apex colpi = tip of the colpus.

C, mb = membrana colpi = colpus membrane: the thinned exine of a colpus.

C, op = operculum colpi = the thicker part of a colpus which is situated within a furrow or suture. The structure of the operculum is of

similar construction as the remaining exine of the pollen grain.

C, med = medianum colpi = colpus median: a line which splits a colpus into two nearly symmetrical halves. One differentiates:

C, med, + = *medianum colpi longitudinalis*: the longitudinal median of the colpus.

C, med, ÷ = *medianum colpi transversalis*: the transverse median of the colpus. [Iversen and Troels-Smith, 1950, p. 33: *Tr.*]

C, cent = centrum colpi = center of the colpus median.

mg = margo: area which surrounds the colpus like a belt and which is differentiated from the remaining exine of the pollen grain by deviations within the ektexine.

mg, 1 = limes marginis = edge of the margo: exterior demarcation line of the margo.

mg, ap = apex marginis = tip of the margo.

cost C = costae colpi: ridges of thickened endexine which surround the colpi.

tr C = colpus transversalis = transversal furrow: a furrow in the endexine which crosses a colpus at an almost right angle.

cost tr = costae transversales: ridges of thickened endexine which surround the transversal furrow.

cost aequ = costae aequatoriales: two parallel, ring-shaped ridges of the thickened endexine which surround the equator.

pseudo C = pseudocolpus: pseudocolpus differs from the true colpi in that it is not the normal place of exitus for the pollen tube.

h. Pokrovskaya, et al., 1950: *Tr.* Furrow (embryonic). Dent in exine, mostly longitudinal, elongate, covered as a rule with a thinner layer of exine (colpus membrane). The furrow has two purposes, it is the place of the growth of the pollen tube and it carries a function of the harmomegathus. The form of furrows may be long (longus) such as in pollen grains of *Magnolia*, short (brevis) as in *Fraxinus*, wide (latus) as in *Platanus*, elliptic (ellipticus) as in *Drimys*.

i. Erdtman, 1952, p. 461: . . . equatorial, usually longitudinal apertures (length: breadth > 2; in transcolpate grains, very rarely met with, the length-breadth ratio is < 0.5."

j. Cranwell, 1953, p. 16: "cf. furrow (germinal)."

k. Thomson and Pflug, 1953, p. 20, 31–41: [see SOLUTION]. cf. germinal apparatus.

l. Ingwersen, 1954, p. 40: "germinating furrow; the area on the grain forming, or surrounding, normal place of emergence of the pollen tube, with a length-breadth index higher than 2."

m. Potonié and Kremp, 1955, p. 12: [see TENUITAS].

n. Pike, 1956, p. 50: "Equatorial, usually meridionally elongated apertures (length : breadth < 2)."

sdb. Takhtadzhyan and Yatsenko-Khmelevskiy, 1945, p. 38; Beug, 1961, p. X.

II. Steeves and Barghoorn, 1959, p. 227: ". . . [Fig. 301–305] In the polar view the [*Ephedra* pollen] grains are polygonal with the alternating ridges and deep concave furrows forming an angular outline. A narrow and serpentine colpus is situated at the base of each furrow.* In most cases the colpi are highly undulate. The undulation of the colpus may be either rounded or highly angular and the frequency of undulation may vary, as well as the degree. The colpus may divide forming lateral branches which extend up the ridge where they may occasionally divide again. In this manner the ektexine forms a reticulate pattern, such as that found in *Ephedra distachya.* In a few cases, as in *E. clokeyi,* the colpus scarcely divides. Also, the width of the colpus may vary as well as the depth to which it cuts into the ektexine." [*see* POLYPLICATE, Erdtman, 1950; Fig. 298–300; and LONGITUDINAL FOLD, Beug, 1961, p. XI].

* Colpus, . . . germinating furrow, the area on the grain forming or surrounding the normal place of emergence of the pollen tube, with a length-breadth ratio higher than 2. The ektexine is reduced, even absent (Faegri and Iversen, 1950; Ingwersen, 1954). Although the terms furrow and colpus may be used interchangeably, in this case a distinction between the two will be made. The term 'furrow' will be used to refer to the region between the ridges; the term 'colpus' will be used to describe the thin, longitudinal, serpentine grooves located in the middle of the furrow area and formed by the absence of or the thinning of the ektexine.

COLPUS EDGE, pl. -s. Demarcation line of the colpus. cf. colpate, colpi, colpus, furrow, furrow rim, limes, margo.

COLPUS MEDIAN, n. Medianum colpi. cf. colpate, colpus, furrow.

COLPUS MEMBRANE, n., pl. -s. Membrana colpi. cf. colpate, colpus, furrow.
a. Iversen and Troels-Smith, 1950, p. 32: *Tr.* Colpus membrane: the thinned exine of a colpus.
b. Pokrovskaya, et al., 1950: *Tr.* Membrane of the furrow. The elastic, more definite part of the exine covering the furrow.

COLPUS TRANSVERSALIS, n., pl. -i, -les. (L. transvertere, to turn or direct across). cf. aperture, colpate, furrow, lalongate os, transversal furrow, transverse furrow.
a. Wodehouse, 1935, p. 544: "Transverse furrow, a short, elliptical or elongate opening in the intine underlying the true furrow and with its long axis crossing that of the latter at right angles."
b. Erdtman, 1943, p. 50: "Transverse furrow; an elliptical or elongated opening in the intexine, underlying the true furrow and with its long

axis crossing that of the latter at right angles. There may be any number of transitions between transverse furrows, and germ pores underlying ektexinous furrows."
c. Selling, 1947, p. 77: "Transverse furrow: an endexinous groove crossing a pore enclosed by a . . . [colpus] or ruga, generally at right angles to the main furrow; often transitional to ± transverse pores. A somewhat arbitrary term."
d. Faegri and Iversen, 1950, p. 22: "In colporate pollen grains the thin exine of the furrows is even thinner in the pore area, and it is characteristic that this thinning is most distinct in the endexine. These thinner parts of the endexine may assume the form of internal *transversal furrows* (*colpi transversales*) which may fuse together to one single internal furrow along the equator of the grain. The internal furrows may be accompanied by thickened edges (*costae transversales,* resp. *costae aequatoriales*)." [*see* Fig. 349–360 for illustration].
e. Pokrovskaya, et al., 1950: *Tr.* Elliptic elongated opening in the endexine, that is lying under the proper furrow and intersecting the furrow at a right angle, e.g., *Castanea dentata* according to Wodehouse.

sdb. Takhtadzhyan and Yatsenko-Khmelevskiy, 1945, p. 38; Iversen and Troels-Smith, 1950, p. 32, [*see* COLPUS]; Beug, 1961, p. X.

COLUMELLA, n., pl. -ae. (L. a small pillar). cf. bacularium, rodlet-layer.
a. Jackson, 1928: "a persistent central axis round which the carpels of some fruits are arranged as in *Geranium;* the axis of the capsule in Mosses; the receptacle bearing the sporangia of *Trichomanes,* and other Ferns;" "the central portion of the anther in Solanaceae (Halsted); a sterile axial body within the sporangium of Fungi; the central column in the pollen-chamber of the apex of the megasporangium of a cycad (Jeffrey)."
b. Faegri and Iversen, 1950, p. 17: "A contrast to the open structure of the above-mentioned exines is formed by those pollen grains in which the outer ends of the granules fuse, forming another, outer sheet which envelops the whole grain covering all the other exine elements. If the granules assume the form of radially placed prismatic elements that are fused along their whole length, the ektexine is compact, but in most cases a very careful examination (preferably of sections) discloses that the granules are fused at their tips, or at their tips and bases, thus forming an outer membrane which is separated from the inner one by a cavity and is borne by small columns (columellae), which may in some cases fuse to form an inner reticulum (*Alisma*).

In some pollen types the outer part of columellae is branched, forming intricate patterns. *Stellaria* [Fig. 693–697]."

sdb. Iversen and Troels-Smith, 1950, p. 35; [*see* STRUCTURE].
  Erdtman, 1952, p. 460; [*see* BACULUM].
  Cranwell, 1953, p. 16.
  Ingwersen, 1954, p. 40.
  Potonié and Kremp, 1955, p. 19; [*see* SPORODERMIS, MESOSPORIUM].
  Kupriyanova, 1956, Fig. 3; [*see* SPORODERMIS and Fig. 540].
  Krutzsch, 1959, p. 40; [*see* ORNAMENTATION].
  Beug, 1961, p. X.

COLUMELLA CONJUNCTA, n., pl. -ae. (L. conjunctirus, joined). cf. bacularium, ramibaculate.
  Iversen and Troels-Smith, 1950, p. 35: *Tr.* Distally united groups of columellae. Also the granula of intectate pollen grains can be called columellae if they occur in combined structures (for instance in a reticulum). [*see* STRUCTURE].

COLUMELLA DIGITATA, n., pl. -ae. (L. digitatus, fingered). cf. bacularium.
  Iversen and Troels-Smith, 1950, p. 35: *Tr.* Distally branched columella. [*see* STRUCTURE].

COLUMELLA SIMPLEX, n., pl. -ae, -ices. (L. simplex, of one piece or series, opposed to compound). cf. bacularium.
  Iversen & Troels-Smith, 1950, p. 35: *Tr.* Simple columella. [*see* STRUCTURE].

COLUMELLATE, adj. cf. tectate.
  Traverse, 1955, p. 95: "Having structure of rods, either in the individual units, for example, in the reticulum of an intectate type, or with rods beneath the tectum of a tectate form. The latter is the common use of the term."

COMB, n., pl. -s. cf. cap-ridge, crest, crista, cristata, fence reticula, marginal ridge. [*see* SACCATE].

COMB-LIKE, adj. cf. cristate, cristo-reticulate, retipilariate.

COMMISSURE, n., pl. -s. (L. commissura, a joint or seam). cf. contact figure, raised commissure.
  I. Jackson, 1928: "the face by which two carpels adhere, as in Umbelliferae."
  II. a. Harris, 1955, p. 25: "the line of dehiscence in the tetrad scar. Compare laesura, which, however, includes the rim (margo) if present."
  b. Couper, 1958, p. 102: "the line of dehiscence in the tetrad scar." [*see* RAISED COMMISSURE].
  sdb. Kosanke, 1950, p. 14; for illustration *see* Fig. 109–110.

CON, prefix. (L. with).
  Jackson, 1928: "modified by euphony frequently into co and com—meaning 'with' in Latin compounds."

CONCAVE, adj. (L. concavus, hollow).
  Couper and Grebe, 1961: [*see* SHAPE and Fig. 176, 180].

CONCLAVE, n. (L. a room that may be locked up).

CONE, cf. conus.

CONE SHAPED, adj. (L. conus, fruit of the pine or fir-tree). cf. conus, echinate, ornamentation.

CONICAL, adj. cf. shape classes.
  I. Jackson, 1928: "having the figure of a cone, as the carrot."
  II. [*see* SHAPE CLASSES]; for outline of disaccate pollen grains [*see* SACCATE, Zaklinskaya, 1957, and Fig. 236–241].

CONIDIOSPORE, n., pl. -s. (Gr. konis, dust). cf. conidium.

CONIDIUM, n., pl. -a. cf. conidiospore.
  a. Jackson, 1928: "= gonidia."
  b. Melchior and Werdermann, 1954, p. 14; [*see* SPORE].
  c. Merriam-Webster, 1960: "a small asexual spore occurring in certain Fungi."

CONSTRICTICOLPATE, adj. (L. constrictus, compact). cf. colpate, furrow.
  Erdtman, 1952, p. 461: "with colpi ± constricted at the equator . . . [*see* Fig. 427–428]."

CONTACT AREA, n., pl. -s. (L. contactus, a touching). cf. contact figure.
  a. Potonié, 1934, p. 15: *Tr.* The contact points are enlarged to considerable contact areas [Fig. 108]. These contact areas are limited proximally by the rays of the Y-mark and distally by the arc-lines or "arc-edges." [*see* APERTURE, area contagionis].
  b. Dijkstra-van V. Trip, 1946, p. 20: *Tr.* Contact area is that part of contact figure which is surrounded by two rays . . . and an arcuate rim . . .
  c. Potonié and Kremp, 1955, p. 10: *Tr.* The position of the spores in the mother-cell is varied. In the case of four spheres which are put together in a tetrahedron, each touches the other three at three points. These three contact points form an isosceles triangle (which lies in the proximal polar hemisphere). The four completely developed spores touch one another, each with three contact areas or contact surfaces (q.v.).
  Potonié and Kremp, 1955, p. 12–13: *Tr.* Considerable contact surfaces sometimes result from the fact that the spores in the mother-cell were arranged as a tetrahedron, rhomb or rectangular parallelepiped. Each of the three contact areas (area contagionis) of the tetrahedral-tetrad spores is proximally limited by two tecta of the dehiscence mark and distally, sometimes, by one of the three curvaturae, arc lines or arc ridges . . . [Fig. 112]. The curvaturae form a bow

open toward the proximal pole. It is to be observed that the curvaturae extend in an equatorial direction with a point in the middle of the arc, therefore it is not exactly bow-shaped.

sdb. Kosanke, 1950, p. 14; [see SPORE and Fig. 109–110].

CONTACT FACE, n., pl. -s. cf. contact figure.

Harris, 1955, p. 25: "the area adjacent to the tetrad scar. In tetrahedral spores the contact area is delimited by the length of the laesurae and equals the proximal surface only when these extend to the equator. In monolete spores the contact area is clearly distinguishable only when much flattened so that the spore is V-shaped proximally."

sdb. Couper, 1958, p. 102.

CONTACT FIGURE, n., pl. -s. cf. alete, aletus, aperture, apertureoid, apex, apex field, arc line, arcuate ridge, arcuate rim, area contagionis, bulla apicalis, commissure, contact area, contact face, contact marking, contact point, contravertex, crest-line, crista arcuata, crista radialis, crista triradiata, curvatura, curvatura imperfecta, dehiscence, dehiscence cone, dehiscence fissure, dehiscence furrow, dehiscence list, dehiscence ridge, discus, extremitas radii, figura contagionis, figura triradiata, fissura dehiscentis, fissura germinativa triradiata, floats, floating body, gula, hilate, hilum, intratectum, kyrtome, laesura, lip, margo arcuata, massa, monolete, monolete mark, monoletoid, monoletus, nozzle, pole field, proximal pole, radial crest, radial extremity, radial ridge, radius, raised commissure, ray, scar, spore, subtectum, sutura, suture, swimming-apparatus, tectum, tetrad scar, torus, top vesicle, trifolium, trilete, trilete aperture, trilete mark, trilete marking, triletoid, triletus, triradiate ridge, vertex, Y-mark, Y-radius, Y-splitmark.

Dijkstra-van V. Trip, 1946, p. 20: Tr. Contact figure is the whole part of the spore surface which is enclosed by the arcuate rims . . . including these rims.

CONTACT MARKING, n., pl. -s. cf. apex field, contact figure, pole field.

CONTACT POINT, n., pl. -s. cf. contact figure.

a. Potonié, 1934, p. 12: Tr. Each of the four spheres, which are joined in a tetrahedral, touch the three other ones at three points . . . [Fig. 2]. These three contact points form an isosceles triangle (that lies in the proximal polar hemisphere). [see APERTURE, under contact point, Potonié, 1934].

b. Potonié and Kremp, 1955, p. 10: [see CONTACT AREA].

CONTOUR, n. (Fr. L. con + tornare, to turn). cf. amb, ambit, ambitus, circumscriptio, equatorial limb, extrema lineamenta, figura, Groups of Sporomorphae, limb, limbus, morphography, optical section, outline, shape classes.

a. Jackson, 1928: "cf. Double-Contour."

b. Thomson and Pflug, 1953, p. 19: Tr. We call the outline of the equatorial plane "contour;" for instance, it can be circular, trilobate, straight, concave, convex, etc. The outline of the sporomorphae in equatorial view we specifically call the "meridional-contour" and the equatorial outline of the pollen-lumen we call "inner contour" . . . [Fig. 162–167]. [see POLE CAP].

CONTRAVERTEX, n. (L. contra, opposite to; vertex, top). cf. contact figure.

Potonié and Kremp, 1955, p. 12: Tr. . . . the beveling of the distal exine . . . into the subtectum. [see Y-MARK and Fig. 141–148].

CONUS, n., pl. -i. (L. a cone). cf. echina, thorn, vestigial spine.

I. Jackson, 1928: "= cone, strobile."

II. Potonié and Kremp, 1955, p. 14 and Fig. 743: Tr. As seen from the side, the cones or coni . . . are pointed, blunt, or rounded cone-shaped. Their height does not exceed twice the diameter of their base; the latter can be extended. [see ORNAMENTATION].

sdb. Couper and Grebe, 1961, p. 7; [see ORNAMENTATION and Fig. 682].

CONVEX, adj. (L. convexus, faulted).

Couper and Grebe, 1961, p. 2: [see SHAPE and Fig. 178, 182].

CONVOLUTE, adj. (L. convolutus, rolled around, coiled).

Jackson, 1928: "when one part is wholly rolled up in another, as the petals of the wallflower; in a spathe when the margins mutually envelop each other."

CORONA, n., pl. -ae. cf. zona, zone.

I. Jackson, 1928: "a coronet, any body which intervenes between the corolla and stamens; the 'eye' of apples or pears, the remains of the calyx limb; the ray of the capitula in Compositae; a whorl of ligules or petals, united or free; a synonym of Cucullus; used by J. Hill for the pericycle, or 'circle of propagation;' the ring of primary wood in the medullary sheath; the medullary-crown or -sheath."

II. a. Potonié and Kremp, 1955, p. 15: Tr. If the equator [of a spore] is covered with a wreath of branched, anastomosed hairs, fimbriae or fringes, we are dealing with a corona (Reinschia, Superbisporites). [see EQUATORIAL REGION and Fig. 156–158].

b. Couper and Grebe, 1961, p. 6: "a surrounding or non-surrounding extension of the exoexine in the form of fimbriate (or hairlike) structural elements in the equatorial region. The outer ends of the structural elements are more or less separated . . . [Fig. 190]."

CORPUS, n., pl. -i. (L., a body). cf. body, central cell.
  a. Pokrovskaya, et al., 1950: *Tr.* Corpus, body. The central part of pollen grains having air sacs.
  b. Erdtman, 1957, pp. 3–4: "The saccate pollen grains in recent gymnosperms . . . consist of a body (corpus) and a varying number of air-sacs or bladders (sacci). [*see* SACCATE; for dimensions of vesiculate grains *see also* DIMENSION].

CORPUS SPORAE, n., pl. -i. cf. body of the spore, sporedisk.

CORRUGATE, adj. (L. corrugatus). cf. corrugatus, costate, ribbed, rugose, rugous, rugulate, rugulose, wrinkled.
  a. Jackson, 1928: "wrinkled."
  b. Erdtman, 1947, p. 106: "with low and irregular sculpturing resembling the creases in a stiff paper tightly wrapped around a globe."
  c. Faegri and Iversen, 1950, p. 160: "corrugatus (Erdtman, 1947) = rugulate p.p."
  d. Thomson and Pflug, 1953, p. 23: *Tr.* Rows of laterally fused warts with a varying crest height.
  sdb. Kupriyanova, 1948, p. 71; [*see* ORNAMENTATION and Fig. 669].

COSTA, n., pl. -ae. (L. a rib). cf. costa aequatorialis, costa colpi, costa pori, costa transversalis, rib, thickening.
  I. Jackson, 1928: "a rib, when single, a midrib or middle-nerve."
  II. a. Faegri and Iversen, 1950, p. 22: "There are also frequently thickenings of the endexine edging the furrows (*costae colpi*), but these are often difficult to observe directly, and they are most easily recognized by their effect on the furrow, the harmomegathis function of which is impeded (*Rumex*). Similar thickening of the endexine may also be found edging the pores (*costae pori*).
    "In colporate pollen grains the thin exine of the furrows is even thinner in the pore area, and it is characteristic that this thinning is most distinct in the endexine. These thinner parts of the endexine may assume the form of internal *transversal furrows* (*colpi transversales*) which may fuse together to one single internal furrow along the equator of the grain. The internal furrows may be accompanied by thickened edges (*costae transversales,* resp. *costae aequatoriales*)." [*see* Fig. 356–357, 360].
  b. Traverse, 1955, p. 95: "Endexinous thickenings under the rims of furrows or germpores. If associated with transverse furrows they are transverse costae, if with longitudinal furrows, they are longitudinal costae."

  c. Oldfield, 1959, p. 21 and Fig. 582–588: "In almost all the pollen grains, whether united into tetrads or not, the endexine thickens around the furrows to form more or less distinct costae . . . The shape in plan and section, width and nature of edges, strength, covering or obscuring by ektexinal ornamentation are noted. The outer edge of the costae are often defined by cracks in the endexine . . . It is obvious that there can be every gradation between a slight 'tuck' around the outer edge and a well marked crack and therefore the character is of limited use for identification purposes . . ."

COSTA AEQUATORIALIS, n., pl. -ae, -les.
  Iversen and Troels-Smith, 1950, p. 32: *Tr.* Two parallel, ring-shaped lists of the thickened endexine which surround the equator. [*see* COLPUS].
  sdb. Faegri and Iversen, 1950, p. 22; [*see* COSTA].

COSTA COLPI, n., pl. -ae. cf. altitude, colpate, colpus, furrow, swelling.
  Iversen and Troels-Smith, 1950, p. 32: *Tr.* Lists of thickened endexine which surround the colpi. [*see* COLPUS].
  sdb. Faegri and Iversen, 1950, p. 22; [*see* COSTA TRANSVERSALIS].

COSTA PORI, n. cf. annulus, anulus, aspis, dissence, endannulus, endanulus, formation of lips, halo, labrum, lip, margo, pore ring, protrudence, oculus, operculum, operculum pori.
  Iversen and Troels-Smith, 1950, p. 33: *Tr.* A thickening of the endexine surrounding the pore. [*see* PORE].
  sdb. Faegri and Iversen, 1950, p. 22; [*see* COSTA TRANSVERSALIS].

COSTATE, adj. (L. costatus, ribbed). cf. caniculate, cicatricose, ribbed, striate.
  I. Jackson, 1928: "ribbed, having one or more primary longitudinal veins."
  II. Harris, 1955, p. 20: "ribbed. Regular, well defined elevations or corrugations, more or less encircling the spore." [*see* ORNAMENTATION].
  sdb. Couper, 1958, p. 102; [*see* ORNAMENTATION].

COSTA TRANSVERSALIS, pl. -ae, -les. cf. colpate, furrow.
  Iversen and Troels-Smith, 1950, p. 32: *Tr.* Lists of thickened endexine which surround the transversale furrow. [*see* COLPUS].
  sdb. Faegri and Iversen, 1950, p. 22; [*see* COSTA].

COVER LID, n., pl., -s. cf. lid, pore lid.
  Kupriyanova, 1948, p. 5: [*see* GROUPS OF SPOROMORPHAE].

CRASS-EXINOUS, adj. (L. crassus, thick).
  a. Erdtman, 1952, p. 461: "with incrassate exine . . . [Fig. 589–590]."

b. Potonié and Kremp, 1955, p. 15: *Tr.* Exines, or exine parts, which are thick throughout are called crass-exinous. [*see* EQUATORIAL REGION].

CRASSIMARGINATE, adj. cf. aperture, pore, tenuimarginate.
Erdtman, 1952, p. 461: "with incrassate aperture margins . . . [Fig. 609–619]."

CRASSINEXINOUS, adj.
Erdtman, 1952, p. 461: "with incrassate nexine (nexine at least twice as thick as sexine) . . . [Fig. 593–596]."

CRASSISEXINOUS, adj.
Erdtman, 1952, p. 461: "with incrassate sexine (sexine at least twice as thick as nexine) . . . [Fig. 591–592]."

CRASSITEGILLATE, adj.
Erdtman, 1952, p. 461: "with incrassate tegillum. Thickness (depth) of tegillum twice as great or greater than length of the supporting bacula or depth of the space between tegillum and nexine . . . [Fig. 602–603]."

CRASSITUDE, CRASSITUDO, n., pl. -s. (L. thickness). cf. arista, cingulum, collar, crassitude, equatorial collar, flange, frassa, ring, zona, zone.
a. Potonié and Kremp, 1955, p. 15: *Tr.* A special development occurs in some cases in the equatorial region of spores (Zonales). In the simplest case, the exine may be a little thicker here than in the remaining area. In this case we are dealing with a crassitude. [*see* EQUATORIAL REGION].
b. Potonié and Kremp, 1956, p. 86: *Tr.* The *crassitudo* consists of a thickening of the exine, which starts more or less in the polar hemisphere and which gets constantly thicker toward the equator, often without equatorial wedging.
sdb. Couper and Grebe, 1961, p. 5; *see* Fig. 191–194.

CRESCENT, adj. Norem, 1958, [*see* SHAPE CLASSES and Fig. 167m].

CREST, n., pl. -s. cf. cap-ridge, comb, crista, fence, marginal ridge, reticula.
Jackson, 1928: "an elevation or ridge upon the summit of an organ; an outgrowth of the funiculus in seeds, a sort of axil."

CRESTLINE, n., pl. -s. cf. contact figure, list, ridge, vertex.

CRIBELLATE, CRIBELLATUS, adj. (L. cribellum, a little sieve). cf. aperture, cribellatus, foraminate, foraminoid, foraminose, forate, fossulate, foveolate, Groups of Sporomorphae, n-stephanoporate, oligoforate, panporate, perforate, periporate, pitted, polyforate, polyporate, porate, pore, scrobiculate, stephanaperturate, stephanoporate.
Wodehouse, 1928, p. 933: "Possessing a number of germinal apertures appearing as rounded and sharply delimited thin areas, equally spaced and generally more or less deeply sunken in the exine. Examples: *Salsola pestifer, Sarcobatus vermiculatus.*"
sdb. Wodehouse, 1935, p. 541; for illustration see Fig. 494.
Erdtman, 1943, p. 50.
Faegri and Iversen, 1950, p. 161.
Pokrovskaya, et al., 1950.

CRISTA, n., pl. -ae. (L. a tuft). cf. cap ridge, comb, crest, crista aequatorialis, crista marginalis, crista proximalis, equatorial crest, fence, marginal ridge, proximal crest, reticula cristata.
I. Jackson, 1928: "a crest of terminal tuft; used by Druce for the ligule of palm-leaves."
II. a. Potonié, 1934, p. 11: *Tr.* A frequent type of arrangement of the above mentioned sculpture elements is that of *cristae*. These are fences or combs that are built up from pila, rodlets, etc., which sometimes are even laterally connected and which can join to form networks (*reticula cristata*) . . . [Fig. 657]. [*see* ORNAMENTATION].
b. Erdtman, 1943, p. 50: "crests; different kinds of sculptural elements may join laterally to form rather intricate ridges or crests."
c. Selling, 1947, p. 76: "crest of reticulate sculpture."
d. Faegri and Iversen, 1950, p. 161: "cristae (Potonié, 1934) = muri p.p."
e. Pokrovskaya, et al., 1950: *Tr.* R. Wodehouse, who studies the pollen of living plants, and G. Erdtman describe the crests as "sculptural elements of different types which form ridges and crests if they are joined together."
f. Potonié and Kremp, 1955, p. 15: *Tr.* A special kind of arrangement of sculptural elements is that of cristae . . . [*Cristatisporites indignabundus* (Loose) in Potonié and Kremp, 1955, Pl. 16, Fig. 295; see also Fig. 742 in this publication]. These are fences or combs which may also join to form networks. They have been built by occasionally laterally fused cones, rodlets, etc. [*see* ORNAMENTATION].
III. Pokrovskaya, et al., 1950: *Tr.* The thickness of a disc which is visible in optical section at the side position of pollen grains with air sacs (Pinaceae and Podocarpaceae).

CRISTA AEQUATORIALIS, n. cf. equatorial crest.

CRISTA ARCUATA, n., pl., -ae. cf. arcuate ridge, arcuate rim, contact figure, curvature, margo arcuata.
Dijkstra-van V. Trip, 1946, p. 20: *Tr.* Arcuate rim, abstract conception, for instance the transition between the finely and coarsely punctuated parts . . . [*Triletes mamillarius* Bartlett]. In *Zonales* the three arcuate rims together form the rim of the spore body,

the so-called equatorial rim ... therefore, the term rim vesical [auricula, valva, ear] is used ... [e.g., *Triletes auritus* Zerndt].

CRISTA MARGINALIS, pl. -ae, -les. cf. cap ridge, crista proximalis, marginal ridge, proximal crest.
Erdtman, 1957, p. 4: "Near the proximal root of the sacci are often found slight, sexinous ridges or frill-like projections (proximal crests, cristae proximales, also referred to as cristae marginales) varying in appearance in different species. At the distal root of the sacci, where these merge into the distal aposaccium, the characteristic pattern of the bladders comes abruptly to an end." [*see* SACCATE].

CRISTA PROXIMALIS, n. cf. cap ridge, crista marginalis, marginal ridge, proximal crest.
Erdtman, 1957, p. 4: [*see* SACCATE].

CRISTA RADIALIS, n., pl. -ae. cf. contact figure, radial crest.

CRISTATE, CRISTATUS, adj. (L. crista, a crest). cf. comb-like, cristatus, cristo-reticulate, retipilariate, sepes a vallis.
a. Pokrovskaya, et al., 1950, p. 2: *Tr.* Cristatus (Cristata)—comb-like.
b. Harris, 1955, p. 20: "Ridges suture-like, as if formed by portions of the outer wall uniting in a seam or crest. Example: *Aspelenium flaccidum.*" [*see* ORNAMENTATION].

CRISTA TRIRADIATA, n., pl. -ae. cf. contact figure, triradiate crest.

CRISTO-RETICULATE, adj. cf. comb-like, cristate, cristatus, retipilariate, sepes a vallis.
Harris, 1955, p. 20: "Crests anastomosing to form a network. Example: *Asplenium richardi.*" [*see* ORNAMENTATION].

CROSS TETRAD, n., pl. -s. cf. aggregation, spore, tetrad.
Cranwell, 1953, p. 17: "a means of designating tetrad shapes." [*see* TETRAD and Fig. 11, 15].
sdb. Selling, 1947, p. 77.

CRUSTATE, adj. (L. crusta, crust).
Erdtman, 1952, p. 461: "aperture membranes or opercula reinforced by densely disposed excrescences are crustate. The excrescences in Dipsacaceae are usually ± spinuloid, those in Didiereaceae [Fig. 431–432] blunt, ± polygonal at the base."

CRYPTOCOLPATE, adj. (Gr. kryptos, hidden). cf. colpate, furrow.
Erdtman, 1952, p. 461–462: "with colpi ± easily seen in optical sections but not (or vaguely only) discernible on examination of the surface of non-transparent grains."

CRYPTOPOLAR, adj. cf. shape classes.
Erdtman, 1952, p. 11: "Cryptopolar spores have much the same appearance as apolar spores, but on closer examination they reveal a more or less distinct polarity (*Larix, Pseudotsuga, Equisetum,* and others)." [*see* POLARITY and SHAPE CLASSES].

CUNEUS, n., pl. -i. (L. a wedge). cf. incidence, platea, solution meridium, solution wedge.
Thomson and Pflug, 1953, p. 33: *Tr.* Remarkable is the stripe-like solution process of the endexine, which takes place in a meridional direction and which proceeds to a different extent in the different groups. We differentiate the following states: the "solution notch" (= incidence), the "solution wedge" (= cuneus), and the "solution channel" (= platea). The overall term is "solution meridium." [*see* APERTURE and Fig. 410–412].

CURVATURA, CURVATURE, n. (L. curvatura, a bending). cf. arc line, arcuate ridge, arcuate rim, contact figure, crista arcuata, margo arcuata, rim.
a. Jackson, 1928: "continued flexure or bending from a right line."
b. Potonié, 1934, p. 15: *Tr.* In many cases one can still find them in this attitude in coal. The contact points have enlarged to considerable *contact areas* ... [Fig. 108]. These *contact areas* are delimited proximally by the rays of the Y-mark; distally by the arc-lines or "arc edges." These *curvaturae* are often nicely arched, concave to the proximal pole (one should also compare *Sporites rugosus* Potonié, Ibrahim and Loose, 1932, Pl. 20, Fig. 59). [*see* APERTURE. cf. Potonié, 1934, p. 15, sub-heading Contact Area].
sdb. Erdtman, 1943, p. 50; [*see* AREA CONTAGIONIS].
Potonié and Kremp, 1955, p. 12; [*see* CONTACT AREA].

CURVATURA IMPERFECTA, n., pl. -ae. cf. contact figure, curvature.
a. Potonié and Kremp, 1955, p. 13: *Tr.* Tecta and curvaturae can be more or less reduced; often the curvaturae are completely lacking. Occasionally, only the rudiments which develop from the ends of the dehiscence ridges are present (Curvaturae imperfectae); these ends are pointed in the direction of the equator. Therefore, the dehiscence ridges look like fork handles with two long fork prongs. This can be true in trilete as well as in monolete types ... [*Lavigatosporites* cf. *maximus* (Loose)].
b. Couper and Grebe, 1961, p. 4: "a curvatura developed only at distal ends of the laesura or laesurae of a tetrad mark ... [Fig. 115, 120–121]."

CURVIMURATE, adj. cf. reticulate.
Erdtman, 1952, p. 462: "with ± curved muri."
sdb. Potonié and Kremp, 1955, p. 15.

CYST, n., pl. -s. (Gr. kystis, a cavity) cf. spore.
Melchior and Werdermann, 1954, p. 16. *Tr.* The protoplast contracts within the cell and forms at its periphery a new membrane which becomes especially thickened . . . [*see* SPORE].

CYSTIDIUM, n., pl. -ia. (Gr. kystis, a cavity).
Jackson, 1928: "large one-celled sometimes inflated bodies, projecting beyond the basidia and paraphyses of the hymenium of Agarics, of unknown function; = utricle."

CYSTOSPORE, n., pl. -s. (Gr.)
a. Jackson, 1928. "= CARPOSPORE (Strassburger)."
b. [*see* Melchior and Werdermann, 1954, p. 14 under SPORE].

D

DEHISCENCE, n., pl. -s. (L. dehiscere, to split open). cf. contact figure, laesura, Y-mark.
Jackson, 1928: "the mode of opening of a fruit capsule or anther by valves, slits, or pores."

DEHISCENCE CONE, n., pl. -s. cf. bulla apicalis, contact figure, gula, nozzle, top vesicle.

DEHISCENCE FISSURE, n., pl. -s. cf. contact figure, fissura dehiscentis, laesura.

DEHISCENCE FURROW, n., pl. -s. cf. contact figure, dehiscence fissure, dehiscence ridge, fissura dehiscentis, germinal furrow, laesura, monolete mark, Y-mark.
Potonié, 1934, p. 12: *Tr.* In contrast to pollen these spores usually show a triradiate mark, the Y-mark, which is formed by the dehiscence-lists and -furrows [Fig. 107–108] and which aids in germination. We find this arrangement in bryophytes too. Comparatively, the uni-radiate dehiscence ridge is a rare type among pteridophytes. [*see* APERTURE].
sdb. Potonié and Kremp, 1955, p. 11; [*see* Y-MARK].

DEHISCENCE LIST, n., pl. -s. cf. dehiscence furrow, dehiscence ridge, monolete mark, trilete mark, trilete marking, Y-mark, Y-radius.

DEHISCENCE RIDGE, n., pl. -s. cf. contact figure, dehiscence list.

DELTOID, adj. Norem, 1958, [*see* SHAPE CLASSES and Fig. 167i].

DEMICOLPORATE, adj. (L. dimidius, half). cf. colpate, colporate, colpus, furrow, Groups of Sporomorphae, porate.
Selling, 1947, p. 76: "[pollen] with colpae every second of which encloses one pore or transverse furrow. Prefix used here: hexa- denoting six colpae."
sdb. Faegri and Iversen, 1950, p. 161.

DEMICOLPUS, pl. -i. cf. aperture, colpate, furrow.
Erdtman, 1952, p. 462: "any apertures with the same shape and arrangement as the double set of apertures which would result from equatorial constriction and splitting of colpi [Fig. 473–474]."

DENSITY OF PATTERN.
Harris, 1955, p. 21: "The appearance of the sculpture pattern depends not only on the relative size of the constituent elements, but also on the distance between them. The practice is to describe the sculpture pattern as *dense* when the space separating the projections is less than one-twentieth the equatorial diameter of the spore and as *sparse* when in the main it is greater than one-tenth the equatorial diameter. Pits or cavities might be similarly treated, except when, together with the intervening walls, they form a compound pattern such as foveo-reticulate or lopho-reticulate. In other cases it is either stated or implied that the pattern is of medium density."

DEPTH, n., pl. -s. cf. dimension, figura. [*see* SACCATE, for depth of disaccate grains].

DESIGN, n., pl. -s. (L. designare, to work out, define). cf. ornamentation, pattern, sculpture, sculpturing, structure, texture.
Merriam-Webster, 1960: "The arrangement of parts, details, form, color, etc., especially so as to produce a complete and artistic unit; artistic invention: as, the design of a rug . . ."

DI, prefix. (Gr. dis, twice, two, or double).

DIAMETER, n. (Gr. diametros). cf. dimension, figura.
Iversen and Troels-Smith, 1950, p. 36: *Tr.* A dimension parallel to the surface of the exine; however, it is used only in connection with sculptures and structures. [*see* DIMENSION].

DIAPORUS, n. cf. aperture, endopore, exitus, germinal apparatus, germ pore, main pore, pore, pseudopore, ulcus.
Iversen and Troels-Smith, 1950, p. 33: *Tr.* Open pore: by absence of the exine (e.g., by casting off an operculum). [*see* PORE].

DICOLPATE, adj. cf. colpate, furrow.
a. Selling, 1947, p. 9: "with two colpae without pores or transverse furrows."
b. Faegri and Iversen, 1950, p. 128: [*see* GROUPS OF SPOROMORPHAE and Fig. 59].

c. Cranwell, 1953, p. 16: "encircled by a single furrow . . . dicolpate or zonate. Zonate seems preferable for this condition, as otherwise both proximal and distal furrows must be assumed."

d. Cranwell, 1953, p. 11: "Continuous or twinned furrows. Aberrant or imperfectly understood types include those with merged (fused) furrows, usually separable into two elements, *zonate,* if running around the equator, and *dicolpate* if running meridionally. The presence of two apertures has frequently been reported for the monocotyledons, as by Fritzsche, Zander, Erdtman and Wodehouse, but little detail has been available. The most penetrating discussion of the condition is that of Money, Bailey, and Swamy (1950), in relation to Ranalian families. They see in *Austrobaileya* and the Laurelieae, of the key family Monimiaceae, a possible transitional series between a primitive monocolpate and a derived dicolpate condition . . . Swamy [1949, cf. Fig. 266] had previously shown that the ends of the single furrow, which are often broadened (as in some monocotyledons), were both exceedingly long and broad in the Degeneriaceae, and that a constriction occurred in the middle of the furrow around the distal pole: further, the pollen tube was seen to emerge from one of the broadened ends, i.e., near the equator. Money, Bailey, and Swamy followed up with the study of the Monimiaceae already mentioned. *Laurelia* was shown to have a furrow even longer than in the Degeneriaceae, and equally constricted over the poles: *Doryphora* [*see* Fig. 267], moreover, with dicolpate grains, had no trace of furrow over the poles."

DICOLPORATE, adj. cf. colpate, colporate, colpus, furrow, Groups of Sporomorphae, porate.

 I. Selling, 1947, p. 13: "with 2 colpae, each with a pore or a transverse furrow. . . . Aberrant grains of numerous genera with 3 (or more) colpae, each having a pore . . .

 II. Cranwell, 1953, p. 16: "With two pores per furrow . . . [Fig. 283]."

DIMENSION, n., pl. -s. cf. altitudo, axis, breadth, depth, diameter, equatorial diameter, figura, height, index, length, longitudo, longitudo transversa, maximal measurement, mensura, morphography, polar area index, P : E ratio, polar area measurement, polar cap index, shape classes, shape class index, size, size classes, size range, width.

a. Iversen and Troels-Smith, 1950, p. 36 and Fig. 351–355, 358–364, 516, 622, 627: *Tr.* As far as the different dimensions are concerned, the following abbreviations are used:

 M = mensura = measurement [dimension]: measurement in general. **M** does not contain

any orientation in respect to the pollen grain. It is used with the following measurement qualifiers:

 1. pollen grain dimensions.
 2. polar area dimension.
 3. interporal and intercolpal dimensions.
 4. pori and colpi dimensions.
 5. dimensions of the exine (thickness of the exine).

Lg = longitudo = length: a dimension parallel to the polar axis in bipolar pollen grains. With this limitation, it is used as **M**.

Lt = longitudo transversa = transverse dimension: a measurement perpendicular to the polar axis of bipolar pollen grains. With this limitation, it is used as **M**.

D = diameter: a dimension parallel to the surface of the exine; however, it is used only in connection with sculptures and structures.

[Iversen and Troels-Smith, 1950, p. 37: *Tr.*]

H = altitudo = height: a dimension perpendicular to the surface of the exine; however, it is used only in connection with sculptures and structures.

For a better definition of the mentioned measurements, the following abbreviations are used:

+ and − indicate respectively the largest and smallest dimension where several possibilities exist.

(−) indicates that the considered measurement stands perpendicularly to the corresponding + measurement.

foc. 0–3: the dimension is given with regard to the observation of the surface of the pollen grain.*

 * Definition of the different focuses . . . [foc. 0-5 *See* LIMES EXTERIOR in this book].

foc. 5: the dimension is given with regard to the observation of the pollen grain in optical cross-section.

Pollen grain dimensions: *practical directions for measuring*—1. All measurements are applied to the outer bounding-surface (limes exterior) of the pollen grain. We call the outer bounding surface the ± rotation-ellipsoidal base which bears the sculpture elements. In those cases where the distal surfaces of the sculpture elements—seen in profile—constitute more than 50 percent of the described outermost boundary line, this line will be perceived to be the exterior border line of the pollen grain, and the corresponding ± rotation-ellipsoidical area is perceived to be the exterior bounding surface.

2. When damaged pollen grains are measured, this must be mentioned particularly, e.g., in the following way:

x = exine fissa: exine is burst.

( ) = exina crispa: exine is crumpled.

(x) = exina fissa et crispa: exine is burst and crumpled.

1. *All pollen types*—M, +: Distance between two parallel planes which are tangent to the pollen grain and which are so situated that the greatest possible distance between the planes is obtained.

M, (−): Distance between two parallel planes which are tangential to the pollen grain and which are perpendicular to both planes indicated by M, +.

In cases where two or more different measurements can be obtained:

M, (÷) + [designates] the largest dimension,

M, (÷) ÷ the smallest measurement.

In certain cases, e.g., when the pollen grain is imbedded in a permanent matrix, one is instructed to measure the dimensions in an arbitrary position. This can be noted in the following way:

M, + (fix) = M, + at arbitrary (fixed) position of the pollen grain.

M, − (fix) = M, − at arbitrary (fixed) position of the pollen grain.

[Iversen and Troels-Smith, 1950, p. 38: *Tr.*]

2. *Bipolar pollen grains*—Lg = longitudo = length : length of the polar axis from pole to pole. Lg designates a dimension parallel to the polar axis if used for colpi, pores, intercolpae and interporae, etc.

Lt = longitudo transversa = transverse dimension at the equator: distance between two parallel planes which are tangent to the pollen grain at the equator and parallel to a plane of symmetry through the polar axes—or a dimension perpendicular to such a dimension also taken in the equatorial plane.

In the case where two or more different measurements can be obtained:

Lt, + [designates] the largest dimension, and

Lt, ÷ the smallest dimension.

In colpi, pores, intercolpae and interporae, Lt designates a transverse measurement in the equatorial plane. In cases where the greatest length and width of bipolar pollen grains are not identical with the length of the polar axis or with the transverse dimension at the equator, respectively, it can be designated in the following way:

Lg, max = *maximum longitudinis* = maximal length: distance between two parallel planes which are perpendicular to the polar axis and tangent to the pollen grain.

Lt, max = *maximum longitudinis transversae* = maximum in the transverse dimension: distance between two parallel planes which are tangent to the pollen grain and are parallel to a plane of symmetry through the polar axes, or a dimension perpendicular to the mentioned dimension and to the polar axis.

In cases where two or more different measurements can be obtained:

Lt, max. + [designates] the largest dimension and

Lt, max. ÷ the smallest dimension.

Colpus dimensions—C-M, + = colpus-length: length of a longitudinal median of a colpus.

C-M, − = colpus transversal dimension: length of a transverse median of a colpus.

C-Lg = colpus length in bipolar pollen grains.

C-Lt = colpus transverse dimension in bipolar pollen grains.

mg-M = margo-width: distance from the edge of the colpus to the edge of the margo.

Iversen and Troels-Smith, 1950, p. 39: *Tr.*

(C, ap-C, ap)-M: distance between two neighboring colpi tips.

(mg, ap-mg, ap)-M: distance between two neighboring margo tips.

Pore dimensions [*see* Fig. 290]—P-M: diameter of the pore. In the case of pores which are not circular the largest and smallest diameter can be designated in the following way:

P-M, +: the greatest diameter.

P-M, −: the smallest diameter.

In bipolar pollen grains P-Lg: [designates] length of the pore . . . P-Lt: width of the pore . . . anl-M: = annulus width: distance from the edge of the pore and the edge of the annulus. (P, 1-P, 1)-M: distance between the edges of two neighboring pores. (anl, 1-anl, 1)-M: distance between the annuli edges of two neighboring pores.

Intercolpium interporium and polar area dimensions—inter C-M: length of the medians of an intercolpium.

inter C-M, +: length of the longest median.

inter C-M, −: length of the shortest median.

inter C-Lg: length of the median which stands perpendicular to the equator (bipolar pollen grains).

inter C-Lt: length of the median which coincides with the equator (bipolar pollen grains).

inter P-M: length of the medians of an interporium.

inter P-M, +: length of the longest median.

inter P-M, −: length of the shortest median.

inter P-Lg: length of the median which stands perpendicular to the equator (bipolar pollen grains).

inter P-Lt: length of the median which coincides with the equator (bipolar pollen grains).

polar-M = *mensura area polaris* = polar area dimension: the longest diagonal or the longest side of a polar area.

Iversen and Troels-Smith, 1950, p. 40: *Tr.*

Dimension of the exine*—The thickness of the exine, endexine, ektexine and tectum is measured in the optical cross-section (foc. 5). In cases where several values can be obtained in the same pollen grain M + designates the largest and M ÷ the smallest dimension.

* All dimensions of the exine refer to the exine outside of annuli, marginae and costae.

ex-M: thickness of the exine.

ex-M, +: maximum thickness of the exine.

ex-M, ÷: minimum thickness of the exine.

end-M: thickness of the endexine.

end-M, +: maximum thickness of the endexine.

end-M, ÷: minimum thickness of the endexine.

ekt-M: thickness of the ektexine.

ekt-M, +: maximum thickness of the ektexine.

ekt-M, ÷: minimum thickness of the ektexine.

tec-M: thickness of the tectum.

tec-M, +: maximum thickness of the tectum.

tec-M, ÷: minimum thickness of the tectum.

One can also give the exine measurements for certain parts of the pollen grain, e.g., at the pole: ex (pol)-M.

Sculpture* and columella dimensions—Example, *clava dimensions*:

* More correctly, sculpture element dimensions, shortened for simplicity.

cla-D, − (foc. 0–3).

cla-D, + (foc. 5).

cla-D, − (foc. 5).

cla-H, (foc. 5).

In the above example "clava" may be replaced by "columella."

Lumina dimensions—Lum-D, +.

Lum-D, −.

Pollen morphological size relationships—Absolute sizes:

The size of the pollen grains is given by the greatest dimension of the grains (M, +: Lg, max or Lt, max, +). According to Erdtman (1945),

Iversen and Troels-Smith, 1950, p. 41: *Tr.* one can differentiate the following size classes [*see* size classes, Erdtman, 1943, p. 48]:

p (<10μ) = pollina perminuta = very small pollen grains.

p (10–25μ) = pollina minuta = small pollen grains.

p (25–50μ) = pollina media = medium-large pollen grains.

p (50–100μ) = pollina magna = large pollen grains.

p (>100μ) = pollina permagna = very large pollen grains.

Exine, sculpture, and lumina dimensions are given according to the largest dimensions within each category (compare pollen morphological measurements . . . .).

One can differentiate the following size classes:

M, + (<1μ) = micro-, e.g., lum-M (<1μ) = micro-reticulate.

M, + (1–4μ) = meso-,

M, + (>4μ) = macro-.

Columella dimensions are given according to the diameter of the thickest columellae; measure the largest diameter of the unbranched part of a columella. The following size classes may be differentiated:

col-D, + (<0.5μ),

col-D, + (0.5–1μ),

col-D, + (1–4μ),

col-D, + (>4μ),

col (incertae) means that no lamellae could be clearly recognized.

b. For dimension of disaccate grains see Fig. 208, *et seq.* For dimensions of trilete spores, Fig. 195; of monolete spores and monocolpate grains, Fig. 248.

DIMORPHIC, DIMORPHOUS, adj. (Gr. di, two or double; morph, shape). cf. figura, Groups of Sporomorphae, morphography, shape classes.

a. Jackson, 1928: "occurring under two forms."

b. Erdtman, 1952, p. 462: "plants with dimorphic pollen have two kinds of sporomorphs. Examples . . . [Fig. 34–37]."

DIORATE, adj. cf. colpate, colporate, colpus, furrow, Groups of Sporomorphae.

Erdtman, 1952, p. 462: "a diorate aperture has two ora. Diorate colpi are shown in . . . [Fig. 284–285, 314]."

DIPLODEMICOLPATE, adj. (Gr. diploos, double). cf. colpate, furrow.

Erdtman, 1952, p. 462: "with two sets of demicolpi, one set in the proximal, the other in the distal face . . . [Fig. 473–474]. If confluent at the equator the two sets would produce a single set of colpi . . . [Fig. 465–466]."

DIPORATE, adj. cf. colpate, colporate, colpus, furrow, Groups of Sporomorphae, porate.

I. Kupriyanova, 1948, p. 10: *Tr.* Having pores which represent simple openings in the exine covered with an exceptionally thin membrane which disintegrates when treated with alkali and acid. [*see* GROUPS OF SPOROMORPHAE].

II. Faegri and Iversen, 1950, p. 128: "two pores restricted to the equatorial area." [*see* GROUPS OF SPOROMORPHAE].

sdb. Traverse, 1955, p. 95.

III. Cranwell, 1953, p. 16: "two pored. Aberrant types only in monocots (cf. Erdtman, 1944a, for cereal grains with more than one pore). Triporate grains unknown outside dicotyledons (but see Cyperaceae)."

DISACCATE, adj. cf. bisaccate. With two air sacs; [*see* Fig. 206].

DISC, DISK, n., pl. -s. cf. cap, cappa, discus.
a. Jackson, 1928: "development of the torus within the calyx or within the corolla and stamens; the central part of a capitulum in Compositae as opposed to the ray; the face of any organ, in contra-distinction to the margin; certain markings in cell walls of circular outline; bordered red pits; the valves of diatoms when circular; the base of a pollinium; the expanded base of the style in Umbelliferae; in a bulb, the solid base of the stem, around which the scales are arranged; disk is the more usual spelling in the case of Compositae . . ."
b. Wodehouse, 1935, p. 541: "disk, see cap." [*see* Fig. 206].

DISCUS, n., pl. -i. cf. cap, cappa, contact, disk, figure, torus.
I. Jackson, 1928: "a flat stroma through which the ostioles of fungi protrude, as in *Valsa*."
II. Pokrovskaya, et al., 1950: *Tr.* Discus, disk or shield. Thickened part of the exine on the proximal surface of pollen grains (*Pinaceae* and *Podocarpaceae*).
III. Thomson and Pflug, 1953, p. 26: *Tr.* At the corners of the equator, the torus either includes an acute angle (e.g., [*Concavisporites*] *Acutus* group) or it surrounds the periphery of a circle-sector and thus forms a "discus" (e.g., [*Concavisporites*] *Rugulatus* group, [*Concavisporites*] *Discites* groups, etc.). [*see* TORUS].

DISSECTION, n., pl. -s. cf. cingulum, zona.
Couper and Grebe, 1961, p. 5 and fig. 15: "rounded to elongated lumen in a cingulum or zona. The position and extension of the dissections should be noted . . . [Fig. 185–187]."

DISSENCE, n., pl. -es. cf. air sac, air sack, annulus, bladder, caverna, oculus, prevestibulum, saccus, tumescence, vesicula.
Thomson and Pflug, 1953, p. 21: *Tr.* Breaks between lamellae (dissences) also usually lie near the germinal apparatus; in a few cases they extend beyond the germinal apparatus but stay close to it in a certain symmetrical relationship. These dissences include a "dissence lumen." This is often empty; in other cases, it is filled with elongated rodlet elements of a lamella.
Partings in the ektexine—1. Anulus . . . : a ring-shaped swelling of the exo-lamella in the pore region. Here the rodlets of this lamella are very much elongated. The lumen anuli is closed toward the pore (in many Brevaxones).
2. Oculus: an analogous swelling of the exolamella b (in some Extratriporates).
3. Praevestibulum: interlamellar parting of the ektexine which is connected with the pore (in the case of some Extratriporates) . . . [for 1–3 *see* Fig. 543–544].
4. Tumescence: the ektexine thickens toward the pore region in a wedge-shaped cross-section (in some Triatriates) . . . [Fig. 376–388].
5. Cavea or saccus (air sac): parting in the ektexine with balloon-like expansion of the split exterior lamellae. No important elongation of the rodlet elements occurs. (Among the *Pityosporites*, *Pityosporites* sp; *P. cedroides* Thomson and Pflug (1953); *P. microalatus* R. Pot.).
6. Pecten (comb): parting in the ektexine by the elongation of the rodlet elements (5 and 6 are not related to a germinal apparatus) . . . [*Pityosporites cedroides* Thomson and Pflug (1953); *P. absolutus* Thierg. (1937)].
Partings between ektexine and endexine—1. Vestibulum: parting between ektexine and endexine which is localized at the germinal apparatus and is connected with the pore . . . [Fig. 379–382].
2. Interloculum: space between the ekt- and endexine which is developed in the whole, or nearly whole exine body. (Among many *Extratriporates*.) . . . [For 1 and 2 *see* Fig. 543–544].
3. Arci: primary garland-like tube-shaped partings between ektexine and endexine that connect neighboring vestibula. (In *Alnus*.) . . . [Fig. 408–409].
4. Endoplicae: small tubes which have been formed by swelling between the ekt- and endexine. They run from atrium to atrium. (Among some *Triatriopoll*.) . . . [Fig. 413–414].
Partings in the endexine—1. Endanulus: inflation of the endexine that is closed toward the pore; analogous to the anulus. (In some *Extratriporates*.) . . . [Fig. 544].
2. Postvestibulum: parting in the endolamellae which is connected with the endopore. (Among the *Intratriporates* and others.) . . . [Fig. 382].
Partings of different positions—Caverna: a meridional parting occurring among nearly all the Tricolpates, Tricolporates, Tetracolporates, and others.
If the parting lies between the lamellae of the ektexine, we speak of a praecaverna; if it lies between the ektexine and endexine, we call it a "caverna" in the strict sense; if it lies between lamellae of the endexine, "postcaverna." Often there are several cavernae arranged behind one another . . . [Fig. 372–375]. [*see* APERTURE].

DISSENCE LUMEN, n., pl. -lumina.
Thomson and Pflug, 1953, p. 20: *Tr.* Breaks be-

tween lamellae (dissences) also usually lie near the germinal apparatus; in a few cases they extend beyond the germinal apparatus but stay close to it in a certain symmetrical relationship. These dissences include a "dissence lumen." This is often empty; in other cases, it is filled with elongated rodlets—elements of a lamella.

DISTAL, DISTALIS, adj. (L. distare, stand apart). cf. exterial, ventral.

a. Jackson, 1928: "remote from the place of attachment; the converse of proximal."

b. Wodehouse, 1935, p. 159–160: "For purposes of discussing the symmetry relations of these spheres it is convenient to speak of their polar axes as lines extending through the centers of the spheres and directed toward the center of the tetrad, where they would all four meet, if so extended, as stated by Fischer (1890). Thus, each sphere comes to have an inner and an outer pole, a proximal and distal polar hemisphere, and the equator is the boundary between the two polar hemispheres."

c. Erdtman, 1943, p. 50: ". . . that part of a pollen grain or spore which is turned outward in its tetrad. In monopored or monocolpate grains, it is the side upon which the pore or furrow is borne. In other grains, the distal and proximal sides are generally not distinguishable from each other after the tetrad has been broken up into individual grains. In spores, the distal side is opposite the tetrad scar (= ventral, Wodehouse)."

d. Pokrovskaya, et al., 1950: *Tr.* Distal (ventral or abdominal, according to Wodehouse, and exterial, according to Kozo-Poliansky). The part of the pollen grain or spore facing outside from the tetrad. Monoporate and monocolpate pollen grains on this side have a pore or a furrow. The distal and proximal sides of other pollen grains are not usually distinguishable from one another after the disintegration of the tetrad.

sdb. Selling, 1947, p. 76.
Kupriyanova, 1948: [*see* GROUPS OF SPOROMORPHAE].
Erdtman, 1952, p. 11; [*see* POLARITY].
Thomson and Pflug, 1953, p. 18; [*see* SYMMETRY].
Potonié and Kremp, 1955, p. 10; [*see* APEX].
Traverse, 1955, p. 91.
Couper, 1958, p. 102.

DISTAL FACE, n., pl. -s. cf. distal hemisphere, distal surface, figura, pars distalis.
Erdtman, 1952, p. 462: "that part of a spore surface which is directed outwards in its tetrad." [*see* POLARITY].

DISTAL HEMISPHERE, n., pl. -s. cf. distal face.
Potonié and Kremp, 1955, p. 10: *Tr.* With respect to the arrangement in the mother-cell, each grain has an inner, or proximal, and an outer, or distal, pole as well as a *proximal* and a *distal* hemisphere. [*see* APEX].

DISTALOAPERTURATE, adj. cf. aperture, colpate, furrow.
Erdtman, 1947, p. 113: "with one aperture on the distal part; exceptionally with one or several additional apertures, more or less symmetrically arranged." [*see* GROUPS OF SPOROMORPHAE].

DISTALO-PROXIMALO-APERTURATE, adj. cf. ambi-aperturate, aperture, colpate, furrow, pore.
Erdtman, 1947, p. 113: "with two apertures, one on the distal, the other on the proximal part."

DISTAL SURFACE, n., pl. -s. cf. distal face.
Ingwersen, 1954, p. 40: "the surface of a pollen grain or spore facing outwards in the tetrad."

DISULCATE, adj. cf. colpate, furrow.
a. Selling, 1947, p. 9: "with two undivided sulci . . ."
sdb. Erdtman, 1947, p. 113; [*see* GROUPS OF SPOROMORPHAE].
Cranwell, 1953, p. 16; [*see* DICOLPATE].
b. Kupriyanova, 1948: *Tr.* Disculcate pollen grains (pollina bisulcata) have two symmetrically arranged furrows, intersecting the polar axis at a right angle. [*see* GROUPS OF SPOROMORPHAE].

DITREME, adj. cf. aperture.
Erdtman and Straka, 1961, p. 65: "(2-treme) spores: with two apertures." [*see* GROUPS OF SPOROMORPHAE].

DIVERSE PORATE, adj. cf. colpate, colporate, colpus, furrow, Groups of Sporomorphae, porate.
Kupriyanova, 1948, p .74: *Tr.* Pollen grains with pores different in form but essentially with homologous furrows, i.e., each pore is homologous to a furrow (sulcus); apparently must be called diversely porus (pollinia diverse porata). [*see* GROUPS OF SPOROMORPHAE].

DORSAL, adj. (L. dorsalis, the back). cf. internal, proximal, proximalis.
a. Jackson, 1928: "relating to the back, or attached thereto; the surface turned away from the axis, which in the case of a leaf is the lower surface (Note: this is reversed by some authors)."
b. Wodehouse, 1935, p. 541: "the side of the grain turned inward in the tetrad and opposite the furrow in monocolpate grains, opposed to ventral."
sdb. Cranwell, 1953, p. 16.

DORSAL ROOT OF THE BLADDERS.
Wodehouse, 1935 [Fig. 206]. [*see* SACCATE].

DUPLIBACULATE, adj. (L. duplex, two-fold, double; baculus, a staff, support). cf. baculate.
Erdtman, 1952, p. 462: "muri, etc., supported by

41

two rows of bacula, are duplibaculate. Examples: . . . [Fig. 633 and Fig. 717–720]." [*see* LO-ANALYSIS].

DYAD, n., pl. dyas. (Gr. two). cf. aggregation, Groups of Sporomorphae, tetrad.

  I. Jackson, 1928: "a subdivision of a tetrad by mitosis, again dividing into single elements (Calkins); a bivalent chromosome."

 II. a. Erdtman, 1943, p. 50: "Dyas: dyad; pollen grains united in pairs . . . [*see* Fig. 47]."
    b. Selling, 1947, p. 76: "a unit of two grains (spores) formed by one mother-cell."

sdb. Faegri and Iversen, 1950, p. 128; [*see* GROUPS OF SPOROMORPHAE].

  Prokrovskaya, et al., 1950.

  Cranwell, 1953, p. 16.

E

EAR, n., pl. -s. cf. auricula, valva.
Jackson, 1928: "the spike of corn."

EARED, adj. cf. auriculate, auricles.
Jackson, 1928: "auriculate."

ECHINATE, adj. (L. echinatus; Gr. echinos, hedgehog, sea urchin). cf. echinatus, echinulate, setose, spinose, thorn-like.
a. Jackson, 1928: "beset with prickles."
b. Wodehouse, 1928, p. 933: "Adorned with prominent sharp spines more or less evenly distributed. Examples: *Silphium perfoliatum, Tanacetum camphoratum.*"
c. Wodehouse, 1935, p. 541: "provided with long or conspicuous and generally sharp, pointed spines, e.g., the grains of *Solidago* [Fig. 737] and *Oxytenia* . . [Fig. 731]."
d. Iversen and Troels-Smith, 1950, p. 46: *Tr.* Pointed sculpture elements. The largest diameter can be larger or smaller than the height.
sdb. Erdtman, 1943, p. 50; [*see* ORNAMENTATION and Fig. 639].
   Kupriyanova, 1948, p. 70; [*see* ORNAMENTATION and Fig. 658].
   Faegri and Iversen, 1950, p. 27; [*see* ORNAMENTATION and Fig. 626].
   Pokrovskaya, et al., 1950.
   Cranwell, 1953, p. 16.
   Thomson and Pflug, 1953, p. 22; [*see* ORNAMENTATION and Fig. 647].
   Harris, 1955, p. 19–21; [*see* ORNAMENTATION].
   Traverse, 1955, p. 93.
   Couper, 1958, p. 102; [*see* ORNAMENTATION].

ECHINOLOPHATE, adj. cf. abporal-lacuna, lophate, spinolophate.
a. Wodehouse, 1928, p. 933: "With the surface thrown into ridges, anastomosing or free, provided with more or less prominent spines. Examples: *Cichorium Intybus, Stokesia laevis.*"
b. Wodehouse, 1935, p. 541: "lophate, with the ridges bearing spines on their crests." [*see* Fig. 764–765].
sdb. Erdtman, 1947, p. 106; [*see* SPINOLOPHATE].

ECHINULATE, adj. cf. echinate, echinatus, spinose, thorn-like.
I. Jackson, 1928: "having diminutive prickles."
II. Harris, 1955, p. 19: "Apex [of the projection] more or less sharp. Trunk tapering with narrow base." [*see* ORNAMENTATION].

ECHINUS, n., pl. -i. (L. hedgehog, sea urchin). cf. cone, conus, spina, spine, thorn, vestigial spine.
Iversen and Troels-Smith, 1950, p. 35: *Tr.* Thorns: pointed sculpture element. The largest diameter can be larger or smaller than the height. [*see* ORNAMENTATION].

ECTEXINA, n. (Gr. ektos, outside; L. exter, outside). cf. ektexine.
Kupriyanova, 1956: [*see* SPORODERMIS and Fig. 540].

ECTONEXINE, n. (Gr. ektos, outside; L. nexus, fastened together). cf. ektonexine.
Erdtman, 1948, p. 387: "The outer, thicker, less refractive nexine layer." [*see* SPORODERMIS].
sdb. Faegri and Iversen, 1950, p. 161.
   Erdtman, 1952, p. 462.
   Erdtman, 1952, p. 19; [*see* SPORODERM].

ECTOSEXINE, n. cf. exolamella, tectum, tegillum.
Erdtman, 1952, p. 462: "the upper (outer, distal) part of the sexine. To the ecto sexine belong, e.g., pila heads, tegilla, tecta (at least partially), etc. . . . Synonyms: Exolamelle (Potonié, 1934b), tectum (Faegri and Iversen, 1950)." [*see* SPORODERMIS].

EDGE OF THE ANNULUS, n. cf. annulus, limes annuli.

EDGE OF THE MARGO, n. cf. limes marginis.

EDGE OF THE PORE, n. cf. limes pori, pore.

EGG, n., pl. -s.
Jackson, 1928: "ovum, ovule."

EKTEXINA, n.
Kupriyanova, 1956: [*see* Fig. 540].

EKTEXINE, n. cf. ectexina, ektexinium, exoexine, sexine.
a. Erdtman, 1943, p. 50: "ektexinium: ektexine, the outer of the two main layers of the exine."
sdb. Selling, 1947, p. 76.

b. Kupriyanova, 1948, p. 70: *Tr.* Thorns (spinae) . . . in staining the preparations in some cases only the spines are tinted, while the membrane-like exine that covers the entire surface of the pollen grain remains colorless, as happens in staining grains having a pore or a furrow membrane. Such staining indicates that the spines are composed of the ektexine element of the membrane, while the surface of the pollen is wanting of this layer and is covered only with a layer of the endexine. [*see* ORNAMENTATION].

sdb. Traverse, 1955, p. 92; [*see* SPORODERMIS].

c. Faegri and Iversen, 1950, p. 16: "Where the exine is more complex, it is possible to distinguish between two layers (Fritzsche, 1837), an inner and an outer, which are called, respectively, *endexine* and *ektexine*. The inner layer forms a continuous homogeneous membrane, corresponding to the simple exine of the above mentioned pollen type.* In contrast the outer layer always seems to consist of small elements (*granula,* Fritzsche l. c.), the development and distribution of which cause the extreme variability of the structure of the exine." [*see* FURROW (Faegri and Iversen, 1950) and Fig. 349–350, and ORNAMENTATION (Faegri and Iversen, 1950) and Fig. 624–626].

* In some grains the pores . . . seem to form true holes in the endexine. However, it is very difficult to ascertain this; nevertheless, the continuous endexine may have thinner areas equivalent to fully perforate pores.

sdb. Iversen and Troels-Smith, 1950, p. 34; [*see* STRUCTURE].
Pokrovskaya, 1950.
Cranwell, 1953, p. 16.

EKTEXINIUM, n. cf. ektexina, ektexine, exoexine, sexine.
Erdtman, 1943, p. 50: [*see* EKTEXINE].

EKTEXOSPORE, n.
Thomson and Pflug, 1953, p. 18: *Tr.* The exospore of spores is more simply constructed. Often two layers can be differentiated which may be structureless or structured and which we would like to call from the exterior to the interior, ektexospore and endexospore. [*see* LAMELLA].

EKTEXOSPORE LAMELLA, n., pl. -ae.
Thomson and Pflug, 1953, p. 18: *Tr.* The lamellae of the ektexospore shall be called, from the exterior to the interior, *ektexospore lamellae* a, b, c, etc.; those of the endexospore, *endexospore lamellae* a, b, c, etc. [*see* LAMELLA].

EKTONEXINE, n. cf. ectonexine.

ELEMENTUM ELONGATUM, n., pl. -a.
Iversen and Troels-Smith, 1950, p. 35: *Tr.* Elongated sculpture elements: the largest diameter is more than twice as large as the smallest. Possibly the sculpture elements can be formed by a close union of dot-like elements. [*see* ORNAMENTATION].

ELEMENTUM PUNCTUALIUM, n., pl. -a.
Iversen and Troels-Smith, 1950, p. 35: *Tr.* Dot-like sculpture element: the greatest diameter is less than twice as large as the smallest one.* [*see* ORNAMENTATION].

* In the case of sculpture and structure elements, the measurement parallel to the surface is designated as diameter, and the measurement perpendicular to the surface is designated as height.

ELEVATION, n. cf. baculum, excrescence, mammilla, process, projection, protrusion, protuberance, tuber, tubercle.
Harris, 1955, p. 20: "The pattern is regarded as comprising elevations of, or projections from, the general surface, these being not less than 1μ in height. The distinction here made between projections and elevations is a matter of convenience and it is not necessarily of structural significance . . . Elevations . . . are long in surface view (length more than twice breadth), except where otherwise specified." [*see* ORNAMENTATION].

ELLIPTICAL, adj. cf. shape classes.
I. Jackson, 1928: "shaped like an ellipse, oblong with regularly rounded ends."
sdb. Norem, 1958, [*see* SHAPE CLASSES and Fig. 167a].
II. Zaklinskaya, 1957: [*see* SACCATE (Fig. 227) for outlines of disaccate pollen grains].

EMBRYO, n., pl. -os. (Gr. embryon fetus, thing newly born).
Jackson, 1928: "the rudimentary plant formed in a seed or within the archegonium of Cryptogams."

EMPHYTIC CHARACTERS, n. (Gr. emphytos, innate). cf. figura, Groups of Sporomorphae, morphography, shape classes.
Wodehouse, 1935, p. 541: those that are the result of a specifically inherited cell form."

ENDANNULUS, ENDANULUS, n., pl. -i. (Gr. endon, within). cf. annulus, anulus, aspis, costa pori, dissence halo, labrum, lip, margo, oculus, operculum, operculum pori, pore ring, protrudence.
Thomson and Pflug, 1953, p. 21: *Tr.* . . . inflation of the endexine that is closed toward the pore; analogous to the anulus. (In some Extratriporates). [*see* APERTURE, DISSENCE, and Fig. 544].

ENDEXINA, n. cf. endexine, intexina, nexine.
Kupriyanova, 1956: [*see* SPORODERMIS and Fig. 540].

ENDEXINE, n. (L. endexina, endexinium). cf. ektexine, endexina, intexine, nexina, nexine, nexinium, sporodermis.
a. Selling, 1947, p. 76: "the inner of the two main layers of the exine."
sdb. Faegri and Iversen, 1950, p. 16; [*see* EXTEXINE (for illustrations see Faegri and Iversen,

1950, Fig. 349–350 and 622–626, FURROW and ORNAMENTATION].

Iversen and Troels-Smith, 1950, p. 34; [see STRUCTURE].

Pokrovskaya, et al., 1950.

Cranwell, 1953, p. 16.

Thomson and Pflug, 1953, p. 18; [see LAMELLA].

Potonié and Kremp, 1955, p. 17–19; [see SPORODERMIS].

b. Traverse, 1955, p. 92: "the distinct inner layer of the exine. The endexine is generally of simple homogenous structure. Sometimes the exine consists only of endexine."

ENDEXOSPORE, n. cf. sporodermis.

Thomson and Pflug, 1953, p. 18: *Tr.* The exospore of spores is more simply constructed. Often two layers can be differentiated which may be structureless or structured and which we would like to call from the exterior to the interior, ektexospore and endexospore. [see LAMELLA and SPORODERMIS].

ENDEXOSPORE LAMELLA, n., pl. -ae. cf. ektexospore lamella, sporodermis.

ENDO-CRACKS, n. cf. sporodermis.

Oldfield, 1959, p. 21: "This term has been coined to characterize certain small cracks observed in the endexine of most of the species considered. The pattern, width, and strength of these cracks are noted. Where they occur within a species they have been observed on pollen from all the herbarium specimens regardless of provenance or method of preparation, though there is intra-specific variation in strength and distribution. They are sometimes useful in distinguishing between types showing superficial resemblances. Close examination shows them to be definitely endexinous and in many cases confined over much of the grain to the inner layer of the endexine (cf. Erdtman, p. 161) . . ."

Oldfield, 1959, p. 22–23: "When this is in fact the case it is usually those cracks bordering the costae which are most likely to be deeper and stronger, breaking into the outer layer of the endexine. They may or may not be related to some of the structures observed in parts of the ektexine, depending on the species, and to a lesser extent on the individual grain. They occur in the following patterns. a) bordering the costae; b) in the polar area; c) from the ends of the pore, especially where it is a transverse gash across the furrow; d) less often, in the intercolpia; e) crossing the costae, more or less parallel to the pore, but nearer to the poles than the pore is. They have frequently been observed under oil immersion on fossil grains, and on fresh material."

ENDOLAMELLA, n., pl. -ae.

Thomson and Pflug, 1953, p. 18: [see LAMELLAE, EXOLAMELLA].

ENDONEXINE, n.

Erdtman, 1952, p. 462: "(Erdtman, 1948 . . . p. 387, [see SPORODERMIS]): the inner, ± thin, more refractive zone of the nexine."

sdb. Erdtman, 1948, p. 387.

Erdtman, 1952, p. 19; [see SPORODERM].

ENDOPLICA, n., pl. -ae. cf. plica.

Thomson and Pflug, 1953, p. 37: *Tr.* From the plicae, the "endoplicae" can probably be differentiated, despite the fact that the latter are similar in shape and that they probably arise by specific swelling differentials. Here, however, only the endexine is folded so that tube-like detachments occur between the ektexine and the endexine. The endoplicae always open into atria and are attached to these because the atria make the endexine an unstable structure. Endoplicae are characteristic of many *Triatriopollenites*. [see APERTURE, DISSENCE, and Fig. 413–414].

ENDOPORE, ENDOPORUS, n., pl. -i. cf. aperture, diaporus, exitus, exopore, germinal apparatus, germ pore, main pore, oris, os, pore, pseudopore, simple pore, ulcus.

Thomson and Pflug, 1953, p. 39: *Tr.* The germinal elements are called, according to their position, exocolpus a (in exolamella a), exoporus b, endoporus a, etc. In cases that are not exactly known, the centrifugal pore is called the "exopore;" the centripetal pore the "endopore." This is without regard to whether they lie in the ektexine or endexine, but it should be emphasized that the exopore of the Tricolporates never lies in the exolamella a. The most distinct pore, along with the pore of the thickest lamella, is called the "main pore." In many cases it lies in the separated endexine or is limited to the a or b lamella of the endexine. [see APERTURE and Fig. 379–382, 408–409, 543–544].

ENDO-RETICULUM, n., pl. -a, cf. intrareticulum.

ENDOSEXINE, n. cf. insulating layer.

Erdtman, 1952, p. 462: "the basal (lower, inner, proximal) part of the sexine (what remains if the upper part of the sexine, the ectosexine, is removed; see Sporodermis for Tab. 2, p. 19). The endosexine in the pollen grains of angiospermous plants is often baculate. Synonym: insulating layer (Potonié, 1934 . . .)." [see SPORODERMIS and MESOSPORIUM].

ENDOSPORE, n., pl. -s. cf. endosporium, intinium, spore intine, sporodermis.

I. Jackson, 1928: "the innermost coat of a spore;" "the intine of a pollen grain;" "the interior membrane of the pollen in Angiosperms."

sdb. Pokrovskaya, et al., 1950.

II. Endogene spore. Melchior and Werdermann, 1954, p. 15, [*see* SPORE].

ENDOSPORIUM, n. cf. endospore, intinium, spore intine.

Erdtman, 1943, p. 50: "intine."

sdb. Dijkstra-van V. Trip, 1946, p. 22; [*see* MESOSPORIUM]. Erdtman, 1952, p. 462; [*see* INTINE].

Thomson and Pflug, 1953, p. 17–18; [*see* LAMELLA OPPRESSA].

Potonié and Kremp, 1955, p. 19; [*see* SPORODERMIS].

EPISPORE, n. cf. episporium, perine, sporodermis, perinium, perispore, perisporium.

Jackson, 1928: "an external coat or perinium formed from the periplasm round the oospore in some Fungi and the spores of certain of the higher Cryptogams."

EPISPORIC, adj.

Jackson, 1928: "connected with the outer coat of a spore."

EPISPORIUM, adj. (Gr. epi, on upon). cf. epispore, perine, perinium, perispore, perisporium.

Erdtman, 1952, p. 462: "(Russow, 1872, etc.): [*see* PERINE]."

EQUATOR, n. (LL. aequator, one who equalizes). cf. amb, extrema, limb, lineamenta, outline.

a. Wodehouse, 1935, p. 541: "the great circle midway between the two poles and dividing the grain into two polar hemispheres."

b. Iversen and Troels-Smith, 1950, p. 31: *Tr.* Aeq = equator: the line of intersection between the surface of a pollen grain and a plane which goes through the centrum of the polar axes and stands at right angles to it.

c. Potonié and Kremp, 1955, p. 10: *Tr.* The shape of the fossil is determined, among other ways, by data on the *outline* of the flattened spore. The outline is often, but not always, identical with the equator . . . or a meridian . . . . One considers the line between the proximal and distal hemispheres to be the *equator* of a spore. The equatorial contour is classified as circular, triangular, etc.

sdb. Erdtman, 1943, p. 49.

Selling, 1947, p. 76.

Pokrovskaya, et al., 1950.

Erdtman, 1952, p. 462.

Traverse, 1955, p. 91.

Couper, 1958, p. 102.

EQUATORIAL AXIS, n. cf. axis aequatorialis, symmetry axis.

Potonié, 1934, p. 8: *Tr.* A line connecting the poles is termed the main axis. The axis perpendicular to it is called the equatorial axis.

sdb. Thomson and Pflug, 1953, p. 18; [*see* SYMMETRY].

Potonié and Kremp, 1955, p. 10; [*see* POLAR AXIS].

EQUATORIAL COLLAR, n., pl. -s. cf. arista, cingulum, collar, crassitudo, flange, frassa, ring, zona, zone.

EQUATORIAL CONTOUR, n., pl. -s. cf. amb, ambit, ambitus, extrema lineamenta, limb, limbus, outline.

Couper, 1958, p. 102: "the shape of the outline of a spore or pollen grain when seen in polar view."

EQUATORIAL CREST, n., pl. -s. cf. crista aequatorialis, equatorial fringe, fimbria aequatorialis.

Wodehouse, 1928, p. 933: "A ridge or interlacunar crest extending from pore to pore along the equator in lophate grains. It may be continuous or interrupted to admit the equatorial lacunae when present. Example: *Vernonia anthelmintica.*"

EQUATORIAL DIAMETER, n., pl. -s. cf. dimension, figura.

EQUATORIAL FORM INDEX, n. cf. equatorial, form index.

EQUATORIAL FRINGE, n., pl. -s. cf. fimbria aequatorialis.

EQUATORIAL LACUNA, n., pl. -ae. cf. lacuna aequatorialis.

a. Wodehouse, 1928, p. 933: "A lacuna in lophate grains, situated midway between the pores and astride the equator when single, or on either side of the equator when double. Examples: *Vernonia wrightii,* single; *V. anthelmintica,* double."

b. Wodehouse, 1935, p. 541: "a lacuna situated on the equator between two germ pores and as much in one polar hemisphere as the other. It may be remote from contact with the pores or poral lacunae, as in the grains of *Vernonia jucunda,* or in contact with one of them, as in those of *Scorzonera* . . . [Fig. 763], or with two of them, as in those of *Tragopogon* . . . [Fig. 762]."

EQUATORIAL LIMB, n., pl. -s. cf. amb, ambit, ambitus, circumscriptio, contour, extrema lineamenta, figura, Groups of Sporomorphae, limb, limbus, morphography, optical section, outline, shape classes.

EQUATORIAL PLANE, n., pl. -s. cf. figura, shape classes, symmetry.

EQUATORIAL PORE, n., pl. -s. cf. oris, os.

EQUATORIAL POSITION, n., pl. -s. cf. face, figura, shape classes.

EQUATORIAL REGION, n., pl. -s. cf. arista auricula, cingulum, corona, crassitude, crassitudo, ear, equatorial rim, flange, margo aequatorialis, ring, valva, zona, zone.

Potonié and Kremp, 1955, p. 15: *Tr.* A special development occurs in some cases in the equatorial region of spores (Zonales). In the simplest case, the exine may be a little thicker here than in the remaining area. In this case we are dealing with a *crassitude*. Exines, or exine parts which are thick throughout are called crassexinous. The swelling may be more pronounced at the "corners" of the equatorial section of the spore than on the rest of the equator. Then we speak of valvae . . . [Fig. 149]. If the valvae continue to swell into cushions or similar structures which stand out from the outline, we speak of *little ears, ears,* or auriculae . . . [Fig. 150–151]. If a massive ridge is present which often is wedge-like in cross section and which distinctly enlarges the equatorial diameter and circles the entire equator, then we have a cingulum, also ring or arista in a more specific sense . . . [Fig. 152–153]. Occasionally, the cingulum does more or less wedge sharply toward the exterior as in *Lycospora* and *Densosporites,* but is in cross section more or less circular with a narrow neck; according to the definition, this is true in *Rotaspora* . . . [Fig. 154]. If the equator has a wide membranous *flange* which is not wedge-shaped in cross section, we are dealing with a *zona* or *frassa* which is also called equatorial collar or simply collar (as in *Cirratriradites* . . . [Fig. 155]). If the equator is covered with a wreath of branched, anastomosed hairs, fimbriae or fringes, we are dealing with a *corona* (*Reinschia, Superbisporites*) . . . [Fig. 157–159].

EQUATORIAL RIDGE, n., pl. -s.
a. Wodehouse, 1935, p. 542: "an interlacunar ridge extending from pore to pore along the equator in lophate grains. It may be continuous . . . as in the grains of *Taraxacum* . . . [Fig. 765], or interrupted to admit the equatorial lacunae when these are present, as in the grains of *Scorzonera hispanica* . . . [Fig. 763] and *Tragopogon pratensis* . . . [Fig. 762]."
b. Faegri and Iversen, 1950, p. 161: "(Wodehouse, 1935), cp. lophate. NB is *not* identical with costae equatoriales."

EQUATORIAL RIM, n., pl. -s. cf. equatorial region, margo aequatorialis.
Dijkstra, 1946, p. 21: *Tr.* Equatorial rim, abstract concept, visible on . . . [*Triletes brasserti* Stach and Zerndt] as the spot where the equatorial collar was attached.

EQUATORIAL RUGA, n., pl. -ae. cf. colpus, pore, ruga aequatorialis, transverse fold.
Potonié, 1934, p. 21: Tr. . . . [Fig. 317, lower part] shows that the double circle bounding the vestibulum produces two small pointed lobes (tips) opposite each other on the equator and [which are] equatorially directed [parallel to it]. These "tips" in *Tilia* are very indistinct and often hardly recognizable. In this we are dealing here with the evolution of an equatorially directed fold of the exine, with a *transverse fold*. All differentiations related to it are to be called *equatorial rugae* . . . [*see* RUGAE (Potonié, 1934)].

EQUIDISTANT, adj. (L. aequus, equal). cf. ornamentation.
Erdtman, 1952, p. 462: "(Wodehouse, 1935: isometric): ± evenly distributed."

EUINTINA, EUINTINE, n. (Gr. eu, good, advantageous).
Kupriyanova, 1956, Fig. 3: *Tr.* Internal layer of the intine that participates in the formation of the pollen tube. [see SPORODERMIS and Fig. 540].

EURYPALYNOUS, adj. (Gr. eurys, wide, broad). cf. figura, Groups of Sporomorphae, morphography, shape classes.
Erdtman, 1952, p. 462: "(Erdtman, 1951 . . . [Statens Naturvet. Forskningsrads Arsb., v. 4]): said of plant families, etc., characterized by a ± great array of spore types (different in apertures, exine stratification, etc.)."

EXCRESCENCE, n., pl. -s. (L. excrescens, growing out). cf. baculum, elevation, mammilla, process, projection, protruberance, protrusion, tuber, tubercle.
I. Jackson, 1928: "a gnaur or wart on the stem of a tree; enation."
II. Erdtman, 1952, p. 462: " 'excrescences' (as understood in the present book) may or may not—as regards origin and growth—be real outgrowths from the exine."

EXINA, EXINE, n., pl. -s. (L. exter, on the outside). cf. exinium, exospore, exosporium, extine, intine, medine, sporodermis.
a. Jackson, 1928: "= extine."
b. Potonié, 1934, p. 8: *Tr.* The wall of fresh pollen grains consists of two parts which lie closely on top of each other and which are strongly chemically differentiated; the exine and the intine (Fritzsche) . . . The *exine* is stronger, less flexible, and strongly cutinized.
c. Ingwersen, 1954, p. 40: "the outer, highly resistant layer of the pollen (or spore) wall. It may be simple, in which case it consists of structureless endexine, or complex, when it is built of structureless endexine (inner) and more or less structured ektexine (outer)."
sdb. Erdtman, 1943, p. 50.
    Cranwell, 1953, p. 16.
    Harris, 1955, p. 25.

Kupriyanova, 1948.

Kupriyanova, 1956; [*see* SPORODERMIS and Fig. 540].

Pike, 1956, p. 50.

Pokrovskaya, et al., 1950.

Ueno, 1958, p. 176.

Van Campo, 1954, p. 253; *see* Fig. 546.

EXINE FISSA, n. cf. burst exine.

EXINE INDEX, n.

Iversen and Troels-Smith, 1950, p. 42: *Tr.* Exine-Index (ex-I). The relative thickness of the exine may be expressed by the ratio of the largest thickness of the exine . . . to the largest transverse dimension of the pollen grain . . .

ex-I (0.05): Exine index small

ex-I (0.05–0.10): Medium

ex-I (0.10–0.25): Large

ex-I (0.25): Very large

sdb. Ingwersen, 1954, p. 40.

EXINE RUGA, n., pl. -ae. cf. colpate, furrow.

Potonié, 1934, p. 24: *Tr.* Also along the line of the simple sulcus the exine can be strongly folded (in the dried living pollen grain as well as in the fossil) . . . [Fig. 286–287] and indeed in such manner that the sulci move inward and the equatorial cross-section of the pollen grain will be trilobate. In this case, we have an *exine ruga.* It must be differentiated from the intexine ruga, a fold mainly of the meridional area . . . [Fig. 329]. [*see* APERTURE (Potonié, 1934)].

EXINIUM, n. cf. exine.

EXINTINA, EXINTINE, n. (L. ex, out, intine). cf. intine.

a. Jackson, 1928: "the middle coat of a pollen grain, that which is next the intine."

b. Kupriyanova, 1956: *Tr.* The outside layer of the intine, frequently having subapertural thickenings (the term proposed by Fritzsche). [*see* SPORODERMIS and Fig. 540].

EXITUS, n., pl. -us. (L. outlet, passage). cf. aperture, colpate, diaporus, furrow, germ pore, germinal aperture, germinal pore, porate, pore.

I. Jackson, 1928: "the inner aperture of the slit of a stoma . . .;"

II. a. Potonié, 1934, p. 17: *Tr.* Where the curvaturae come closest to the equator, close to their out side lie the exitus or germinal points . . . [Fig. 4], which are those particular spots of the often rather expanded germinal apparatus, where germination will be effected.

b. Erdtman, 1943, p. 50: "exit; germinal aperture, a hole in the furrow membrane through which the germ pore—the place of emergence of the pollen tube—protrudes."

sdb. Wodehouse, 1928, p. 933; [*see* GERMINAL APERTURE].

Wodehouse, 1935, p. 542; [*see* GERMINAL APERTURE].

Potonié, 1934, p. 12–24; [*see* APERTURE].

Faegri and Iversen, 1950, p. 161.

Cranwell, 1953, p. 16; [*see* APERTURE].

Thomson and Pflug, 1953, p. 24, 31–41; [*see* APERTURE].

EXITUS DIGITATUS, n. cf. colpate, furrow.

Potonié, 1934, p. 22: *Tr.* Let us now further consider the fossils in which the tenuitas is not protrudent, but is indented. Here the indentation can deepen more and more in such manner that the upper and lower edges of the tenuitas approach each other more and more covering the hollow forming below them . . . [Fig. 322–323]. The *exitus digitatus* is developed. The enclosure is accomplished in such a way that the parts of the exoexine above and below the tenuitas and right and left of the sulcus approach one another like four fingers or papillae. The vestibulum which the four fingers enclose can then to a certain degree be compared with the vestibulum of *Tilia* and other species; only, in *Tilia* the equatorial ruga is practically not yet developed. Conditions very similar to those in *Pollenites megadolium* are found in *P. pompeckji* . . . [Fig. 319]; exactly the same conditions are found in many other forms as, e.g., *P. exactus,* to be compared with *Castanea* . . . [Fig. 318, 324]. [*see* APERTURE, Potonié, 1934].

EXITUS OPENING, n. cf. exitus.

EXITUS PAPILLA, n. cf. ligula [*see* Fig. 212].

Faegri and Iversen, 1950, second revised edition, 1964, p. 173: "Taxodiaceae—with a well-marked exitus papilla."

EXOCONIDIUM, n., pl. -a. [*see* SPORE, Melchior and Werdermann, 1954, p. 14, under Exospores].

EXOEXINE, n. (Gr. exo, outside). cf. ektexina, ektexine, ektexinium, sexine, sporodermis.

a. Potonié, 1934, p. 8: *Tr.* In fossil pollen, two layers of the exine are frequently well differentiated and separated from each other by sharper lines: the *exoexine* and *intexine.*

b. Potonié, 1934, p. 15: *Tr.* The main agreement with the contact areas of Carboniferous spores is that by omitting the external exine parts (namely the exoexine) the meridional areas lack surface sculpture (or structure of the exine) which is preserved on the remaining parts of the pollen grain. [*see* SPORODERMIS and APERTURE; Potonié, 1934, p. 8].

c. Faegri and Iversen, 1950, p. 161: "exoexine = ektexine."

EXOLAMELLA, n., pl. -ae. (L. lamella, a thin plate or scale). cf. ectosexine, sporodermis, tectum, tegillum.

I. Potonié, 1934, p. 9–10: *Tr.* This [exine of

fossil Pinaceae pollen] shows a construction in which at a cursory glance, one would say that it is here (as an exception) the intexine which shows the just-mentioned rodlet construction. On a rather thin innermost lamella (which sits atop the intine which is not preserved in fossils) there is namely a rather strong layer of rodlets which are separated from the outside by a strong *exolamella*. Because of the strong frame of the exolamella, one is tempted to take this alone as the exoexine and to interpret the layer of rodlets, together with the innermost lamella, as endoexine. That this interpretation is incorrect can be shown in the construction of the air-sacks of Pinaceae pollen which were hitherto not yet clearly understood. These air-sacks originate in the following way; the exolamella together with the shrinking rodlet-layer separates from the innermost lamella and inflates . . . [Fig. 547]. [*see* SPORODERMIS and SACCATE; Potonié, 1934, p. 8–11].

II. Erdtman, 1943, p. 50: *see* ORNAMENTATION." [*see* ORNAMENTATION; Erdtman, 1943, p. 51–52].

III. Faegri and Iversen, 1950, p. 161: "exolamella (Potonié, 1934) = tectum."

IV. Thomson and Pflug, 1953, p. 18: *Tr.* One often differentiates an exterior group of lamellae, the "ektexine" (exoexine) and an inner lamellae group, the "endexine" (indexine) (Iversen) . . . Ekt- and endexine are usually formed by alternating structureless and structured lamellae . . . If eight lamellae can be noticed, five of which are located in the ektexine and three in the endexine, then we designate the lamellae of the ektexine from the outside to the inside, as exolamellae a, b, c, d, e; those of the endexine, as endolamellae a, b, c. [*see* LAMELLA and Fig. 544].

EXOPORE, EXOPORUS, n., pl. -i. cf. aperture, diaporus, endoporus, exitus, germinal apparatus, germ pore, oris, os, pore, pseudopore, simple pore, ulcus. Thomson and Pflug, 1953, p. 39: *Tr.* The germinal elements are called, according to their position, exocolpus a (in exolamella a), exoporus b, endoporus a, etc. In cases that are not exactly known, the centrifugal pore is called the "exopore;" the centripetal pore the "endopore." This is without regard to whether they lie in the ektexine or endexine, but it should be emphasized that the exopore of the Tricolporates never lies in the exolamella a. The most distinct pore, along with the pore of the thickest lamella, is called the "main pore." In many cases it lies in the separated endexine or is limited to the a or b lamella of the endexine. [*see* ENDOPORUS and Fig. 379–382].

EXOSPORE, n., pl. -s.
I. cf. exina, exine, exinium, exosporinium, exosporium, extine, sclerine, spore exine. Jackson, 1928: "the outer covering of the spore;" "a thick coat developed from the periplasm round the oospore in Peronosporeae;" "the three outer layers of the spores of *Isoetes* (Fitting)."
II. exogene spore. Melchior and Werdermann, 1954, p. 14; [*see* SPORE].

EXOSPORINIUM, n. cf. exosporium.
Jackson, 1928: "the outer integument of a pollen grain, or microspore of flowering plants (Fitting)."

EXOSPORIUM, n. cf. exospore, exosporinium.
I. Erdtman, 1943, p. 50: "exine (fern spores excepted)."
sdb. Erdtman, 1952, p. 18; [*see* SPORODERMIS].
II. a. Pokrovskaya, et al., 1950: *Tr.* Outside membrane of a spore.
b. Potonié and Kremp, 1955, p. 19: *Tr.* In using the term exospore, one has to remember that this term does not include the conception of the perispore; the perispore lies on the exospore. [*see* SPORODERMIS].
sdb. Potonié, 1934, p. 15; [*see* APERTURE].
Dijkstra- van V. Trip, 1946, p. 22; [*see* MESOSPORIUM];
Thomson and Pflug, 1953, p. 18; [*see* LAMELLA].

EXPANSION FOLD, n., pl. -s.
Wodehouse, 1928, p. 933: "A longitudinal thin-walled area, generally present in heavy-walled grains and serving to accommodate changes in volume. Example: *Inula helenium.*"
sdb. Thomson and Pflug, 1953; [*see* APERTURE (swelling fold) and Fig. 413–414].

EXTERIAL, adj. (L.). cf. distal, distalis, exterial, ventral.

EXTERIOR, n. (L. exterus, on the outside). cf. distal, distalis, exterial, external, ventral.
a. Faegri and Iversen, 1950, p. 15: "The terms *interior* and *exterior* refer to the distance from the center of the individual grain."
b. Traverse, 1955, p. 91: "Refers to parts away from the center of a pollen grain or spore. A part is exterior to another if farther from the center of the pollen grain."

EXTERNAL, adj. (L. externus, outward). cf. exterial.
Jackson, 1928: "outward."

EXTINE, n., pl. -s. (L. extimus, outside). cf. exina, exine, exinium, exospore, exosporium, sclerine.
Jackson, 1928: "the outer coat of a pollen grain."

EXTRAPORATE, adj. cf. colpate, colporate, colpus, furrow, Groups of Sporomorphae, porate.
Faegri and Iversen, 1950, p. 129: "pollen grains free from each other . . . two or more distinct apertures . . . Lacunae (pseudocolpi) present . . .

Free pores present." [*see* GROUPS OF SPOROMOR-PHAE and Fig. 74].

EXTRARETICULATE, adj. cf. reticulate.

Potonié and Kremp, 1955, p. 15: *Tr.* The "reticula" . . . [Fig. 773–774] usually consist of smooth, low *stripes* or higher *ridges* (muri) which join to form a net. In both cases the exine is designated as *extrareticulate* . . . We distinguish the reticulum = *extrareticulum* from the infrareticulum and from the reticuloid or negative reticulum. [*see* ORNAMENTATION].

EXTRARETICULUM, n. cf. extrareticulate, reticulum.

Potonié and Kremp, 1955, p. 14: [*see* ORNAMENTATION].

EXTREMA LINEAMENTA, n. (L. outline). cf. amb, ambit, ambitus, circumscripto, contour, equatorial contour, equatorial limb, limb, limbus, optical section, outline.

Potonié and Kremp, 1955, p. 15: [*see* ORNAMENTATION].

F

FABIFORMIS, adj. (L. faba, a bean). cf. bean-shaped, fabaeformis, shape classes.
Jackson, 1928: "applied to Lichen spores which are bean-shaped."

FACE, n., pl. -s. (L. facies, form, shape, face). cf. distal, equatorial position, figura, proximal, shape classes.
I. Jackson, 1928: "that surface of an organ which is opposed to the back, usually the upper or inner side."
II. Erdtman, 1952, p. 463: "a polar spore has two faces (sometimes spoken of as hemispheres or halves), one directed outward, the other inward (i.e., toward to center of a tetrad). The two faces meet at the equator."

FENCE, n., pl. -s. cf. comb, crest, cristata, reticulum cristatum.
Jackson, 1928: "Withering's word for INVOLUCRE."

FENESTRATE, adj. (L. fenestra, a window).
I. Jackson, 1928: "pierced with holes, as the septum in some Cruciferae."
II. a. Faegri and Iversen, 1950, p. 17–18: "In the *Liguiflorae* the tectum is broken by a limited number of rather large openings (*lacunae,* Wodehouse, 1928), which are arranged in a symmetric pattern. As this highly developed pollen type derives from types with a complete tectum, we class it as a tectate pollen of a special type (*fenestrate,* i.e., equipped with windows), even if the *lacunae* occupy a greater area of the total surface of the grain than the tectate parts of the exine." [*see* GROUPS OF SPOROMORPHAE and Fig. 72; Faegri and Iversen, 1950].
b. Cranwell, 1953, p. 16: "Faegri and Iversen, p. 17. Used tentatively here for larger meshes of *Phormium* sp."

FIGURA, pl. -ae. (L. shape). cf. altitudo, amb, area basalis, axis, axis equatorialis, axis pole, breadth, circumscriptio, contour, depth, diameter, dimension, distal, equatorial axis, equatorial diameter, equatorial limb, equatorial plane, equatorial position, equidistant, extrema lineamenta, face, figure, focus 0–5, form, form element, height, index, interior, interior boundary, isometric, length, limb, limbus, limes exterior, limes interior, longitudinal, longitudo, lune, main axis, marginal, maximal measurement, mensura, meridional, optical section, outline, P:E ratio, polar, polar area, polar area index, polar area measurement, polar axis, polar cap index, polar hemisphere, polar position, polaris, polarity, pole, pole axis, polus, positio aequatorialis, positio polaris, proximal surface, shape, shape classes, shape class index, size, size classes, size range, subequatorial symmetry, symmetry axis, symmetry element, transverse.
Thomson and Pflug, 1953, p. 38: *Tr.* After the usual treatment with alkalies, the sporomorphae are in a swollen condition; this condition is assumed in the definition of "figura." The "figure" (figura) we define by approximate comparison with geometrical bodies: balls, spindles, lenses, etc. Differences in the specific swelling ability of the different parts of the sporomorph-body influence the "figura." These are significant for single morphologic units.

FIGURA CONTAGIONIS, n. cf. contact figure, contact marking, trilete marking, triradiate marking.

FIGURA TRIRADIATA, n. cf. contact figure, trilete mark, triradiate figure.

FIGURE, n., pl. -s. cf. figura, form, shape classes.

FIMBRIA, n., pl. -iae. (L. fringe). cf. capillus, hair.
I. Jackson, 1928: "a fringe;" "an elastic-toothed membrane beneath the operculum of mosses."
II. Potonié and Kremp, 1955, p. 14 and Fig. 159: *Tr.* The hairs, capilli, or fimbriae . . . are cylindrical or almost ribbon-like, especially they are not straight but more or less winding or coiled, knobbed, forked, branched, or anastomozed . . . and perhaps also somewhat laterally expanded. Even the cylindrical, flexuose hairs are probably almost never completely cylindrical. Their base may be thickened and the narrowing in the main part of the hair, if it occurs, is very small. Sometimes they are rather abruptly pointed. [*see* ORNAMENTATION].

FIMBRIA EQUATORIALIS, adj. (L. fimbriatus). cf. equatorial fringe, fimbria.

FIMBRIATE, adj. cf. fimbria, fimbriatus.

Jackson, 1928: "with the margin bordered by long slender processes."

FISSURA, n., pl. -ae. cf. fissure.

Potonié, 1934, p. 23–24: *Tr.* We refer to such species as *Taxus, Larix, Sequoia, Taxodium, Cupressus,* and *Juniperus* as plant pollen without germinal apparatus. When the pollen of these genera are examined more closely, it appears that the exine, through stronger swelling of the fresh pollen grain does not burst altogether irregularly. On the contrary, there arises a quite sharp, straight rent, a *fissura* . . . [Fig. 213]. A corresponding illustration of *Cupressus* was already given by Fritzsche (1832, Pl. 1, Fig. 16). This fissure is very characteristic for certain fossil pollen; we can refer to *Pollenites hiatus,* which is probably related to one of these genera. Weakening of the exoexine at a stripe corresponding to the fissure may also have led to the formation of the primary meridional area. [*see* APERTURE (fissure and sulcus, simplex) and Y-MARK; Potonié and Kremp, 1955, p. 12].

FISSURA DEHISCENTIS, n., pl. -ae, -es. (L. fissura, split; dehiscene, to gape). cf. contact figure, dehiscence, fissure, laesura, monolete, mark.

Erdtman, 1943, p. 50: "fissure of dehiscence; a central longitudinal fissure in the scar of monolete spores; also the three-armed fissure in the tetrad scar of trilete spores."

sdb. Pokrovskaya, et al., 1950.

FISSURA GERMINATIVA TRIRADIATA. (L. iriradiate germinal cleft). cf. contact figure, trilete aperture.

FISSURE, n., pl. -s. cf. fissura.

FIXIFORM, adj. (fr. L. fixus, past part. of figere, to fix; forma, form). cf. figura, Groups of Sporomorphae, morphography, shape classes.

Erdtman, 1947, p. 112: "with fixed shape." [*see* GROUPS OF SPOROMORPHAE].

sdb. Erdtman, 1952, p. 463.

FLAGELLUM, n., pl. -a. (L. a whip).

Jackson, 1928: "a runner or sarmentum, branchlets in Mosses;" "the whip-like process of the protoplasm of a swarmspore;" "similar organs in the cells of some Schizomycetes."

FLANGE, n., pl. -s. cf. arista, cingulum, collar, crassitude, crassitudo, frassa, ring, spore, zona, zone.

I. Jackson, 1928: "A ring-like projection of the integumental lining of the micropyle of certain fossil seeds;" "Bower's term for the apparent margin of the pinnae in *Blechnum*."

II. Kosanke, 1950, p. 14: *see* Fig. 109.

FLATTENED, adj. cf. polar cap, shape classes.

FLECKED, adj. cf. blotched, infrapunctate, intrapunctate, maculate, maculatus, maculose, papillate, planitegillate, psilotegillate, punctate, punctitegillate, scabrate, subpsilate.

a. Potonié, 1934, p. 10: *Tr.* In all these cases, when in the cross-section no rodlets are recognizable and where, in the plan view of the exine, punctate to maculate patterns become noticeable without the existence of sculpture, one speaks of punctation or flecks (maculation). [*see* ORNAMENTATION].

b. Cranwell, 1953, p. 16: "W., 1943, a useful term where detail is obscure at ordinary working magnifications (about 400X)."

c. Harris, 1955, p. 18: "With minute pits or elevations, the psilate in part and scabrate of Faegri and Iversen (1950). Example: *Paesia scaberula*." [*see* ORNAMENTATION].

FLOATING BODY, n., pl. -ies. cf. contact figure, massa, swimming-apparatus.

Potonié, 1956, p. 83: *Tr.* On the contact areas three or nine floating bodies are found. If only three are present, then we have a form similar to that of *Cystosporites,* where the contact areas bear three rudimentary megaspores [compare Fig. 134–135].

In *Azolla,* as well as in *Cystosporites,* only one fertile megaspore forms in the megasporangium. The other megaspores become disorganized and together with the dissolved tapetum cells of the sporangium form a mucus which is deposited as an exospore on the fertile megaspore. In this way, the remains of the three rudimentary spores which belong to the tetrad of the fertile megaspore, by rearrangement of the substance, form into 3 or 3 x 3 floating bodies. If one detaches the three floating bodies form the megaspore of *Azolla filiculoides* Lam. (compare Maedler, 1954, Pl. 5, Fig. 6, 7), then the Y-rays, contact areas, and curvaturae remain on the amputated area as in *Cystosporites* (compare Potonié and Kremp, 1955 I, Pl. 10, Fig. 77).

The floating-body group consists partly of a foamy substance, corresponding to that in *Cystosporites varius* (compare Potonié and Kremp, 1955 I, Pl. 10, Fig. 81, 82, and the text 1956 II, p. 149 above, p. 152 center). [*see* MASSA].

FLOATS, n. cf. contact figure, swimming-apparatus.

FOCUS 0–5. (L. hearth). Critical optical sections of spores and pollen under the microscope. cf. figura, LO-analysis, optical section, ornamentation. [*see* Fig. 636].

Iversen and Troels-Smith, 1950, p. 43: [*see* LIMES EXTERIOR].

FOLD, n., pl. -s. cf. colpate, colpus, furrow, ruga, sulcus.

FOOT LAYER, n., pl. -s. cf. ektonexine.
Larson, Skvarla, and Lewis, 1963. [see SPORODERMIS and Fig. 545].

FORAMEN, FORAMINA, n. cf. fovea, foveola, pit, scrobiculus.
I. Jackson, 1928: "an aperture, especially in the outer integuments of the ovule; cf. MICROPYLE."
II. Erdtman, 1952, p. 463: "global (q.v.) ± circular apertures. Example: . . . [Fig. 489, 495, 502]."

FORAMINATE, adj. cf. aperture, cribellate, foramen, forate, Groups of Sporomorphae, porate, pore.
Krutzsch, 1959, p. 40: Tr. The wall shows small roundish apertures, certainly no pores in germinal sense. Noticed in combination with structure. [see ORNAMENTATION and Fig. 772].

FORAMINOID, adj. cf. foraminate.
Erdtman, 1952, p. 463: "global apertures, ± similar to foramina."

FORAMINOSE, adj. (L. foraminosus). cf. foraminate.
Jackson, 1928: "perforated by holes."

FORATE, adj. (L. forara, to bore). cf. foraminate, Groups of Sporomorphae, oligoforate, oligoperiporate, panporate, perforate, periporate, polyforate, polyperiporate.
Erdtman, 1952, p. 463: "with foramina . . . . [see Fig. 493; see also FORAMEN, Fig. 489, 502, BROCHAL, Fig. 757]."

FORKED Y-MARK, n., pl. -s. cf. Y-split mark, Fig. 113.

FORM, n., pl. -en. (L. forma, form). cf. figura, figure, Groups of Sporomorphae, morphology, shape classes.
Potonié, 1934, p. 5: Tr. In the simplest case, the form of a pollen grain is that of a sphere. This applies especially clearly to the systematically lower types. At the same time, there are forms which more or less approach the ellipsoid. Pollen of many conifers are, for example, ball-like (apart from specializations). Those of many monocotyledons are ball-like or ellipsoidal. Here, filiform pollen (Zostera) constitutes a peculiarity. In dicotyledons the many variations of form which are present are by far in most cases based on the six-sided (trigonal) bipyramid (. . . [Fig. 517], compare Pollenites euphorii). This is a consequence of the arrangement of the pollen in the mother cell . . . [Fig. 2]. Rarely, eight-sided (tetragonal, etc.) bipyramids appear. Pollen with this basic bipyramidal construction in large part again approach the original sphere or ellipsoidal form through a more or less marked rounding of the edges and corners as well as by increasing the number of sides . . . [Fig. 518]. There are, however, also spindle-, cylinder-, and rod-shaped forms. Frequently, rather flat lenses are found among the ellipsoids. Where the form of pollen of dicotyledons is not derived from that of the double pyramid, we almost always find the sphere form. [see SACCATE for outlines of disaccate pollen grains; also SHAPE and SHAPE CLASSES].

FORMATION OF LIPS, n. cf. anulus, labrum.

FORM-ELEMENT, n. (L. elementum). cf. figura, ornamentation.

FORM INDEX, EQUATORIAL, n.
Beug, 1961, p. 11: Tr. Quotient of greatest and smallest diameter of equator, measured in polar view.

FORMULA OF SPORES AND POLLEN SYMBOLS. (L. dim. of forma, form). cf. figura, Groups of Sporomorphae, morphography, palynogram, pollen and spore formula, shape classes.
I. Potonié, 1934, p. 32–34 (translation of German version in parentheses):

O	Extrema lineamenta orbiculata (outline roundish)
△	Extr. lineam. triangulata (outline triangular)
□	Extr. lineam. polygonalia (outline multilateral)
⊙	Corpus cum aequatoriali corona cuticulari (equatorial zona present)
( )	Extr. lineam. aut ovalia aut elliptica. poli in lateribus angustis (outline oval or elliptic, poles at the short sides)
⌣	Extrem. lineam. aut ovalia aut elliptica, poli in lateribus latis (outline oval or elliptic, poles at the long sides)
⌒	Extr. lineam. fabaeformis (outline bean shaped)
⌂	Extr. lineam. trilobata (outline trilobate)
∧	Poli fastigiati (poles pointed)
⌒	Poli rotundati (poles roundish)
⌐	Poli plani (poles flattened)
::	Spora aut pollen ex quattuor cellis aequalibus composita (spore or pollen composed of four equal cells)
∴	Spora aut pollen ex tribus cellis aequalibus composita (spore or pollen composed of three equal cells)
{	Spora aut pollen cum sacculis (spore or pollen with air sacs)
l	Superficies levis (surface smooth)
p	Superf. punctata (surface punctate)
m	Superf. maculata (surface maculate)
gr	Superf. granifera (surface granulate)
v	Superf. verrucosa (surface verrucate)
h	Superf. hirsuta (surface with hairs)
c	Superf. canaliculata (surface striate, the grooves are smaller than the lists between them)

cic   Superf. cicatricosa (surface covered with somewhat longer scars, stripes, or lists; the grooves between the lists are wider or of equal width)

r   Superf. reticulata (surface reticulate)

rg   Superf. rugosa (surface rugose)

sp   Superf. spinosa (surface spinose)

r+r   Superf. bireticulata (surface bireticulate)

n   Superf. nexa (surface interwoven)

s   Superf. saetosa (surface saetose)

f   Superf. fibrosa (surface fibrose)

t   Superf. tuberosa (surface tuberous)

●   Exitus ± orbiculati (germinal points ± circular)

○   Rugae aequatoriales ovales (equatorial rugae oval)

■   Rugae aequatoriales rectangulatae (equatorial rugae ± square)

⟩⟩   Exitus non multum/multum prominentes (exitus somewhat/distinctly protruding)

⟨⟨   Exitus non multum/multum insculpti (exitus somewhat/distinctly indented)

〜   Ruga (= area meridionalis aut sulcus aut linea dehiscens simplex) adest (ruga [meridional area, sulcus, or simple dehiscent furrow] present)

〰   Ruga etc. perexpressa (ruga very distinct)

(〜)   Ruga etc. brevis (ruga short)

((〜))   Ruga etc. perbrevis (ruga very short)

=   Ruga etc. cum altitudine (ruga with altitude or lists)

≡   Ruga etc. cum altitudine lata (ruga with broad altitude or broad lists)

y   Y-radii usque ad extrema lineamenta pertinent (Y-mark rays about length of radius)

y (〜)   Y-radii usque ad 2/3 radii corporis pertinent (Y-mark rays about 2/3 of radius)

y ((〜))   Y-radii usque ad 1/3 radii corporis pertinent (Y-mark rays about 1/3 of radius)

( )   Parentheses around symbol when feature not distinct

!   Exclamation mark after symbol when feature very distinct

The symbols of the Formula of Spores and Pollen must be arranged in the sequence as given above to be well understood . . .

The formula ○ — △, ○ — ◗, ∧!, gr 2μ, 3 ○ 4μ, 3 (=), 30.4μ, 30μ thus means:

In aspectu polari extrema lineamenta aut orbiculata aut triangulata sunt. In aspectu laterali extrema lineamenta aut orbiculata aut ovalia sunt. Poli in lateribus latis, multi fastig-

iati sunt. Superficies granifera est. Grana 2 μ in diametro. Tres rugae aequatoriales ovales, 4 μ longae. Tres rugae meridionales breves, cum altitudinibus. Axis aequatorialis 30.4 μ. Axis polaris 30 μ.

II. Erdtman, 1943, p. 53: "Pollen diagnoses:—The pollen morphology of many genera and even of some families is still almost entirely unknown. In many cases, pollen grains of our most common plants are but imperfectly known. Several exhaustive plant monographs dismiss the morphology of the male gametophyte in a few lines if they consider the pollen at all. Therefore, it seems logical to insist that a pollen or spore diagnosis would be given in descriptions of every species. A short pollen diagnosis would be somewhat as follows:

"Diagnosis pollinaria (imaginata):—Pollina subprolata, tricolpata; exinium reticulatum. Axis polaris circiter 38–42 μ; diameter aequatorialis c. 31–35 μ; colpae longae (c. quattuor quintas continentes distantiae interpolaris), angustae, extremitatibus acutis, nihil seu parum immersae, singulae cum pori germinali aequatoriali circulari instructae diametri c. 3 μ; lumina reticuli hexagonalia, in medio areae intercolparis maxima (ad 1.8 μ), deinde paulatim decrescentia adversus polos et margines colparum; altitudo maxima murorum c. 0.9 μ; crassitudo exinii (muribus exceptis) c. 2.5 μ; cuius tertiam partem endexinium explet."

III. Erdtman, 1946, p. 73–75: "The definitions and abbreviations . . . may be used in pollen and spore formulae in the following way . . .

". . . [Fig. 24]: monosulcate, plano-convex pollen grain. Formula: *Pollina monosulcata, media* (20–46–30); abbreviated: 1-su—ME. '20' is the length of the polar axis (the depth of the grain), '46' the length of the 'sulciferous' diameter (the length of the grain), and '30' the length of the 'not-sulciferous' diameter (the breadth of the grain).

". . . [Fig. 25]: monosulcate, concavo-convex pollen grain. Formula: *P. monosulcata, magna* (20–56–20), *concavo-convexa* (—4); abbreviated: 1-su-ME-cccv.

"The actual length of the polar axis is obtained by subtracting the figure (4) in the second set of brackets from the 'depth' figure (20) in the first set.

". . . [Fig. 26]: monosulcate, biconvex grain. Formula: *P. monosulcata, media* (20–46–40), *biconvexa* (9 + 11); abbreviated: 1-su-ME—2cv.

"The figures in the second set of brackets indicate that the polar axis and the longitudinal, 'sulciferous' diameter cross each other 9 μ

within the contour line of the distal and 11 μ within the contour line of the proximal part of the grain. The value $(10 + 10)$ would indicate a biconvex grain with a contour line displaying the same convexities on both sides of the 'sulciferous' diameter, whilst $(18 + 2)$ would indicate a nearly convexoplane grain with the sulcus on the convex side, etc.

"... [Fig. 27]: monolete, concavo-convex spinose spore. Formula: *Sporae monoletae, mediae* $(20\text{–}46\text{–}22)$, *concavo-convexae* $(\text{—}4)$, *spinosae* (<5); abbreviated: 1-let—ME—cccv; spin.

"IF —4 in the second set of brackets be exchanged for $2 + 18$, the corresponding spore would be nearly planoconvex with the laesura on the flat side.

"... [Fig. 28]: nonaperturate (asulcate, acolpate, arugate, or aporate) pollen grain. Formula: *Pollina nonaperturata, minuta, sphaeroidea* (20.0); abbreviated: nap—MI—sph.

"... [Fig. 29]: dodekacolpate, peroblate pollen grain. Formula: *P. dodekacolpata, media* $(20\text{—}42)$, *peroblata* (0.48); abbreviated: 12-c-ME-po.

"... [Fig. 30]: tricolporate, prolate-perprolate grain. Formula: *P. tricolporata, minuta* $(20.0\text{—}10.0)$, *prolata-perprolata* (2.00); abbreviated: 3-cp—MI—p-pp.

"This pollen grain differs from the other pollen grains and spores in ... [Fig. 24–33] by reason of the possession of apertures consisting of two parts, one outer, ektexinous, and one inner, endexinous. The outer part of the aperture has the shape of a colpa, the inner part that of a porus. Pollen grains provided with such apertures are termed *colporate* (cp; *Pollina colporata*). The endexinous pori of colporate grains may be of various shape: extended meridionally, rounded, or equatorially elongated, sometimes so far as to join laterally to a continuous ring (cf. Erdtman, 1945, p. 188, Figs. 1, 2). Frequently the endexine is severed by meridional fissures underlying the ektexinous colpae. In such cases the pores are often represented by lateral semicircular notches in the endexine on each side of the fissure.

"... [Fig. 31]: hexaporate grain. Formula: *P. hexaporata, minuta* $(20.0\text{–}22.0)$, *oblatosphaeroidea* (0.91); abbreviated: 6-po—MI—o-sph.

"... [Fig. 32]: polyporate grain. Formula: *P. polyporata, minuta* (20.0), *sphaeroidea*; abbreviated: po-po—MI—sph.

"... [Fig. 33]: polyrugate grain. Formula: *P. polyrugata, minuta* (20.0), *sphaeroidea*; abbreviated: po-ru—MI—sph.

"Formulae of trilete spores may, e.g., be arranged as follows: *Sporae triletae, mediae* $(25(11 + 14)\text{–}20)$; abbreviated: 3-let—ME; or *Sporae triletae, mediae* $(25(5 + 8 + 12)\text{—}20)$; abbreviated: 3-let—ME.

"In these formulae '25' is the length of the polar axis, '20' the length of the longest (equatorial) diameter. '11' is the proximal part of the polar axis up to that point of the axis which would be cut by the longest equatorial diameter. (It should be understood that the polar axis would be cut at the same point, or nearly so, by a plane passing through the distal ends of the three laesurae.) '14' is the length of the remaining, distal part of the polar axis. The second formula would apply to spores of a shape similar to that of the megaspores of *Lagenicula*. '5' is the length of the proximal part of the polar axis, from the proximal apex of the spore to that point of the polar axis which would be cut by a plane laid through the distal ends of the three laesurae. '8' is that part of the polar axis stretching from the point just mentioned to that point of the polar axis which would be cut by the longest equatorial diameter. '12' is the remaining, most distal part of the polar axis.

"The thickness of the exine should be given in μ; a special scale (cf. Cranwell, 1942) does not seem to be necessary.

"The sculpture and texture, etc., of the grains and spores are not dealt with here, since it is felt that still much research work, particularly in the palynoanatomical line, should be done before any classification of the different sculptures and textures could be attempted." [*see* POLLEN CLASSES, Faegri and Iversen, 1950, and Erdtman, 1947].

FOSSAPERTURATE, adj. (L. fossa, ditch). cf. aperture, colpate, furrow.
Erdtman, 1952, p. 459: "... with apertures situated in the ditch-like indentations between the lobes (of the amb)." [*see* AMB and Fig. 496–501; 507–508; 82, 88].

FOSSULA, n., pl. -ae. (L. a little ditch). cf. negative reticulum, ornate, reticuloid, reticulum, reticulum fossulare.
I. Jackson, 1928: "a small groove in some Diatomvalves;" "a space between the ridges of an oospore of Charads; sulcus."
II. Kupriyanova, 1948, p. 70: *Tr.* If the tubercles have the form of areas (areolata) surrounded by narrow ditches (fossula) a negative reticulate structure is created (areolata, reticulum fos-

sulare). *Dracaena aurea* . . . *Lysichiton kamt-schatkense* . . . , and *Chamaedorea concolor* . . . as well as some others can serve as examples of the negative reticulate structure. [*see* ORNAMENTATION and Fig. 662].

FOSSULATE, adj. (L. fossulatus). cf. aperture, crilellate, fossulatus, grooved, ornamentation, pore, vermiculate.
a. Faegri and Iversen, 1950, p. 27: "sculpturing elements are absent. Surface with grooves." [*see* ORNAMENTATION].
b. Iversen and Troels-Smith, 1950, p. 46: *Tr.* True sculpture elements absent . . . With scattered, elongated indentations. The fossulate type presupposes that the fossulae do not anastomise to form sculpture elements (verrucae or vallae). [*see* ORNAMENTATION and Fig. 692].
c. Cranwell, 1953, p. 16: "E.—surface grooved, e.g., Centrolepidaceae."
d. Thomson and Pflug, 1953, p. 23: *Tr.* Irregular depressions. [*see* ORNAMENTATION and Fig. 675].
e. Harris, 1955, p. 19: "cavities, elongate, regular or irregular, but not anastomising." [*see* ORNAMENTATION].
f. Traverse, 1955, p. 93: "sculpturing of grooves."
sdb. Beug, 1961, p. 11.

FOVEA, pl. -ae. (L. a pit). cf. foramen, foveola, pit, scrobiculus.
I. Jackson, 1928: "a depression or pit as (a) in the upper surface of the leaf-base in *Isoetes,* which contains the sporangium; (b) the seat of the pollinium in Orchids."
II. Potonié, 1934, p. 16: *Tr.* The relations are also obvious in those cases where the pollen does not produce three large meridionally-stretched areas but only three small, generally equatorially situated foveae . . . [Fig. 491], which are circular zones of weakened exine resulting from removal of the exoexine. [Potonié, 1934, p. 18: *Tr.*] The simplest case of the strongly localized germinal apparatus is after all, the fovea.

The fovea originates in this way; the exoexine develops a small circular pit whose floor is probably mostly composed of the intexine . . . [Fig. 491]. Thus, for example, the germinal apparatus of *Sagittaria* sp. (from sapropel from sediments of Lake Ahlbeck) consists of many small shallow circular depressions regularly distributed over the surface of the spherical pollen cell and which originate by the weakening of the exoexine . . . [Fig. 490]. Sometimes it is not recognizable if the exoexine is really weakened at the foveae sites (compare *Pollenites. multistigmosus,* Cecilie Mine). The foveae then appear merely as small depressions over the entire exine. [*see* APERTURE].
III. Faegri and Iversen, 1950, p. 161: "badly defined pore without annulus."

FOVEATE, adj. cf. pitted.
I. Jackson, 1928: "pitted."
II. Krutzsch, 1959, p. 39: *Tr.* In contrast to foveaolate sculpture, where small lacunae of different sizes are distributed more or less irregularly, here exist more or less regularly big, bigger or small roundish lacunae between elevated parts of the wall. In the habitual end result, neither the lacunae nor the net-like positive reticulate parts of the wall predominate, but they counterbalance each other. A special linear arrangement does not exist. Transitions to fovealate, reticulate, corrugate, and other structures are possible. [*see* ORNAMENTATION and Fig. 771].

FOVEOLA, n., pl. -ae. cf. pit, scrobiculus.
I. Jackson, 1928: 3A "a small pit;" 1 "the perithecium of certain Fungals (Lindley);" 2 "in *Isoetes,* a small depression above the fovea, from which the ligule springs."
II. Erdtman, 1952, p. 463: "± rounded lumina (cf. . . . [Fig. 780–781]). Very small foveolae are termed scrobiculi (q.v.)."
sdb. Potonié, 1934, p. 12; [*see* SCROBICULUS].
Erdtman, 1947, p. 106; [*see* FOVEOLATE].
Potonié and Kremp, 1955, p. 15; [*see* ORNAMENTATION].
Couper and Grebe, 1961, p. 8; [*see* ORNAMENTATION and Fig. 687].

FOVEOLATE, adj. (L. foveolatus). cf. aperture, cribellate, ornamentation, pitted, pore, scrobiculate.
a. Jackson, 1928: "marked with small pitting."
b. Erdtman, 1947, p. 106: "Foveolatus: foveolate, pitted; provided with pits ('foveola')." [*see* FOVEOLA; Erdtman, 1952, p. 463].
c. Kupriyanova, 1948, p. 71: *Tr.* Pitted – (foveolate) – Sculpture in which the pits are represented by narrow indentations in the upper layer of the exine; such sculpture is found in the representatives of the family *Restionaceae* . . . An example of the coarse pitted sculpture is the pollen of some genera of the family *Centrolepidaceae.* [*see* Fig. 665].
d. Ingwersen, 1954, p. 40: "sculpturing type with holes or pits whose diameter >1 μ but simultaneously $<$ the shortest distance to a neighboring hole or pit."
e. Harris, 1955, p. 18: "The cavities or lumina may be up to 2 μ in diameter or, if larger, are too widely separated to form a reticulum (*Lycopodium varium*)."
sdb. Faegri and Iversen, 1950, p. 27; [*see* ORNAMENTATION].

Iversen and Troels-Smith, 1950, p. 46; [*see* ORNAMENTATION and Fig. 691].

Cranwell, 1953, p. 16.

Thomson and Pflug, 1953, p. 23; [*see* ORNAMENTATION and Fig. 674].

Couper, 1958, p. 102; [*see* ORNAMENTATION].

FOVEO-RETICULATE, adj.

Harris, 1955, p. 19–21: "The pits are large enough and close enough together to form a reticulum comprising the cavities (lumina or lacunae) and the intervening walls (muri) which separate them. Example: *Phylloglossum drummondii.*" [*see* ORNAMENTATION].

sdb. Couper, 1958, p. 102; [*see* ORNAMENTATION].

FRAGMENTIMURATE, adj. (L. fragmentum; Fr. grangere, to break). cf. reticulate.

Erdtman, 1952, p. 463: "with ± broken muri."

sdb. Potonié and Kremp, 1955, p. 15; [*see* ORNAMENTATION].

FRASSA, n., pl. -ae. cf. arista, cingulum, collar, crassitude, crassitudo, equatorial collar, flange, ring, zona, zone.

Potonié and Kremp, 1955, p. 15: *Tr.* If the equator (of a spore) has a wide membraneous *flange* which is not wedge-shaped in cross-section, we are dealing with a *zona* or *frassa* which is also called equatorial collar or simply collar (as in *Cirratriradites*). [*see* EQUATORIAL REGION and Fig. 155].

sdb. Potonié and Kremp, 1956, p. 86; [*see* ZONA].

FRINGE, n., pl. -s. (Fr. L. fimbria, fringe). cf. capillus, fimbria, hair.

Jackson, 1928: "used by Sir W. J. Hooker for the peristome of Mosses."

FRINGED, adj. cf. fimbriate.

Jackson, 1928: "margined with hair-like appendages, fimbriate."

FURROW, n., pl. -s. cf. altitude, apertura, aperture, centrum colpi, clinocolpate, colpate, colpus, colpus edge, colpus median, colpus membrane, colpus transversalis, costa colpi, costa transversalis, demicolpus, equatorial ruga, exine ruga, exit, exitus, exitus digitatus, expansion fold, fold, furrow, furrow membrane, furrow rim, geniculus, germinal, germinal aperture, germinal apparatus, germinal furrow, germinal lid, germ pore, hilum, intexine ruga, knee, lalongate, limes colpi, limes marginis, linea dehiscence, lolongate, loxocolpate, margo, margo colpae, margo equatorialis, medianum colpi, medianum colpi longitudinalis, medianum colpi transversalis, membrana colpi, meridional area, meridional furrow, meridional pleat, meridional ruga, operculum colpi, oris, orthocolpate, os, plica, primary germinal, primary meridional area, pseudocolpus, rimula, ruga, ruga aequatorialis, ruga compressa, rupus, secondary fold, secondary germinal, secondary meridional area, sulcoid groove, sulculus, sulcus, sulcus simplex, su-

tura, suture, tenuitas, tenuity, tip of the colpus, transversal furrow, transverse furrow, transverse fold, transverse median of the colpus, zoned furrow.

a. Potonié, 1934, p. 24: *Tr.* The simple sulcus, i.e., the ordinary furrow, shows no differentiation in its equatorial regions, it lacks any special apparatus, and in some cases could have been a transitional stage leading to the meridional area. [*see* APERTURA, under simple sulcus and geniculus].

b. Wodehouse, 1935, p. 542: [*see* GERMINAL FURROW and HARMOMEGATHUS, and Fig. 206].

c. Erdtman, 1945, p. 190–191: "As is well known, the furrow of a monocolpate pollen grain is not strictly homologous to the furrows of a tricolpate pollen. In monocolpate grains the furrow is borne on the distal part of the grain whilst in tricolpate grains the furrows are developed as meridional furrows crossing the equator of the grain at right angles. If it be considered that 'colpa' should not denote two not strictly homologous features, grains of the typical monocotyledonous monocolpate type may be termed 'monosulcate,' or 'trichotomosulcate,' provided the sulcus presents a three-slit opening.

"Terms such as 'polycolpate' ('nonacolpate,' etc.) may indicate pollen grains of two types, either grains with many equatorial (meridional) furrows, or grains with many furrows equally distributed over the general surface of the exine. Misinterpretations are avoided if the term 'colpae' is used to indicate furrows arranged more or less meridionally, and confined to the equatorial, more or less extended zone of the grain. Although originally applied by Potonié (1934) in another sense, the term 'ruga' may be used to denote furrows equally distributed over the surface of the grains. Of the three terms suggested 'ruga' is shortest, 'colpa' intermediate, and 'sulcus' longest. As to the features, which these terms are supposed to denote, the rugae in a polyrugate grain, as a rule, are shorter than the colpae in, e.g., a grain of the common tricolpate type. The colpae, in their turn, are as a rule comparatively shorter than the sulcus in a typical monocotyledonous monosulcate pollen grain."

d. Kupriyanova, 1945, p. 6: *Tr.* . . . The furrow (sulcus) intersects the polar axis at right angles . . . [*see* GROUPS OF SPOROMORPHAE].

e. Selling, 1947, p. 76: "a markedly longitudinal groove or (preformed) opening in the exine. Includes colpa, ruga, and sulcus. cf. pore."

f. Faegri and Iversen, 1950, p. 20: "Most pollen grains possess openings in, or thin parts of the exine, through which the pollen tube emerges. Two different types of apertures can be recognized, which are generally called *pores* and *fur-*

rows (*colpi*) (Wodehouse, 1935). In many cases the furrows also function as *harmomegathi* (Wodehouse, 1935: mechanism which accommodates the changes in volume of the semi-rigid exine).

"Furrows are with few exceptions boat-shaped parts of the exine where this membrane is much thinner than in the rest of the grain. The ektexine is reduced, even to absence, whereas the endexine is less affected. The furrows thus form no holes through the exine, and if no pores are present, the pollen tube must force its way through the thin membrane. The volume of the dry, living grain is comparatively small, the exine is contracted and the furrows appear as narrow slits. When the grain is moistened, it expands and the thin membrane of the furrows bulges out (this is the state generally depicted by Wodehouse, 1935). When the cell contents have been removed—by fossilization or by chemical action—the furrows generally appear as rather open grooves in the exine." *see* Fig. 349–360.

g. Cranwell, 1953, p. 16: "a very old term, covering active and inactive germinal grooves and their membranes. Normally elongate (length about twice width) but with many transitions to short or even ragged porelike areas (cf. colpi, sulci, rugae)."

h. Traverse, 1955, p. 93: "an opening in the ektexine, endexine, or both that has a slit or straplike shape."

i. Oldfield, 1959, p. 21: "Regular grains in regular tetrads are, with very few exceptions, more or less clearly tricolporate. This is also the form in most of the species considered which do not produce tetrads. In the tetrads, all regular and many irregular furrows are concurrent at the grain junctions. The double furrow thus formed is considered as a single unit for practical purposes. Shape, width, nature of edge, regularity or otherwise of disposition, number, and presence or absence of opercular materials are noted."

j. Steeves and Barghoorn, 1959, p. 227: "In *Ephedra* . . . the term 'furrow' will be used to refer to the region between the ridges, the term 'colpus' will be used to describe the thin, longitudinal, serpentine grooves located in the middle of the furrow area and formed by absence of or the thinning of the ektexine."

FURROW-LESS, adj. cf. acolpate, alete, colpate, furrow, inaperturate, non-aperturate.

FURROW MEMBRANA, FURROW MEMBRANE, n.

a. Wodehouse, 1935, p. 542: "the area of the exine enclosed by the germinal furrow, generally a delicate elastic membrane which stretches as the furrow opens, e.g., in the grains of *Solidago speciosa*." [*see* Fig. 737].

b. Kupriyanova, 1948, p. 70: *Tr.* Sculpture on the furrow is frequently wanting, and its surface is covered only with a thin transparent and a quite smooth film-membrane . . . Quite frequently only the fragments of the sculpture are observed on the surface of the furrow membrane . . . [*see* GROUPS OF SPOROMORPHAE].

sdb. Faegri and Iversen, 1950, p. 20; [*see* FURROW].
Cranwell, 1953, p. 542.
Traverse, 1955, p. 93.

FURROW RIM, n., pl. -s. cf. colpate, colpus edge, furrow, margo.

Wodehouse, 1935, p. 542: "the lip of the furrow, the edge or fold of exine bounding the furrow, sometimes thickened and in the winged grains of the Podocarpineae, bearing the ventral roots of the bladders."

G

GAMETE, n., pl. -s. (Gr. gamete, a wife).
Jackson, 1928: "a unisexual protoplasmic body, incapable of giving rise to another individual until after conjugation with another gamete, and the joint production of a zygote."

GAMETOPHYTE, n., pl. -s. (Gr. phyton, a plant).
Jackson, 1928: "the generation which bears the sexual organs, producing gametes, in turn giving rise to the sporophyte."

GEMINICOLPATE, adj. (L. twins). cf. colpate, furrow.
Erdtman, 1952, p. 463: "with colpi arranged in pairs."

GEMMA, n., pl. -ae. (L. a bud). cf. gemmate, microgemma.
  I. a. Jackson, 1928: "a young bud, either of flower or leaf, as used by Ray;" "an asexual product of some Cryptograms, as in the Hepticae, analogous to leaf-buds."
    b. Melchior and Werdermann, 1954, p. 14; [*see* SPORE].
  II. Iversen and Troels-Smith, 1950, p. 35: *Tr.* . . . like verrucae, but with proximal constriction. [*see* ORNAMENTATION].
  III. Beug, 1961, p. 11: *Tr.* In plan view ± circular sculpture element with proximal constriction: greatest diameter (measured parallel to surface of exine) is as great or greater than the height and 1 μ.

GEMMATE, GEMMATUS, adj. cf. microgemmate.
  I. Faegri and Iversen, 1950, p. 27: see Fig. 626.
  II. Cranwell, 1953, p. 16: ". . . Diameter of base of granule greater than height, constricted at base."
  sdb. Iversen and Troels-Smith, 1950, p. 46; [*see* ORNAMENTATION].
    Thomson and Pflug, 1953, p. 23; [*see* ORNAMENTATION and Fig. 645].
    Harris, 1955, p. 19; [*see* ORNAMENTATION].
    Beug, 1961, p. 11.

GENERAL EXINE, n. cf. sporoderm, sporodermis, synexine.
Kupriyanova, 1956: *Tr.* It is impossible not to mention here another layer, covering not only individual pollen grains but the whole tetrad, and which can be observed in pollen tetrads of *Juncaceae, Cyperaceae, Thurniaceae, Orchidaceae, Ericaceae,* and in certain other families. The layer mentioned is nothing but the membrane of the maternal cells. To distinguish it from the exine of the singular pollen grain it can be called the general exine or synexine (synexina mihi). [*see* SPORODERMIS and Fig. 540].

GENICULATE, adj. (L. geniculatus, with bent knees).
Jackson, 1928: "abruptly bent so as to resemble the knee-joint."

GENICULUM, n. (L. a little knee; dim. of genu, knee). cf. geniculus.
Jackson, 1928: "a node of a stem (Lindley);" "the junction of the articuli of Corraline Algae, which is destitute of crustation."

GENICULUS, n., pl. -i. cf. aperture, colpate, furrow, geniculum, knee, pore.
  I. Potonié, 1934, p. 24: *Tr.* In some cases the sulcus simplex shows a small peculiarity, in so far as it produces equatorially, an in-or-outwardly turned knee, a *geniculus.* [*see* APERTURE, Potonié, 1934].
  II. Stanley and Kremp, 1959, p. 351–353: "Among the European workers, it has been generally accepted that the small equatorial peculiarity found in the pollen of the genus *Quercus* was what Potonié meant by the term 'geniculus.' Consequently, this study was concentrated upon pollen of one species of the genus *Quercus,* with the hope that, with caution, the results would be projected to the pollen of other species. Cursory examination of pollen grains from several species assigned to the genus *Quercus* showed that the geniculus is best developed in *Quercus prinoides* (Chestnut scrub oak). However, it should be mentioned that the geniculus was observed to be present in the other species, but is developed to a lesser extent.
    "The pollen grains were acetylated and chlo-

rinated according to Erdtman's method, and then stained with Bismarck brown. A new technique was used in this investigation, which involved taking photographs at every half micron through the pollen grains with the new Leitz 100x Plano flat-field objective. This objective was designed to give a very small depth of focus, thereby permitting a photograph representing a section of very limited thickness to be obtained at every half micron. From these photographs, it was possible to reproduce the construction of the grain in great detail. In this study, both the equatorial and polar views were photographed, inasmuch as they were both readily available. However, the study could have been made with either one of the views presented . . .

"The interpretations are summarized in [Fig. 341–344] . . . which are schematic drawings of the grain under investigation. The polar view of *Quercus prinoides* pollen showing the location of section A is shown in . . . [Fig. 342] . . . [Fig. 343] illustrates how the geniculus is believed to be constructed, and also indicates the locations of sections B and C, which are shown in . . . [Fig. 343–344] . . . . Three well-developed colpi are illustrated in . . . [Fig. 343]. However, in . . . [Fig. 344] it can be seen that the colpi become shallow, indicating the presence of the geniculus at this location.

"Conclusions: From the illustrations shown, the geniculus in *Quercus prinoides* appears to be an outwardly turned equatorial bulge. The splitting and consequent thinning of the exine in this region suggest that the geniculus functions as an aid in germination. It is also interesting to speculate on the possibility that this equatorially located thin area is intermediate, in an evolutionary sense, between the tricolpate and the tricolporate pollen grains. Approximately 90 percent of the grains examined had a geniculus present. Among this 90 percent, the geniculus was found to vary from specimens with only a slight trace of a bulge to specimens with a marked or well-developed bulge. The mode for the species is thought to lie somewhere in the region of the latter extremity. Although no statistical tests were run to determine the exact amount of variation it is felt that the variation is so great that the height of the geniculus alone cannot be used as a criterion for species determination in *Quercus*, except possibly on a statistical basis. However, the geniculus does appear to be characteristic for at least some species of the genus *Quercus*. Finally, it should be mentioned that at least thirteen families other than the Fagaceae were observed in Erdtman's volume on the angiosperms to have genera within them that are

thought to contain geniculi. These families are the Caprifoliaceae, Clethraceae, Compositae, Cyrillaceae, Elaeagnaceae, Eriocaulaceae, Flacourtiaceae, Marcgraviaceae, Nyssaceae, Ranunculaceae, Rubiaceae, and Turneraceae. Within the Fagaceae, however, it appears that the geniculus is confined to *Quercus,* and it is suggested that it may be characteristic of the genus both in Recent and in fossil pollen grains. However, in order to substantiate this hypothesis, much more work other than this preliminary investigation will have to be done."

GERM CELL, n., pl. -s. (L. germen, a bud).
a. Jackson, 1928: "a female reproductive cell; a spore of the simplest character, a sporidium. . . ."
b. Melchior and Werdermann, 1954, [see SPORE].

GERMINAL, n., pl. -s. cf. aperture, furrow, pore. Jackson, 1928: "relating to a bud."

GERMINAL APERTURE, n., pl. -s. cf. aperture, colpate, exit, exitus, furrow, germinal pore, germ pore.
a. Wodehouse, 1928, p. 933: "The point of emergence of the pollen tube."
b. Wodehouse, 1935, p. 542: "a hole in the furrow membrane through which the germ pore protrudes. The term is also used to designate the rounded apertures which frequently occur in the general surface of the exine in the absence of germinal furrows, e.g., in the grains of *Salsola pestifer,* though these should probably be regarded as germinal furrows which are rounded in form and coinciding in extent with their enclosed germ pores."
c. Erdtman, 1943, p. 50: "exit; germinal aperture, a hole in the furrow membrane through which the germ pore [protrudes] . . . the place of emergence of the pollen tube . . . ."

GERMINAL APPARATUS, n. cf. aperture, colpate, furrow, porate, pore.
I. Jackson, 1928: "= egg apparatus."
II. Thomson and Pflug, 1953, p. 24: *Tr.* We designate as the "germinal apparatus" the totality of all differentiations of a sporomorph in which we assume, on the basis of extant examples, that they function, or may function, during the development of the pollen tube. [see APERTURE].

GERMINAL FURROW, n., pl. -s. cf. aperture, colpate, colpus, sulcus.
a. Wodehouse, 1928, p. 933: "A longitudinal area surrounding the germinal pore, differing from the remainder of the surface generally in the unadorned and thinner character of the exine; usually serving as an expansion fold. Example: *Inula Helenium.*"
b. Wodehouse, 1935, p. 542: "A longitudinal groove or opening in the exine, either enclosing a

germ pore or serving directly as the place of emission of the pollen tube, also generally serving as a harmomegathus (q.v.)."
sdb. Cranwell, 1953, p. 16.

GERMINAL LACUNA, n., pl. -ae. cf. poral lacuna.
Potonié, 1934, p. 12: *Tr.* . . . in the *poral lacuna* lies the germinal; we . . . prefer to use the general term *germinal lacuna* . . . (cf. Wodehouse, 1928, p. 933 . . .). [*see* ORNAMENTATION].

GERMINAL LID, n., pl. -s. cf. furrow, pore.
Jackson, 1928: "a separable area of a pollen-grain, breaking away to permit a pollen-tube to issue."

GERMINAL PAPILLA, n., pl. -ae.
Beug, 1961, p. 12: [*see* PAPILLA].

GERMINAL POINT, n., pl. -s. cf. aperture, exit, exitus, germinal pore.

GERMINAL PORE, n., pl. -s. cf. aperture, diaporus, endopore, exitus, exoporus, germinal apparatus, germ pore, open pore, oris, os, pore, pseudopore, simple pore, ulcus.
Wodehouse, 1928, p. 933: "The point of emergence of the pollen-tube." [*see* APERTURA].

GERMINATION, n. (L. germinatio, a sprouting).
Jackson, 1928: "the first act of growth in a seed; sprouting."

GERM PORE, n., pl. -s. cf. aperture, diaporus, endopore, exitus, exitus opening, exopore, germinal apparatus, germinal points, germinal pore, open pore, oris, pore, pseudopore, ulcus.
a. Wodehouse, 1928, p. 933: "the point of emergence of the pollen-tube." [*see* GERMINAL APERTURE].
b. Wodehouse, 1935, p. 542: "a pollen-tube anlage or the place of emergence of the pollen tube, generally denoted by a rounded papilla, e.g., in the grains of *Eriogonum gracile* . . . [Fig. 336]. Germ pores are generally enclosed in a germinal furrow as in the above example, but they may penetrate the exine directly, e.g., in the grains of *Salsola pestifer* . . . [Fig. 494]."
c. Traverse, 1955, p. 93: "(equals pore): A circular opening, usually more or less circular in ektexine, endexine, or both. It may or may not be associated with a furrow."

GLOBAL, adj. (Fr., L. globus, sphere). cf. shape classes.
Erdtman, 1952, p. 463: "said of apertures ± uniformly spread over the surface of a spore."

GLOBOID, adj. Norem, 1958. [*see* SHAPE CLASSES and Fig. 176b].

GRANIFER, adj. (L. fera, ferum, grain-bearing). cf. granulate, granulatus, granulose.

GRANULA, granule, n., pl. -a, -s. cf. granulum, granum, punctum.
Jackson, 1928: "Any small particles, as pollen, chloroplasts, etc.;" "the Naviculae of *Schizonema* (fide Lindley);" "sporangia in Fungi (Lindley);" "by Frommann used for the nucleolus-like structure in the nucleus of the terminal cells of the glandular hairs of *Pelargonium zonale,* Ait.;" "a minute particle, the assemblage of such being held to constitute protoplasm (Oltmanns)."

GRANULAR, adj.
Jackson, 1928: "Composed of grains;" "divided into little knots or tubercles, as the roots of *Saxifraga granulata,* Linn."

GRANULATE, granulatus, adj. cf. granifer, granulose.
I. Potonié, 1934, p. 11: *Tr.* The grana are generally more or less round to globular . . . [*see* GRANUM].
II. Erdtman, 1947, p. 106: "with granules ('granula'). The basal diameter of a granule is frequently equal to, or shorter than, any other tangential diameter of the granule (cf. Verrucatus)."
III. Kupriyanova, 1948, p. 71: *Tr.* The smallest sculpture should be considered as granulate sculpture (granulata). Erdtman defines it as follows: having small grains, the basal diameter of which does not exceed their height. This sculpture differs from the tubersulate structure by being much smaller. [*see* ORNAMENTATION].
IV. Faegri and Iversen, 1950, p. 161: "(Erdtman, 1947) = gemmate."
V. Pokrovskaya, et al., 1950, p. 2: *Tr. Granulatus* (granulata)—grained.
VI. Harris, 1955, p. 19: "More or less isodiametric not less than 1 μ or more than 1/20 equatorial diameter of spore if the latter is over 20 μ (*Lygodium articulatum*)."
VII. Couper, 1950, p. 102: "Granulate—more or less isodiametric, not more than 1/20th of the equatorial diameter of the spore if the latter is over 20 μ." [*see* ORNAMENTATION].
VIII. Krutzsch, 1959, p. 39: *Tr.* Single elements 1–2 μ, no longer dotlike, but look flat. (Transitions to sculpture and sculpture-species with little knobs and warts, etc.). [*see* ORNAMENTATION].
IX. Potonié and Kremp, 1955: [*see* GRANUM].

GRANULOSE, adj. cf. granifer, granulate, granulatus, punctate.
I. Jackson, 1928: "composed of grains."
II. Kosanke, 1950, pl. 11: for illustration [*see* Fig. 640 and ORNAMENTATION].

GRANULUM, n., pl. -a. (L. dim. of granum, grain). cf. granulate, granule, granum, punctum.
I. a. Potonié, 1934, p. 11: *Tr.* The grana are

generally more or less round to globular . . . (*see* ORNAMENTATION].

II. a. Faegri and Iversen, 1950, p. 16: "In contrast the outer layer [the ektexine compared with the endexine] always seems to consist of small elements (*granula,* Fritzsche, 1.c.), the development and distribution of which cause the extreme variability of the structure of the exine." [*see* EKTEXINE].

b. Faegri and Iversen, 1950, p. 161: "grana (Potonié, 1934) = gemmae + verrucae."

III. Iversen and Troels-Smith, 1950, p. 34: *Tr.* Sharply limited grains, rodlets and so on (structure elements) which are imbedded in the homogeneous ground substance or are deposited on it.

IV. Erdtman, 1952, p. 463: "granules, often small and ± rounded sexinous excrescences (cf. Erdtman, 1947c, p. 106). Examples: . . . [Fig. 733–734].

"Faegri and Iversen (1950) and Iversen and Troels-Smith (1950) define granula as 'scharf begrenzte Körner, Stäbchen u. dgl. (Strukturelemente), die der homogenen Grundsubstanz eingelagert oder aufgelagert sind' and quote Fritzsche (1837) as having suggested the term 'Granula' in this sense. Fritzsche's definition of granula is however as follows (1. c., p. 24): 'Die kleinen Oltröpfchen und Amylumkörner machen nun die sogenannten Granula des Pollen aus . . .' Nor does Fritzsche speak of a 'matrix' in contradistinction to 'granula.' The term granule was also used by Guillemin (1824, p. 106): 'Le grain de Pollen est un utricule renfermant une multitude de grains globuleux d'une extrême ténuité. Ceux-ci que, pour éviter des périphrases, je désignerai sous le nom de *Granules* . . .'"

V. Ingwersen 1954, p. 40: "sharply delimited granules, rods or similar structure elements embedded in or on the homogeneous basal substance of the exine (the endexine)."

VI. Potonié and Kremp, 1955, p. 14 and Fig. 732: *Tr.* The *granula,* or *grana,* are more or less circular in plane view; as seen from the side, they are more or less circular, semi-circular, or semi-oval. We speak of grana only if their size does not change too much on the individual spore which they decorate, or if it only changes gradually from one area of the spore surface to the other.

GRANUM, n., pl. -grana. (L. grain). cf. granule, granulum, punctum.

GROUND SUBSTANCE, n., pl. -s. cf. groundmass, matrix.

Potonié and Kremp, 1955, p. 18: *Tr.* As porphyritic structure [Mörtelstruktur] we call (along with Haberlandt) a phenomenon where micelle nuclei or puncta are imbedded in the groundmass or matrix of the exine. [*see* SPORODERMIS].

GROUPS OF SPOROMORPHAE, n. cf. aggregation, amb, dimension, formula of spores and pollen, pollen classes, shape classes, shape class index, size classes, tetrad.

I. Erdtman, 1943, p. 44: "Pollen grains may broadly be grouped in three classes according to their general shape: tricolpate radiosymmetrical grains with three furrows; monocolpate bilateral grains with one furrow; and alcolpate grains without furrows. Tricolpate grains are found mainly in dicotyledons, monocolpate grains in monocotyledons and gymnosperms. Acolpate are rarer than monocolpate grains; they occur in gymnosperms and angiosperms, both monocotyledons and dicotyledons.

"The spores of bryophytes and pteridophytes may also be grouped into three classes: trilete radiosymmetrical spores with a triradiate tetrad scar; monolete bilateral spores with a single unbranched scar; and alete spores without scars. Spores belonging to different classes may occur in the same family (e.g., *Polypodiaceae*) or even in the same genus or species (e.g., *Isoëtes,* where the megaspores are trilete and the microspores monolete).

"Pollen grains of the dicotyledonous type—tricolpate grains—are produced in fours by pollen mother-cells. The two necessary nuclear divisions take place in rapid succession, almost simultaneously, at right angles to each other. The daughter nuclei usually tend to take up positions as far from each other as possible within the confines of the pollen mother-cell. This results in their being arranged tetrahedrally. Subsequently, ridges grow inwardly from the wall of the pollen mother-cell, dividing it into four spaces, each of which corresponds to a pollen grain cell. The polar axes of pollen grains of this type are defined as lines extending through the centers of the grains and directed toward the center of the tetrad, where all four would meet if extended. Thus, each pollen grain has an inner, or proximal, and an outer, or distal, pole and a proximal and distal half, meeting at the equator of the grain. In the tetrad with the daughter-cells still in close contact with one another, the proximal part of each grain has three contact areas with its neighbors. Each contact area with its extension to the equator and further on to the distal pole is provided with a central furrow or colpa. These furrows cross the equator at right angles and are, therefore, called meridional furrows. Tricolpate grains have four planes of symmetry, one transverse, coincident with the equatorial plane, and three vertical (longitudinal), extending from the furrows past the polar axes to the wall opposite the furrows." [Erdtman, 1943, p. 45:] "In single tricolpate grains it is almost always impossible to make any distinction between the two poles. However . . . [Fig. 50] shows a young pollen grain of *Trapa natans,* where it was possible to make a distinction because of a distinct triradiate scar denoting the last place of contact between the four daughter-cells of a tetrad . . .

"Diagrammatic outlines of tricolpate grains in four different positions are presented in . . . [Fig. 42] . . .

"Pollen grains of the monocotyledonous type—monocolpate grains—are also produced in fours by pollen mother-cells. The two nuclear divisions take place successively and the resulting grains are usually arranged in a single plane and not in tetrahedral tetrads. The grains are typically bilateral, boat-shaped, provided with two planes of symmetry, one of which extends from 'prow to stern,' the other from 'port to starboard.' The keel corresponds to the proximal part of [Erdtman, 1943, p. 46:] "the grain. The opposite distal part is provided with a single longitudinal furrow, usually dividing the deck in equal portions. The water-line represents the equator of the grain and a mast shipped on deck at the intersection of the two symmetry planes would form an elongation of the polar axis of the grain. (However, the terms equator and polar axis are but seldom used in connection with monocolpate grains.) In descriptions of monocolpate grains, it is customary to speak of their distal and proximal part, their outline in lateral (apical or transversal) view, etc. (compare . . . [Fig. 45]).

"In addition to tri- and monocolpate grains—already described—and alcolpate grains (which need no special description), many other pollen types have been distinguished: di-, tetra-, hexa-, octo-, nona-, dodeca-, pentadeca-, and triacontacolpate grains. In hexacolpate grains, the colpae are equivalent in number and orientation to the six lines of contact between the four triangles corresponding to a tetrahedron; those of a dodecacolpate grain to the lines of contact between the six squares corresponding to a cube; and those of a triacontacolpate grain equivalent to the lines of contact between the twelve pentagons corresponding to a pentagonal dodecahedron, etc. (Wodehouse, 1. c.).

"The pores in cribellate grains (i.e., grains with a varying number of pores more or less uniformly scattered over the surface) are generally

considered to be short germinal furrows, rounded in form and coinciding in size with their enclosed germ pores (exits). Further information regarding these and other types is given by Fischer (1890) and Wodehouse (1.c.).

"The pollen grains of many *Cyperaceae* are quite aberrant. Here [Erdtman, 1943, p. 47:] "such terms as proximal, distal, etc. are not applicable, at least not in the same sense as used above, since these pollen grains are not homologous with ordinary pollen grains but with pollen mother-cells instead. After the nuclear divisions within the pollen mother-cell, three of four daughter nuclei are pressed down into the thick intine of the apex of the more or less tetrahedral grains and finally degenerate, a feature which was observed earlier by Elfving (1878). Thus the wall of the mature grain is nothing but the wall of the pollen mother-cell. . . . [Fig. 39] shows the probable correlation between this type of sedge pollen grain and a tetrahedral pollen tetrad.

"The trilete spores of mosses and ferns are produced in fours by spore mother-cells and, as the nuclear divisions are similar to those in the pollen mother-cells of plants and tricolpate pollen, the daughter-cells are arranged in tetrahedral tetrads. The proximal part of every spore has three contact areas (*areae contagionis,* Wicher, 1934). These areas touch each other laterally along three lines, which meet at the proximal pole, forming a triradiate streak which is usually known as the tetrad scar. Each arm of the scar is severed longitudinally by a dehiscence fissure. Trilete spores have three planes of symmetry, all meridional, one for each dehiscence fissure . . . [Fig. 40].

"Monolete spores are similar to monocolpate pollen grains in several respects. However, monolete spores resemble closely boats turned upside down; the keel corresponds to the distal part, while the proximal part is provided with a straight longitudinal streak which is severed by a dehiscence fissure. When describing monolete spores, as well as [Erdtman, 1943, p. 48:] "trilete, it is customary to speak, just as in monocolpate pollen grains, of their distal and their proximal part, of their shape in lateral (apical or transverse) view, etc. . . . [Fig. 46].

"Not only spores but also pollen grains may be trilete. Some pollen grains (e.g., in *Trapa natans*) may, at a certain stage of development, present a triradiate scar, which later disappears. But, on the other hand, some pollen grains (especially among the pteridosperms and possibly also among other classes of extinct spermatophytes) were provided with a permanent triradiate scar and did not develop any colpae at all. For this reason, it is impossible in many cases—at least at the present stage of our knowledge—to decide whether a spore *sensu lat.* is a pollen grain or a spore *sensu str.*

"It should also be emphasized that it sometimes, particularly when dealing with old and poorly preserved material, may be difficult to make a distinction between monocolpate pollen grains and monolete spores, or to decide whether a certain grain be a pollen of *Nuphar*-type . . . [Fig. 49], or a spore of the *Dryopteris thelypteris*-type . . . [Fig. 48], or again if it be a palm or a lily pollen grain with a three-slit opening or a fern spore with a triradiate scar . . . [Fig. 40, 44] . . ."

II. Erdtman, 1947, p. 112–114:

I. Nonfixiform (without fixed shape).

II. Fixiform (with fixed shape).

  A. Apolar (without distinct polarity).

    a. Nonaperturate (without apertures or aperturoid areas).

      1. Psilate (smooth, without sculpturing).

      2. Apsilate (not smooth, with sculpturing).

    b. Aperturate (with apertures).

      1. Rugate (with rugae without pori or poroid apertures).

      2. Rugoporate (with poriferous or poridiferous rugae).

      3. Porate (with nonaspidate pores; as to the term non-aspidate, see below under "aspidoporate").

      4. Aspidoporate (with aspidate pores; the term aspidate (Wodehouse, 1935) is replaced, here, by the terms 'aspidorate' and 'aspidoporate.' The first term designs ora, the second pori with a more or less well-defined shield-shaped area, or 'aspis,' surrounding the ektexinous part of every germinal aperture (the longest diameter of the ektexinous openings are thus always shorter than that of the aspides). The application of the terms postulates that a clear distinction between ektexinous and endexinous layers can be made. In aspides of the *Betula*-type (cf. Erdtman, 1943, . . . [Fig. 407, 406], there is no contact between ektexine and endexine; in aspides of the *Carya*-type (cf. . . . [Fig. 400–401]) there is apparently no endexine at all.

      5. Irregular (apertures irregular in shape or departing from the regular distributional patterns).

    c. Aperturidate (with aperturoid areas).

  B. Polar (with distinct polarity).

    a. Isopolar (hemispheres not different).

      1. Radiosymmetric (with more than two vertical planes of symmetry, or, if provided with two such planes, always with equilong equatorial axes).

        (a) Colpate (with colpae without ora or oroid openings).

        (b) Colporate (with oriferous or oridiferous colpae).

        (c) Colpato-colporate (with plain colpae—as in colpate grains—and oriferous or oridiferous colpae alternately).

        (d) Orate (with nonaspidate ora).

        (e) Aspidorate (with aspidate ora).

      2. Bilateral (with two vertical planes of symmetry; equatorial axes not equilong).

      3. Colpate, colporate, etc., as above.

    b. Subisopolar (hemispheres slightly different; example: tricolpate spores with colpae meeting in one pole).

    c. Anisopolar (hemispheres dissimilar).

      1. Distaloaperturate (with one aperture on the distal part; exceptionally with one or several additional apertures, more or less symmetrically arranged).

(a)		Radiosymmetric.
(1) +		Monoporate (with one pore).
(2) ++		Trichotomosulcate (with a three-split aperture).
(3) +++		Tetrachotomosulcate (with a four-split aperture).
(b)		Bilateral.
(1) +		Monosulcate (with one sulcus; aberrant forms: Zonosulcate, with one continuous sulcus-zone; subzonosulcate, as in zonosulcate, but with the sulcus-zone interrupted at one or more places; saccate, with air sacs).
(2) ++		Disulcate (with two sulci).
(3) +++		Polysulcate (with many sulci).

      2. Ambiaperturate (distalo-proximalo-aperturate; with two apertures, one on the distal, the other on the proximal part).

      3. Proximaloaperturate (with an aperture, simple or composite, on the proximal part).

(a)		Radiosymmetric.
(1) +		Trilete (with three laesurae, forming a triradiate scar).
(2) ++		Triletoid (with a three-slit aperture, not as distinct as a triradiate scar).
(3) +++		Poroletoid with one poroid aperture (scar).
(b)		Bilateral.
(1) +		Monolete (with one laesura).
(2) ++		Monoletoid (with one laesuroid-sulcoid aperture, not as distinct as a laesura)."

III. Kupriyanova, 1948, p. 174–179 and Fig. 242:
Tr. The pollen grains with a three-slit opening (*Pollina trichotoma fissurata*) have a nearly ball-like shape, sometimes in outline more or less triangular, with the three-slit opening being located on the distal side of the grains and its center coincides with the distal pores, three slits diverge in the meridial direction in a 120° angle [Fig. 242–1]. The three-slit opening in swollen grains opens in the shape of flaps of an envelope. . . .

Grains with three-slit openings are rare among the monocotyledonous plants. They were discovered in the palm, lily, and amarillo families.

Some species have four-slit grains along with those of three-slit grains. . . .

Monosulcate pollen grains (*Pollina monosulcata*) are bilateral in structure with two planes of symmetry [Fig. 242–2]. Their proximal side is furrowless . . . , the opposite distal side is provided with a longitudinal furrow; looking from the side, i.e., from the lateral side, a typical monocolpate grain has the form of a boat.

Erdtman (1946) proposes to call the furrow of this type sulcus, in order to distinguish it from the three furrows of dicotyledons, called colpa.

The furrow (sulcus) intersects the polar axis at a right angle, while the colpa proceeds in the meridial direction and intersects the equator of the pollen grain at a right angle. The proximal and distal poles in single-furrowed grains can be well distinguished when the grains are still in the maternal tetrad, with the polar axis arranged in the direction toward the center of the maternal tetrad . . .

Monocolpate pollen grains have several derivative types, such varied types of the monocolpate pollen include: pollen grains furrowless or asulcate (*asculcata*), monoporous with a simple pore in the shape of an opening; monoporous with a complex pore possessing a ring of thickened exine, an operculum; and finally with a furrow rolled spirally in the shape of a band bending around the grain, and some others.

Furrows in monocolpate pollen grains habitually differ in form, degree, and the character of the sculpture. Sculpture on the furrow is frequently wanting, and its surface is covered only with a thin, transparent, and quite smooth film —membrane—that is not stained with fuchsin. In other instances, for example, the pollen of *Aponogetion desertorum*, the furrow is covered with a thinner sculpture than on the surface of the remaining exine.

Quite frequently only the fragments of the sculpture are observed on the surface of the furrow membrane; examples are: *Lilium martagon, Pitcairnia leycalema* and certain species of *Iris* and others. These fragments can be stained with fuchsin to an intensive red color. In some species, the monocolpate pollen is provided with a lanceolate-lid (tectum) on the bottom of the furrow; such as in species *Tulipa, Ixia, Veratrum, Chamaerops humilis* and some others. In treating with alkali or acid this lid falls off very easily. The lanceolate lid has not yet been described for the monocotyledons, but its presence apparently is more frequent than we now indicate.

Pollen of *Calestesia cyanantha* has three thickenings of the exine band along the furrow, with the middle one being wider than the two lateral ones; they can be considered as a formation analogous to the lanceolate lid.

Pollen grains having a zoned furrow (*Pollina zonosulcata* [Fig. 242–5]); the furrow in this case envelopes the grain with a more or less wide band, dividing it into two hemispheres. Pollen grains of *Nymphaea capensis* . . . can serve as an example. In some cases the furrow happens to be broken up in one or more places.

Pollen grains of specimens *Costus igneus* and *C. nepalensis* (Fischer, 1890) are an example of a zone furrow illustrating the relation of this grain type to the pore grains.

The pollen grains with a spiral furrow (*Pollina spirali sulcata* [Fig. 242–4]); the furrow has a shape of ribbon-like band making two or four turns; such pollen is characteristic of the family Eriocaulaceae.

However, grains are found where the furrow has an appearance of parallel bands on one side of the grain; while on the other side, the bands are bent in the shape of closed rings. Such a "false" spiral distribution of the furrow is shown in grains of *Aphyllanthus monospellensis* and pollen of specimens of the genus *Crocus*. . . .

Disulcate pollen grains (*Pollina bisulcata*) have two symmetrically arranged furrows, intersecting the polar axis at a right angle [Fig. 242–3]. This type of grain is relatively rare; such grains occur in representatives of families Dioscareaceae, Pontederiaceae, Liliaceae, Amarillydaceae, Araceae and some others.

Monoporate pollen grains (*Pollina monoporata*) can be constructed in two ways: in one case the pollen grains have a pore surrounded by an annulus (porus annularis [Fig. 242–7]), that is, they have a ring-like thickening of the exine around the pore opening; usually the grains of this type also have in the middle of the pore membrane several fragments of the exine, isolated or grouped together in the form of a cover lid (operculum).

The pollen grains with an operculum pore occur in representatives of families Restionaceae, Flagellariaceae, and in grasses. But in the Cyperaceae family they are found very seldom . . .

The pollen grains with one simple pore (porus simplex) have a simple opening in the exine [Fig. 242–6], usually round and without the rim and lid. Such pollen grains may be found in the families Sparganiaceae, Centrolepidaceae, Pandanaceae, and Palmaceae.

Diporate pollen grains (Pollina diporata) have pores which represent simple openings in the exine covered with an exceptionally thin membrane which disintegrates when treated with alkali or acid [Fig. 242–8]. Such pollen grains occur in the representatives of the families Bromeliaceae *(Achamaea, Bromellea)* and Liliaceae *(Colchicum).*

Triporate pollen grains (Pollina triporata) have three round pores [Fig. 242–9], the formation of which, according to Erdtman, is related to the contraction of the central part of the three-slit opening. The grains of this type have been discovered so far only in the pollen of genus *Vanilla* of the orchid family.

Polyporate pollen grains (Pollina polyporata) have very many pores distributed on the entire surface of the pollen [Fig. 242–10]. They are exceedingly rare and are not characteristic of the pollen of monocotyledonous plants. This type of pollen has been known only in families Alismataceae, Butomaceae, Iridaceae, and Zingiberaceae.

Pollen grains with pores different in form but with essentially homologous furrows, i.e., each pore is homologous to a furrow (sulcus), apparently must be called diversely porus *(Pollina diverse porata)*. Three pores arranged on the lateral surface of the grains are slit-like and in certain cases are hardly visible, but the fourth pore is located at the broad end of the grain and can be seen distinctly and is round in form. Both the large pore and the lateral pores can serve equally as an outlet for the pollen tube. The grains of such structure occur only in the representatives of family Cyperaceae. Erdtman (1947) proposes to call this grain-type "Trisulcido-monoporites" (three-furrowed monoporous); we feel this name is unsuitable because all these pores are evidently of furrow origin.

Pollen grain without outlets (Pollina nonaperturata), in addition to the absence of outlets (furrows or pores), is characterized by a relatively thin exine and usually by a layer of intine that is thick and swollen when moist [Fig. 242–12]. The difference in grains of this type becomes apparent when they are stained with fuchsin. In some cases, both layers-of-exine and intine omit the stain without getting tinted and remain colorless; this takes place when the intine is uniformly thick.

But in other cases the exine becomes stained and it is possible to see the colorless intine irregularly distributed or in the form of a thinner layer.

Examples of pollen with a thickened intine can be found in the pollen grains of families Araceae, Cannaceae, Musaceae, and some others; grains having stained exine and a nonuniformly distributed and thinner intine are known in genera *Lilaea, Triglochin, Potamogeton,* and also in certain lily-genera, *Trillium, Lapageria* (Fischer, 1890), and *Smilax* (Erdtman, 1947).

Pollen grains without exine (Pollina nuda) do not have exine and are covered only with a layer of intine at the same time having an indefinite form [Fig. 242–13]. An example may be the pollen of the genus *Najas,* the form of which is definitely elongate-elliptic. Such exineless pollen occurs usually in plants that bloom in water *(Zannichelta, Cymodocea, Posidonia, Zostera).*

The inclosed pollen grains (Pollina occulusa) are enclosed within a membrane of the maternal pollen cell and are arranged in a form of a tetrahedral tetrad [Fig. 242–14]. The pollen of this type is distinctive only of the families Junaceae and Thurniaceae.

We feel it necessary to separate this type of pollen as an independent one, because it represents a special line in the development of pollen grains. Precisely, this type should be considered as initial type for diversally porus pollen which is characteristic of the sedge-family.

In addition to our review of types of pollen grains of the monocotyledons it should be noted that the growth of pollen grains into tetrads is characteristic of some families of the monocotyledons. The grains of the tetrads of the monocotyledons are usually arranged in one plane, while in the dicotyledonous, as we have pointed out, the pollen grains are arranged in the tetrahedral manner. The tetrahedral tetrads in the Monocotyledons are found only in the representatives of families Orchiaceae, Juncaceae,

Thurniaceae; Tetragonal and cross-tetrads are widely distributed among the monocotyledons. However, it must be said that transition stages exist between them.

IV. Faegri and Iversen, 1950, p. 128: For illustration see Fig. 6–13; 51–74.

"A. Pollen grains united in groups
  B. More than 4 grains in each group .... polyads
  BB. Groups of 4 .... tetrads
  BBB. Groups of 2 .... dyads
"AA. Pollen grains free from each other
  "B. Apertures 1 or none
    C. With air-sacks .... vesiculate
    CC. No air-sacks
    D. No furrow
      E. Pore rudimentary or absent .... inaperturate
      EE. One distinct pore present .... monoporate
    DD. One furrow present. monocolpate
  "BB. Two or more distinct apertures
    C. Lacunae (pseudocolpi or pseudopores) absent
    D. Furrows present, no free pores
      E. Furrows fused to spirals, rings, etc...... syncolpate
      EE. Furrows not fused
        F. Two furrows ...... dicolpate
        FF. More than two furrows
          G. Furrows without distinct pores or transversal furrows
          H. All furrows meridional
            I. Three furrows.. tricolpate
            II. More than 3 furrows .... stephanocolpate
          HH. Some or all furrows not meridional ...... pericolpate
        GG. Furrows with pores or transversal furrows (sometimes missing in one or two furrows)
          H. All furrows meridional
            I. 3 furrows ..... tricolporate
            II. More than 3 furrows .... stephanocolporate

          HH. Some or all furrows not meridional .... pericolporate
  "DD. Free pores present, no furrows
    E. Pores restricted to the equatorial area
      F. 2 pores .... diporate
      FF. 3 pores .... triporate
      FFF. More than 3 pores .... stephanoporate
    "EE. Pores outside the equatorial area... periporate
[Faegri and Iversen, 1950, p. 129]:
  "CC. Lacunae present
    D. Pseudopores present . fenestrate
    DD. Pseudocolpi present
      E. Some furrows with, others without pores, free pores absent ... heterocolpate
      EE. Free pores present. extraporate"

V. Erdtman, 1958, p. 4:
  "No aperture .... inaperturate
  One aperture 1-aperturate
  Two apertures 2-aperturate
  "More than two apertures:
    Apertures situated within a usually circular zone (zonaperturate):
    Three apertures 3-aperturate
    More than three apertures. stephanaperturate
    Apertures ± uniformly distributed over the surface . panaperturate
  "This scheme gives six main groups of very unequal size. Probably all of them are so large as to be in need of further subdivision.

"Within the inaperturate group a primary subdivision can be accomplished by observing whether or not there is a tenuitas (sensu Erdtman, 1952).

"Within the 1-aperturate group there are several possibilities for a subdivision. One may consider the shape of the aperture according to the scheme given below, or its position (cf. the ana and cata positions as defined by Erdtman and Vishnu-Mittre).

"Within the remaining groups the basis for subdivision is furnished by the shape of apertures and whether they are simple or composite. That gives the following scheme, which is the simplest of several possible ones:
  Outer and inner parts of apertures congruent:
    Apertures circular .... n-porate
    Apertures elongate .... n-colpate
  Outer and inner parts of apertures not congruent:
    Outer part of aperture circular... n-pororate
    Outer part of aperture elongate.. n-colporate

"By combining this key with the first one a useful net-work for a primary classification may be obtained."

VI. Erdtman and Straka, 1961, p. 65:

"Chiefly according to the number of the apertures the following classes can be established . . .

N 0. Atreme spores: without aperture(s).

"N 1. Monotreme (1-treme) spores: with one aperture.

N 2. Ditreme (2-treme) spores: with two apertures.

N 3. Tritreme (3-treme) spores: with three apertures.

N 4. Tetratreme (4-treme) spores: with four apertures.

N 5. Pentatreme (5-treme) spores: with five apertures.

N 6. Hexatreme (6-treme) spores: with six apertures.

N 7. Polytreme (po-treme) spores: with more than six apertures.

N 8. Anomotreme spores: with irregular or irregularly placed apertures (the spores in N 1–7 are nomotreme, i.e., provided with regular apertures).

"Pleotreme spores have more than one aperture. Spores with four to many zonally (monozonally or pleozonally) distributed apertures may be referred to as stephanotreme (monozono-stephanotreme or pleozono-stephanotreme; in the latter there are four or more than four apertures in each zone). N.B.: 'Spore' here and in the following means pollen grain or spore or both according to the context.

"A more detailed classification can be made if, besides number, the character (shape, etc.) of the apertures is known. The following classes may be distinguished . . . :

"C 0. Character unknown (ignote).

C 1. Monolept, dilept, etc. ('tenuitatiferous') spores: with one or several leptomata (Gr. leptoma, plur. leptomata; from leptos, thin), i.e., thin-walled areas which may be engaged in forming an opening in connection with the normal exit of substance from the inner part of the spores.

[Erdtman and Straka, 1961, p. 66]:

"C 2. Trichotomocolpate spores: with one (or exceptionally two) three-slit (three-armed) colpus (colpi).

C 3. Colpate (atomocolpate) spores: with one or several unbranched colpi.

C 4. Porate spores: with one or several pori.

C 5. Colporate (colp-orate) spores: with apertures consisting of a distal (or lateral-marginal) colpoid part with a proximal (or central) part known as an os (gen. oris, plur. ora). Colporate spores are with but few

exceptions pleotreme. They form a morphologically heterogeneous group in need of revision.

C 6. Pororate (por-orate) spores: with apertures consisting of a distal, poroid part and a proximal part (os). Pororate spores are practically always pleotreme.

"A more definite classification can be made if, besides number and character, the exact position of the apertures is known. The following position classes may be distinguished . . . :

P 0. Position unknown (ignote).

P 1. Catatreme spores: 1-treme spores with the center of the aperture at the proximal pole.

P 2. Anacatatreme spores: 2-treme spores with one aperture with its center at the distal and the other with its center at the proximal pole.

P 3. Anatreme spores: 1-treme spores with the center of the aperture at the distal pole.

P 4. Zonotreme (monozonotreme) spores: usually pleotreme spores with the centers of the apertures situated at the middle line of a zone coinciding with or parallel to the equatorial belt.—N.B. Zonocolpate spores are either orthocolpate (central line of the zone cutting the colpi at right angles) or loxocolpate (colpi converging in pairs), rarely clinocolpate (longest axis of the colpi coincident with the middle line of the zone).

P 5. Pleozonotreme (2-zonotreme, 3-zonotreme, etc.) spores: pleotreme spores with the centers of the apertures situated at the middle lines of two (or exceptionally more than two) mutually parallel zones.

P 6. Pantotreme spores: spores with the apertures more or less uniformly distributed over the spore surface."

GULA, n., pl. -ae. (L. throat). cf. bulla apicalis, contact figure, dehiscence cone, nozzle, top vesicle.

a. Potonié and Kremp, 1955, p. 12: *Tr.* If the tecta become very high (namely near the area of the apex), a dehiscence cone is formed which is called a *gula*. One should compare the diagnosis of the second subdivision Lagenotriletes to which the genus *Lagenicula* belongs. [*see* Y-MARK and Fig. 131].

b. Potonié, 1956, p. 49–50: *Tr.* Trilete spores, equatorial axis often shorter than the polar axis. The apex rises to form a "gula," that is, a higher dehiscence cone which elongates the polar axis.

The dehiscence cone originates in the following way: the tecta, at least in the immediate environs of the apex, become higher and in this way they are also more or less widened. The tecta considerably exceed their usual height and, in the typical Lagenotriletes, combine in this way to form a cone. Sometimes this is so distinct that the vertex of the tecta is not only steep or perpendicular, but is first bent backwards and then runs in an arc-like line toward the apex. In this case, a constriction originates in the dehiscence cone.

c. Couper and Grebe, 1961, p. 4: "A marked extension of the tectum at the proximal pole. This term is used mainly in the description of megaspores."

H

HAIR, n., pl. -s. cf. capillus, fimbria.
Jackson, 1928: "an outgrowth of the epidermis, a single elongated cell, or row of cells."

HALO, n. (L. halos, acc. halo; Fr. Gr. halos, disk of the sun or moon). cf. annulus, aspis, costa pori, dissence, endannulus, formation of lips, labrum, lip, margo, oculus, operculum, operculum pori, pore ring, protrudence.
Erdtman, 1952, p. 463: "(Bischoff, 1833): ±narrow area surrounding an aperture. Corresponds to 'annulus' and 'margo' of Faegri and Iversen (1950)."

HALONATE, adj. cf. annular, annulate, aperture, porate, pore.
a. Jackson, 1928: "when a coloured circle surrounds a spot."
b. Erdtman, 1952, p. 463: "(Bischoff, 1833): with halo."
c. Cranwell, 1953, p. 16: "Purkinje, 1830, annular.

HAMULATE, adj. (L. hamus, a hook).
Krutzsch, 1959, p. 39: *Tr.* Hitherto found only as sculpture form: little hook elements in irregular, disorderly arrangement; no distinct relief of an organized net-like character. [*see* ORNAMENTATION and Fig. 735].

HAPTOTYPIC CHARACTERS, n. (Gr. touch, and make an impression). cf. figura, Groups of Sporomorphae, morphography, shape classes.
Wodehouse, 1935, p. 542: "those which are due to internal or prenatal environment, such as the stimuli received by a developing pollen grain from contacts with its neighbors."

HARMOMEGATHUS, n., pl. -i. cf. colpus, figura, furrow, Groups of Sporomorphae, morphography.
a. Wodehouse, 1935, p. 542: "an organ or mechanism which accommodates a semirigid exine to changes in volume, e.g., the three germinal furrows of the grains of *Solidago speciosa* . . . [Fig. 737]."
b. Pokrovskaya, et al., 1950: *Tr.* The organ which regulates the volume of microspores depending on the amount of moisture in them (e.g., in the spores of *Salvinia* and in pollen grains of *Juglans* and *Carya*).

HARMOMEGATHY, n. (Gr. armozo, accommodate; megathos, size). cf. colpus, figura, furrow, Groups of Sporomorphae, morphography.
Wodehouse, 1935, p. 542: "volume-change accommodation . . ."

HEIGHT, n., pl. -s. cf. altitudo, dimension, figura.
Means of measurement of pollen; for height of disaccate grains; [*see* SACCATE and DIMENSION].

HEMISPHERICAL, adj. cf. polar cap, shape classes.
For outlines of disaccate pollen grains [*see* SACCATE, Zaklinskaya, 1957, Fig. 208-210].

HETEROBROCHATE, adj. (Gr. heteros, other). cf. reticulate.
Erdtman, 1952, p. 463: "with brochi of ± distinctly different sizes. Cf. . . . [Fig. 807–810]."
sdb. Beug, 1961, p. 11.

HETEROCOLPATE, adj. cf. colpate, colpato-colporate, colporate, colpus, furrow, Groups of Sporomorphae, porate.
a. Kupriyanova, 1948, p. 70: *Tr.* In contrast with the dicotyledons, the pore of the monocotyledons is never located in the furrow. Therefore both furrows and pores are not present on the surface of the grain simultaneously. [*see* PORE].
b. Faegri and Iversen, 1950, p. 129: "Some furrows with, others without pores, free pores absent." [*see* GROUPS OF SPOROMORPHAE and Fig. 73].

HETEROCYST, n., pl. -s. (Gr. hetero-, other; kystis, sack, bladder).
Jackson, 1928: "large inert cells in the filaments of certain Algae, separating contiguous hormogonia."

HETEROPOLAR, adj. cf. shape classes, subisopolar.
I. Jackson, 1928: "for the axis of Diatomaceae when the extremities differ."
II. Erdtman, 1952, p. 463: "in heteropolar spores the distal and proximal faces are ± distinctly different as regards apertures, etc. Cf. also 'Subisopolar'." [*see* POLARITY].
sdb. Beug, 1961, p. 11.

HETERORUGATE, adj.
  Erdtman, 1952, p. 463: "rugate; all rugae not of the same size."

HETEROSPORE, n., pl. -s.
  Melchior and Werdermann, 1954, p. 16. *Tr.* Formation of two kinds of spores which are of different size and of different sex: heterospory. [*see* SPORE].

HETEROSPOROUS, adj. (Gr. spora, seed).
  Jackson, 1928: "with spores of two kinds, as in *Selaginella*."

HETEROTASITHYNIC, n. (Gr. other, a straining, in a straight line). cf. figura, furrow, Groups of Sporomorphae, isotasithynic, morphography, tasithynic.
  Wodehouse, 1935, p. 542: "due to unequal lateral stresses, i.e., bilateral stresses, the forces which produce vertical cracking in a wall. This is due to a lateral shrinking at right angles to the vertical thrust of gravity. Such an effect is encountered in the oblong pollen grains of *Impatiens,* with four furrows, one at each corner of the grain and not arranged in the trischistioclasic system."

HEXATREME, adj. cf. aperture.
  Erdtman and Straka, 1961, p. 65: "(6-treme) spores: with six apertures." [*see* GROUPS OF SPOROMORPHAE].

HILATE, adj. cf. aperture, porate, pore.
  Erdtman, 1952, p. 12: "In hilate spores the laesura(e) is (are) reduced to a ± circular, indistinctly delimited aperture." [*see* HILUM].

HILUM, n., pl. -a. (L. a trifle). cf. aperture, laesura, pore, tetrad mark.
  I. Jackson, 1928: "the scar left on a seed where formerly attached to the funicle or placenta;"
  "the central point in a starch granule which the ring-like markings seem to surround;" "any kind of attachment;" "an aperture in pollen grains."
  II. Erdtman, 1952, p. 12: "In hilate spores . . . [Fig. 75] the laesura(e) is (are) reduced to a ± circular, indistinctly delimited aperture (L. hilum; cf. certain moss spores)."

HOMOBROCHATE, adj. (Gr. homos, one and the same, like). cf. polybrochate, reticulate.
  Erdtman, 1952, p. 464: "with brochi of ± the same size. Cf. [Fig. 751–753]."
  sdb. Beug, 1961, p. 11.

HOMOSPORE, n., pl. -s. cf. isopore.
  Dijkstra-van V. Trip, 1946, p. 22: *Tr.* Isospores, i.e. either spores of heterosporic plants whose heterosporic character is not externally discernible (homospores) . . . or spores which deliver only one kind of prothallium, that means a prothallium with both male as well as female archegonia (monospore). [*see* MEGASPORE].
  sdb. Schopf, 1938, p. 13; [*see* SPORE].

HOMOSPORIC, adj. cf. homosporous, isosporous.
  Jackson, 1928: "derived from one kind only of spore (Blakeslee)."

HOMOSPOROUS, adj. cf. homosporic, isosporous.
  Jackson, 1928: "similar-seeded, in opposition to heterosporous;" "neutral-spored."

HYALINE SMOOTH, adj. (Gr. hyalos, glass). cf. laevigate, psilate.
  Thomson and Pflug, 1953, p. 22: *Tr.* Completely transparent, strongly reflective. [*see* ORNAMENTATION].

HYPHA, n., pl. -ae. (Gr. kyphe, a web).
  Jackson, 1928: "element of the thallus in Fungi, a cylindric thread-like branched body developing by apical growth and usually septate."

I

INAPERTURATE, adj. (L. in-, not). cf. alete, aletus, colpate, currow, furrowless, non-aperturate.
  a. Faegri and Iversen, 1950, p. 128: "Pore rudimentary or absent." For complete definition [*see* Groups of Sporomorphae]. Non-aperturate (Erdtman, 1947) = inaperturate.
  b. Cranwell, 1953, p. 16: "Inaperturate: Faegri and Iversen (replacing non-aperturate, Erdtman and Straka), e.g., *Ruppia*."
  c. Traverse, 1955, p. 95: "Having no openings whatsoever—equivalent to but preferable to acolpate, used by others, because acolpate suggests that furrows are absent, whereas actually all openings are absent, including pores as well as furrows."
  sdb. Beug, 1961, p. 11.

INCIDENCE, n., pl. -en. cf. cuneus, platea, solution meridium, solution notch.
  Thomson and Pflug, 1953, p. 33: *Tr.* Remarkable is the stripe-like solution process of the endexine, which takes place in a meridional direction and which proceeds to a different extent in the different groups: We differentiate the following states: "solution notch" (=incidence), "solution wedge" (=cuneus), and "solution channel" (=platea). The overall term is "solution meridium." [*see* Aperture and Fig. 410–412].

INDEX, n., pl. indices (L. Fr. indicare, to point out). cf. dimension, exine index, figura, polar area index, shape classes.
  Ingwersen, 1954, p. 40: "figure indicating the ratio of a measurement to the maximum equatorial diameter (cf. exine index, polar area index, shape class index)."

INFRA-, (L. below, lower). cf. intra.
  Merriam-Webster, 1960: "a prefix meaning below, beneath, as in infrared, infracostal."

INFRA-GRANULATE, adj. cf. intra-granulate.

INFRA-PUNCTATE, adj. cf. intrapunctate, planitegillate, psilotegillate, punctate, punctitegillate.

INFRA-RETICULATE, adj. cf. intrareticulate, reticulate.

INFRA-RETICULUM, n. cf. intrareticulum, reticulum.

INFRA-RUGULATE, adj. cf. intrarugulate.

INFRA-SCULPTURE, n., pl. -s. cf. intrasculpture.
  Potonié and Kremp, 1955, p. 13: *Tr.* We speak of infra-sculpture (also intra-sculpture) if, as in the case of saccus formation, the exine lamellae detach from each other and a relief remains on the inner saccus wall. This relief is formed by the elements which stood as "structure" of the insulation layer between the exolamella and the intexine and originally joined them together. [*see* Fig. 198–199 and diagram, Potonié and Kremp, 1955, p. 18; placed in this text under Sporodermis].
  Definitions such as intra- or infra-punctate, -granulate, -reticulate have been used (the last term in Iversen and Troels-Smith, 1950) to describe structural elements that are between unseparated lamellae of the exine, and are not ornamentation features added as sculpture to the exine. They were often mistakenly called granulation, reticulation, etc.
  It must be said that structure and sculpture can grade into each other. One reason for this, among others, is that structural elements (that is, elements of the insulation layer) can project, even though smooth lamellae are draped over them.

INFRATECTAL, adj. cf. intratectal.
  Erdtman, 1952, p. 464: "infratectal details or patterns are confined to the space between the nexine and the overlying tectum."

INFRATEGILLAR, adj. cf. intrategillar.
  Erdtman, 1952, p. 464: "infrategillar details or patterns are confined to the space between the nexine and the overlying tegillum."

INFUNDIBULIFORMIS, adj. (L. infundibulu, a funnel; forma, shape). cf. aperture, cone-shaped, obinfundibuliformis, porate, pore.
  a. Jackson, 1928: "shaped like a funnel."
  b. Pokrovoskaya, et al., 1950: *Tr.* To clarify cone shaped pore canals. [*see* Pore Canal].

INORDINATUS, INORDINATE, adj. (L. disordered). cf. ordinatus.

Jackson, 1928: "when spores in an ascus show no regular arrangement."

sdb. Iversen and Troels-Smith, 1950, p. 47 and Fig. 620.

INSULATING LAYER, n., pl. -s. (L. insulatus, insulated). cf. endosexine, sporodermis.

I. a. Potonié, 1934, p. 9: *Tr*. . . . structures can be seen in the cross section of the exine as many rodlets which stand closely together and which are oriented radially. In opposition to the space between them, they have the ability to take on a dye-material like fuchsin, gentianviolet, etc., in strong intensity. H. Fischer and others recognized these rodlets as minute pillars which stand between the lamella and which separate them. An air interspersed insulating layer originates in the exine.

b. Potonié and Kremp, 1955, p. 19: *Tr*. The sum of the more or less loosely standing, radially directed rodlets, or *columellae,* which support the exolamellae, and their interspaces is to be called the insulation layer. [*see* SPORODERMIS].

II. Faegri and Iversen, 1950, p. 161: "(Isolierschicht Potonié, 1934) = cavity between endexine and tectum."

INTECTATE, INTECTATUS, adj.

a. Faegri and Iversen, 1950, p. 17: "Pollen grains in which the ektexine elements, if present, are free and isolated (*Ilex*) or form an open pattern (*Armeria*) are called *intectate* (Iversen and Troels-Smith, 1950)." [*see* ORNAMENTATION and TECTUM and Fig. 623, 626].

b. Harris, 1955, p. 26: "means that the wall is considered to consist of a single complete layer corresponding to the endexine in tectate spores. There may be surface projections forming a sculpture pattern, but these do not fuse to form either a continuous or a minutely perforate outer surface veiling the inner layer." [*see* TECTATE].

sdb. Iversen and Troels-Smith, 1950, p. 46; [*see* ORNAMENTATION].

INTECTATE-RETICULATE, adj. cf. extrareticulate, reticulate.

Erdtman, 1952, p. 470: "Simply reticulate ('intectate-reticulate' according to Iversen and Troels-Smith) with isolated bacula rising from the bottom of the lumina as in *Catopheria chiapensis* . . . [Fig. 754–756], *Cobaea spp.,* and other plants."

INTERCOLPAR, INTERCOLPARIS, adj. (L. inter, among, between).

Wodehouse, 1935, p. 543: "between the furrows."

sdb. Erdtman, 1943, p. 50.

Selling, 1947, p. 76.

Pokrovskaya, et al., 1950.

INTERCOLPAR THICKENING, n., pl. -s.

Wodehouse, 1935, p. 543: "thickened areas in the exine, e.g., in the grains of *Chorizanthe pungens*." see Fig. 608.

INTERCOLPIUM, n., pl. -ia. cf mesocolpium, mesorium.

a. Faegri and Iversen, 1950, p. 161: "the part of the exine between neighboring furrows."

b. Iversen and Troels-Smith, 1950, p. 34: *Tr*. These terms are used only for bipolar pollen grains.

inter C = intercolpium: area which is defined by the edges of the colpi (or the edges of the margo where well-limited margins exist) and the connecting lines of neighboring colpi tips (or tips of the margo).

inter C, med = medianum intercolpii: line which separates an intercolpium into two symmetrical halves.

inter C, cent = centrum intercolpii: point of intersection of the medians of an intercolpium.

sdb. Ingwersen, 1954, p. 40.

Beug, 1961, p. 11.

INTERIOR, n. (L. inter, between). cf. figura.

Traverse, 1955, p. 91: "Refers to parts toward the center of a pollen grain or spore. A part is interior to another if nearer the center of the pollen grain."

INTERIOR BOUNDARY, n., pl. -ies. cf. figura, focus 0–5, limes interior.

INTERLACUNAR CREST, pl. -s. INTERLACUNAR CRISTA, pl. -ae, n., cf. interlacunar crista, interlacunar margina, interlacunar ridge.

Wodehouse, 1928, p. 933: "A ridge separating lacunae from each other in lophate grains. Examples: *Cichorium intybus, Pacourina edulis*." [*see* ORNAMENTATION (Interlacunar lists) Potonié, 1934, p. 12].

INTERLACUNAR MARGINA, n., pl. -ae. cf. interlacunar muri, interlacunar ridge.

Potonié, 1934, p. 12: *Tr*. The lists which separate the lacunae are called by Wodehouse *interlacunar lists* (interlacunar cristae were here to be distinguished from interlacunar marginae). [*see* ORNAMENTATION (Interlacunar lists)].

INTERLACUNAR MURUS, n., pl. -i. cf. interlacunar margina, interlacunar ridge.

Erdtman, 1943, p. 51: "If the meshes (lamina) of a reticulum are large and arranged regularly, they may be termed lacunae. The lacunae are separated by interlacunar ridges (muri) or crests (cristae)." [*see* LACUNA].

INTERLACUNAR RIDGE, n., pl. -s. cf. interlacunar margina, interlacunar muri.

Wodehouse, 1935, p. 543: "one separating lacunae from each other in lophate grains, e.g., those of *Taraxacum officinale* . . . [*see* Fig. 765]."

INTERLOCULUM, n., pl. -a. (L. loculus, a little place).

a. Thomson and Pflug, 1953, p. 21: *Tr.* Space between the ekt- and endexine which is developed in the whole, or nearly whole exine body (among many Extratriporates).

b. Thomson and Pflug, 1953, p. 31–32: *Tr.* The distinct interspace between ektexine and endexine is called interloculum. It is especially large at the equator and widens here toward the pores. A direct connection between the ekt- and endexine is often not at all recognizable and is probably usually localized to small areas. [*see* DISSENCE and Fig. 544].

sdb. Krutzsch, 1959, p. 37; [*see* SPORODERMIS].

INTERPOLAR LACUNA, n., pl. -ae. cf. lacuna.

a. Wodehouse, 1928, p. 933: "The circumpolar lacunae, generally three in each hemisphere, which are alternate with the germinal pores in lophate pollen grains. Examples: *Vernonia jucunda, V. gracilis.*"

b. Wodehouse, 1935, p. 543: "a lacuna situated between, and bounded on one or two sides by aboral lacunae and wholly within one polar hemisphere in lophate grains, e.g., those of *Scorzonera hispanica* . . ." [*see* EQUATORIAL RIDGE and Fig. 763].

INTERPORIUM, n., pl. -ia. cf. mesoporium.

Iversen and Troels-Smith, 1950, p. 34: *Tr.* These terms are used only for bipolar pollen grains.

inter P = interporium: the area which is defined by the edges of the pores (or the annulus edges when well defined annuli are present) and the two lines which are tangential to neighboring pores (or their annuli).

inter P, med = medianum interporii: line which divides an interporium into two symmetrical halves.

inter P, cent = centrum interporii: point of intersection of the medians of an interporium.

[*see* MESOPORIUM, Erdtman, 1952].

INTERRADIAL POSITION, n., pl. -s. cf. spore.

Couper and Grebe, 1961, Fig. 1. [General term used in orientation and description of spores. *see* Fig. 160].

INTEXINA, INTEXINE, n. (L. intus, within). cf. endexine, intexine, nexina, nexine, nexinium.

a. Jackson, 1928: "the inner membrane when two exist in the extine, or outer covering of a pollen grain."

b. Potonié, 1934, p. 8: *Tr.* In fossil pollen, two layers of the exine are frequently well differentiated and separated from each other by sharper lines: the *exoexine* and the *intexine*. [*see* EXINE].

c. Faegri and Iversen, 1950, p. 161: "= endexine."

sdb. Pokrovskaya, et al., 1950.

INTEXINE RUGA, n., pl. -ae.

Potonié, 1934, p. 23: *Tr.* Primary meridional areas are already present as a broader surface in the recent freshly-dried pollen grain. They are here folded inward completely so that particularly this feature merits the name fold or ruga; more precisely we are dealing with an *intexine ruga.* [*see* APERTURE (Intexine ruga and exine ruga), Potonié, 1934].

INTEXTINE, n. cf. intexine.

INTINE, INTINIUM, n., cf. exine, medine, sporodermis.

a. Jackson, 1928: "the innermost coat of a pollen grain."

b. Potonié, 1934, p. 8: *Tr.* The intine is fragile and pliable and consists predominantly of pectin-rich cellulose. Hereby the pectin is especially enriched under the germinal apparatus (which see): it is strongly capable of swelling. The exine is stronger, less flexible, and strongly cutinized. In fossil pollen which have been prepared from sediments in the described manner, the intine is no longer identifiable; also, attempts to indicate its presence with chloro-zinc iodide prove unsuccessful. [*see* SPORODERMIS and APERTURE].

c. Erdtman, 1943, p. 50: "the inner, slightly resistant layer of a pollen or spore wall. At the germination of a pollen grain the intine protrudes, forming the membrane of the pollen tube."

d. Kupriyanova, 1948, p. 71: *Tr.* As we have already noted, the intine also has a certain role in morphology of pollen grains.

The intine (intinium) is a colorless membrane tightly attached to the cell plasma, which is not tinted by iodine or analin stains. It decomposes when boiled in alkali or acids. The intine is related to the thickness of the exine reversely, in the place where the exine gets thinner the intine layer becomes thicker and reverse. Naturally the thickened intine is located under pores. The intine of some genera (*Canna, Musa, Hedychium, Aphyllanthes, Crocus*) covers the whole grain uniformly with a thick layer. [*see* SPORODERMIS and Fig. 540].

e. Erdtman, 1952, p. 464: "(Fritzsche, 1837; intinium (Strasburger, 1882, p. 135): endosporium): the inner, usually not very resistant ('malacodermatous') sporoderm layer (= sporoderm except for sclerine)."

f. Cranwell, 1953, p. 16: "essential membrane of a pollen grain: occurring without exine covering in *Zostera;* acts as temporal furrow in *Potamogeton* (inaperturate), . . ."

sdb. Thomson and Pflug, 1953, p. 18; [*see* LAMELLA].

Hyde, 1954, p. 255; [*see* ONCUS].

Potonié and Kremp, 1955, p. 17; [*see* SPORODERMIS].

Beug, 1961, p. 11.

INTRA- (L. within, on the inside). cf. infra.

INTRA-BACULATE, INTRA-BACULATUS, adj. cf. baculate, infra-baculate.

a. Iversen and Troels-Smith, 1950, p. 46: *Tr.* [A structure type based on] distribution of granula beneath the tectum. [*see* ORNAMENTATION].

b. Cranwell, 1953, p. 16: "with rods (columellae) under surface of the exine, e.g., *Astelia*."

sdb. Thomson and Pflug, 1953, p. 21–23; [*see* ORNAMENTATION].

INTRA-GRANULATE, adj. cf. infra-granulate (low carriage).

Thomson and Pflug, 1953, p. 22: *Tr.* Round grains with a diameter of less than 1 micron. [*see* ORNAMENTATION].

sdb. Potonié and Kremp, 1955, p. 13; [*see* ORNAMENTATION (Infrapunctate)].

INTRA-PUNCTATE, adj. cf. infra-punctate, punctitegillate.

Thomson and Pflug, 1953, p. 22: *Tr.* Inhomogeneities in the form of dot-like bodies. [*see* ORNAMENTATION].

sdb. Potonié and Kremp, 1955, p. 13; [*see* ORNAMENTATION (Infrapuctate)].

INTRA-RETICULATE, INTRA-RETICULATUS, adj. cf. infra-reticulate, reticulate.

a. Faegri and Iversen, 1950, p. 161: "structure, refers to the arrangement of granules *below* the tectum. Ref. the significance of terms, cp. the corresponding sculpturing terms."

b. Erdtman, 1952, p. 464: "(Iversen and Troels-Smith, 1950: intrareticulate): an infrareticulate sexine is tegillate and has its endosexinous elements arranged ± as in an ordinary ('nontegillate') reticulum."

sdb. Iversen and Troels-Smith, 1950, p. 46; [*see* ORNAMENTATION].

Thomson and Pflug, 1953, p. 21–23; [*see* ORNAMENTATION].

Cranwell, 1953, p. 16.

Potonié and Kremp, 1955, p. 13–14; [*see* ORNAMENTATION].

INTRARETICULUM, n., pl. -a. cf. infrareticulum.

Beug, 1961, p. 11: *Tr.* (J. Iversen and J. Troels-Smith 1950) structure type: The reticulum is covered by a tectum.

INTRA-RUGULATE, INTRA-RUGULATUS, adj. cf. infra-rugulate.

Thomson and Pflug, 1953, p. 22: *Tr.* With irregularly arranged elongated structure elements. [*see* ORNAMENTATION].

sdb. Iversen and Troels-Smith, 1950, p. 46; [*see* ORNAMENTATION].

INTRA-SCULPTURE, n., pl. -s. cf. infra-sculpture.

INTRA-STRIATE, INTRA-STRIATUS, adj.

Thomson and Pflug, 1953, p. 22: *Tr.* With elongated structure elements arranged parallel, radially, etc. [*see* ORNAMENTATION].

sdb. Iversen and Troels-Smith, 1950, p. 46; [*see* ORNAMENTATION].

INTRATECTAL, adj. cf. infratectal.

INTRATECTUM, n., pl. -a. cf. contact figure.

Potonié and Kremp, 1955, p. 11: *Tr.* The more or less narrow interior of the tectum. [*see* Y-MARK and Fig. 141–148].

INTRATEGILLAR, adj. cf. infrategillar.

I-PATTERN, n. (Infrategillar pattern).

Erdtman, 1952, p. 464: "designation used in descriptions of tegillate (or tectate) spores; connotes, e.g., the 'cryptopattern' produced by infrategillar (infratectal) elements."

ISOMETRIC, adj. (Gr. isos, equal; metron, measure). cf. figura, Groups of Sporomorphae, morphography, shape classes.

Wodehouse, 1935, p. 543: "equal space appropriation, used here in a sense slightly modified from the usual meaning, characterized by equal measure, to describe the arrangement of spines and, occasionally, of pores which tend to be arranged at equal distances in all directions from each other." [*see* EQUIDISTANT of Erdtman, 1952].

ISOPOLAR, adj.

a. Jackson, 1928: "an axis of Diatom frustules is so termed when its extremities are similar (O. Mueller)."

b. Erdtman, 1947, p. 112: "hemispheres not different." [*see* GROUPS OF SPOROMORPHAE].

c. Erdtman, 1952, p. 464: "In isopolar spores there are no (or at least no sclerodermal) differences between the proximal and the distal faces." [*see* POLARITY].

sdb. Pike, 1956, p. 51.

ISOSPORE, n., pl. -s. cf. homospore, megaspore, spore.

a. Jackson, 1928: "a spore produced by one of the isosporeae."

b. Schopf, 1938, p. 13: "= homospore." [*see* SPORE].

c. Dijkstra-van V. Trip, 1946, p. 22: *Tr.* Isospores, i.e. either spores of heterosporic plants whose heterosporic character is not externally discernible (homospores) . . . or spores which deliver only one kind of prothallium, that means a prothallium with both male as well as female archegonia (monospores). [*see* MEGASPORE].

ISOSPOREAE, n.

Jackson, 1928: "plants having one kind of spore, as in Ferns, opposed to heterosporous."

ISOSPOROUS, adj. cf. homosporic, homosporous. Jackson, 1928: "homosporous, or having one kind of spore only."

ISOTASITHYNIC, n. (Gr. isos, equal; tasis, a straining; ithuneto, in a straight line). cf. figura, Groups of Spromorphae, heterotasithynic, morphography, tasithynic.

Wodehouse, 1935, p. 543: "due to equal lateral stresses, the forces which produce trischistoclasis, tending to form hexagons on a plane surface of hexagons, pentagons, squares, and triangles on a spherical surface, e.g., cracks in a plaster wall, caused by the shrinking of a plaster equally in all directions. Stands in contrast to heterotasithynic."

J-K

JUGATE, adj. (L. jugatus, connected or yoked together). cf. aggregation, Groups of Sporomorphae, spore, tetrad.

Jackson, 1928: "used in composition as conjugate, bijugate, etc."

KNEE, n., pl. -s. cf. aperture, colpate, colpus, furrow, geniculus.

I. Jackson, 1928: "an abrupt bend in a stem or tree-trunk;" "an outgrowth of some tree-roots."

II. [see GENICULUS].

KYRTOME, n., pl. -s. (Gr. kyrtos, arched, curved). cf. contact figure, torus.

Potonié and Kremp, 1955, p. 13: *Tr.* Tecta and curvaturae can be more or less reduced; often the curvaturae are completely lacking. Occasionally, only the rudiments which develop from the ends of the dehiscence ridges are present (Curvaturae imperfectae); these ends are pointed in the direction of the equator. Therefore, the dehiscence ridges look like fork-handles with two long fork prongs. This can be true in the case of trilete as well as in the case of monolete types . . . [cf. *Lavigatosporites* cf. *maximus* (Loose)].

The occasionally developed kyrtomes or tori . . . [Fig. 122] are bow-shaped, open toward the equator. They nestle in the angles which have been formed by the Y-rays, running more or less parallel to these rays. The kyrtomes are consistently formed exine folds which, according to Thomson and Pflug, 1952 (there called tori), are supposed to consist only of certain lamellae(?). Two kyrtomes can join at the equator. The name tori must be avoided since it has already been used. That which is called arcus in *Pollenites,* that is the bow-fold which run from aperture to aperture, cannot be equated with kyrtomes.

L

LABRATE LAESURA, n., pl. -ae. cf. labrum, laesura.

Couper and Grebe, 1961, p. 3: "spores exhibiting a comissure clearly bordered by a margo can be described as having a *labrate laesura* (monolete spores) or *labrate laesurae* (trilete spores). The labra can be developed as a rooflike extension over the suture . . . [Fig. 115–121]. Spores with such a feature are frequently referred to in the literature as having a 'raised' or 'crested' tetrad mark."

LABRUM, n., pl. -a. (L. lip). cf. annulus, aspis, costa pori, dissence, endanulus, formation of lips, halo, lip, operculum, pore ring, protrudence, swelling.

I. Thomson and Pflug, 1953, p. 36: *Tr.* Some triporate types do not develop an anulus, but evaginate the unthickened ektexine at the pores as a nozzle. This feature is called "labrum." [*see* APERTURE].

II. a. Potonié and Kremp, 1955, p. 11: *Tr.* In cross section the intratectum is surrounded by the walls of the tectum, that is, by the labra, lips, "swellings" or "dehiscence-ridges." The name "lips" would actually be proper only when the suture has opened . . . [Fig. 141–148]. However, it is also applied with certain justification to the walls of the tectum that are still connected. [*see* Y-MARK].

b. Couper and Grebe, 1961, p. 3: "a transition zone between the suture(s) of the tetrad scar and the remainder of the exine of the proximal surface. It is distinguishable by an increase of thickness of the exine or modification of the sculpture pattern or both. It is fundamentally a structural and/or sculptural differentiation bordering the suture when seen in polar view . . . [Fig. 115–121]."

LACUNA, LACUNE, n., pl. -ae, -s. (L. ditch, pit). cf. abporal lacuna, circumpolar lacuna, equatorial lacuna, germinal lacuna, interlacunar crest, interpolar lacuna, lacuna aequatorialis, lumen, paraporal lacuna, polar lacuna, poral lacuna, pseudocolpus, pseudospore.

a. Jackson, 1928: "an air-space in the midst of tissue; a depression on the thallus of a Lichen; applied to the vallecular canals of *Equisetum*."

b. Wodehouse, 1928, p. 933: "Areas in lophate pollen-grains, bounded by crests and generally smooth. Examples: *Barnadesia spinosa*, *Vernonia jucunda*."

c. Potonié, 1934, p. 11: *Tr.* The meshes of the network are to be called *lumina*. If the meshes or lumina of a network become larger and if, moreover, they are very regularly arranged, then Wodehouse (1928) speaks of *lacunae;* according to their position he gives the lacunae special names. [*see* ORNAMENTATION (Lacunae)].

d. Wodehouse, 1935, p. 543 and Fig. 761–765: "a large pit or depressed space in the exine of lophate or reticulate grains. Lacunae are never germ pores or furrows but may be occupied by one or the other of them."

e. Erdtman, 1943, p. 51: "if the meshes (*lumina*) of a reticulum are large and arranged regularly, they may be termed *lacunae*. The *lacunae* are separated by interlacunar ridges (*muri*) or crests (*cristae*)."

f. Faegri and Iversen, 1950, p. 162: "regularly distributed large openings in the ektexine. Wodehouse (1928) originally uses the term in a somewhat wider sense, comprising also great lumina in coarsely reticulate grains . . . ."

g. Pokrovskaya, et al., 1950: *Tr.* The uniformly arranged large clearances in the net (reticulum). Lacunae are separated from one another by interlacunar partitions. [*see* LUMEN].

h. Cranwell, 1953, p. 16: " 'regularly distributed large openings in the ektexine,' (Faegri and Iversen), e.g., *Phormium*. Wodehouse uses the term in a wider sense, e.g., for mesh of reticulate Liliaceae generally."

i. Harris, 1955, p. 26: "the cavity or space between the walls of a reticulum formed by anastamosing ridges (i.e., a lophoreticulate pattern)." [*see* ORNAMENTATION].

sdb. Iversen and Troels-Smith, 1950, p. 34; [*see* PORE].

LACUNA AEQUATORIALIS, n. cf. equatorial lacuna.

LACUNAR, adj.
Jackson, 1928: "pertaining to or arising from lacunae."

LAESURA, n., pl. -ae. (L. a hurting, injuring). cf. contact figure, fissura dehiscentis, monolete aperture.
a. Erdtman, 1946, p. 72: "Spores of mosses and ferns are, as a rule, provided with dehiscence fissures. Monolete spores (1-let) have one fissure severing the straight longitudinal streak, or 'laesura,' borne on the proximal part of the spore. Trilete spores (3-let) have three fissures longitudinally severing each arm (laesura) of the triradiate streak usually known as the tetrad scar (*cicatrix tetradica*). Alete spores (alet) have no preformed dehiscence fissures."
b. Harris, 1955, p. 26: "dehiscence fissure, including the commissure, and also the margo when this is distinguishable."
c. Couper and Grebe, 1961, p. 3: "consists of the suture (dehiscence line) together with the bordering labrum." [*see* Fig. 115–121].
sdb. Potonié and Kremp, 1955, p. 10; [*see* Y-MARK].
Couper, 1958, p. 102.

LAEVIGATE, adj. (L. laevigatus, smooth, slippery). cf. chagrenate, hyaline smooth, infra-punctate, laevigatus, laevis, levigate, levis, planitegillate, psilate, psilotegillate, smooth, translucent.
Jackson, 1928: "smooth, as if polished."

LAEVIS, adj. (L. smooth). cf. laevigate, levigate, levis, psilate, smooth.
Jackson, 1928: "smooth, in the sense of not being rough."

LALONGATE, adj. (L. latus, broad; elongatus, extended). cf. aperture, furrow, shape classes.
Erdtman, 1952, p. 464: "lalongate ora are transversely elongated. Examples: . . . [Fig. 348, 415, 425]."
sdb. Pike, 1956, p. 51.

LAMELLA, n., pl. -ae. (L. a thin plate or scale). cf lamella conspicua, lamella oppressa, membrane, sublayer.
I. Jackson, 1928: "a thin plate;" "Lamellae, the gills of Agaries."
II. Potonié, 1934, p. 8: *Tr.* If one of the two principle layers (in fossil material without the usual further preparation) can occasionally be further distinguished as two distinctly composite layers, then the line separating these *sublayers* or lamellae from each other is always much less distinct. [*see* SPORODERMIS and PILUM, Potonié, 1934].
III. Thomson and Pflug, 1953, p. 17 and Fig. 543–544: *Tr.* The wall of the sporomorphae is usually constructed of several concentric layers, the "lamellae." In the body of the pollen, the lamellae are usually arranged into two layers which are different in their composition. Because of resistant pollenins, the outer "exine" is preserved in the fossil; the inner cellulose-rich "intine" is destroyed during fossilization.

In spores, one correspondingly differentiates an "exospore" and an "endospore;" some spore groups have an additional outermost, usually sculptured covering which can easily tear open and fall off, the "perispore."

The number of lamellae which compose the exine varies—eight or more lamellae are not unusual.

It must be supposed that the exine (and probably also the exospore) consists in addition to the visible lamellae as seen through the ordinary microscope (N. A. 1.32)—the "lamellae conspicuae"—of still finer laminations which should be designated as "lamellae oppressae;" a lamella conspicua therefore, may consist of several lamellae oppressae. [Thomson and Pflug, 1953, p. 18: *Tr.*] By thickening, the latter can become lamellae conspicuae in one or the other types (N. A. 1.32) or, on the other hand, by thinning of the exine, the lamellae conspicuae develop into a lamellae oppressae. This is the only explanation for the fact that closely related groups show partly more and partly less lamellae. If in the following, the discussion is about "lamellae," then it means precisely "lamellae conspicuae."

One often differentiates an exterior group of lamellae, the "ektexine" (exoexine) and an inner lamellae group, the "endexine" (intexine) (Iversen); both are often separated by an intervening space. If this intervening space is clearly developed everywhere on the equator, then we speak of an "interloculum." Ekt- and endexine are usually formed by alternating structureless and structured lamellae. The structured lamellae usually consist of radially directed individual elements which often have the shape of small rods. Therefore, these lamellae are also called "rodlet-layers." Their thickness varies between fractions of a micron, but can also reach five microns and more in the anulus structure. The structureless lamellae are always thinner-walled than the rodlet layers and form the base and roof surfaces of the rodlets which stand like small pillars between the two surfaces. This sequence of layers of radial and concentric anisotropy gives the exine its high mechanical firmness and its important swelling capacity.

The number, strength, construction, and arrangement of the lamellae are very variable. If eight lamellae can be noticed, five of which are located in the ektexine and three in the endexine, then we designate the lamellae of the ektexine from the outside to the inside, as exolamellae a, b, c, d, e; those of the endexine as endolamellae a, b, c. Of these, exolamellae b and d and endolamella b are structured; the rest are unstructured. Phylogenetic changes can lead to an apparent single layering of the exine, e.g., as a result of the lamellae becoming submicroscopically small (lamellae oppressae) and intimately fusing. Also, the exolamellae a and b frequently break up into individual sculptured elements.

Also systematically important is the thickness ratio of the end- and ektexine and of the individual rodlet layers as well as the magnitude of the interloculum, etc. The shape and the arrangement of the rodlets play a role . . . [Fig. 544].

The exospore of spores is more simply constructed. Often two layers can be differentiated which may be structureless or structured and which we would like to call from the exterior to the interior, ektexospore and endexospore. Very probably each is a bundle of thin lamellae. This question, however, cannot always be decided in detail with the light-microscope. A distinct space between the lamellae that is limited to the equatorial region is called "zone." Such "zones" are especially commonly found in Paleozoic spores. The lamellae of the ektexospore shall be called from the exterior to the interior, ektexospore lamellae a, b, c, etc.; those of the endexospore, endexospore lamellae a, b, c, etc.

sdb. Krutzsch, 1959, p. 36; [see SPORODERMIS].

LAMELLA CONSPICUA, n., pl. -ae.
Thomson and Pflug, 1953, p. 17: *Tr.* Visible lamella . . . a lamella conspicua may consist of several lamellae oppressae. [see LAMELLA].

LAMELLA OPPRESSA, n., pl. -ae.
Thomson and Pflug, 1953, p. 17: *Tr.* It must be supposed that the exine . . . consists in addition to the visible lamellae . . . of still finer laminations which should be designated as "lamellae oppressae;" . . . [see LAMELLA].

LAMELLAR, adj. (L. lamelliatus, made up of thin plates).
Jackson, 1928: "composed of thin plates."

LAMELLIFORM, adj. (L. forma, shape).
Jackson, 1928: "in the shape of a plate or scale."

LATIMURATE, adj. (L. latus, broad). cf. reticulate.
Erdtman, 1952, p. 464: "with broad muri (mural

cross-sections as long as average luminal diameters or longer)."

LATIPORATE, adj. (L.) cf. trilatiporate.
Norem, 1958, p. 670: "with pores in one hemisphere only . . . [Fig. 421–1]."

LENGTH, n. cf. dimension, figura, longitudo.
For length of disaccate grain [see SACCATE].

LEPTOMA, n., pl. leptomata. (Gr. leptos, thin). cf. aperture.
Erdtman and Straka, 1961, p. 65: "thin walled areas which may be engaged in forming an opening in connection with the normal exit of substance from the inner part of the spores."

LEVIGATE, adj. cf. laevigate, psilate, smooth.
a. Kosanke, 1950, p. 11: "Ornamentation types." see Fig. 640.
b. Harris, 1955, p. 18: "Smooth, Example, *Sticherus cunninghamii.*" [see ORNAMENTATION].

LEVIS, adj. cf. laevigate, laevis, psilate, smooth.
Jackson, 1928: "smooth, in the sense of not rough; from the time of Linnaeus downward this has been spelled botanically as 'laevis'."

LID, n., pl. -s. cf. operculum, pore lid, pore plug.
I. Jackson, 1928: "the operculum of moss-capsules (W. J. Hooker);" "the distal extremity of the ascidium of *Nepenthes* which forms a lid-like appendage to the pitcher;" "the areas of pollen grains which are detached to permit the pollen-tubes to pass."
II. Kupriyanova, 1948: *Tr.* In some species the single furrowed pollen is provided with a lanacolate lid (tectum) on the bottom of the furrow . . . [see GROUPS OF SPOROMORPHAE].

LIGULA, LIGULE, n. (L. ligula, a little tongue).
I. Jackson, 1928: "a strap-shaped body, such as the limb of the ray florets in Compositae;" "a lobe of the outer corona in *Stapelia* (N. E. Brown);" "the thin, scarious projection from the top of the leaf-sheath in grasses;" "a narrow membranous, acuminate structure, internal to the leaf-base in *Isoëtes* and *Selaginella*;" "an appendage to certain petals, as those of *Silene* and *Cuscuta* (A. Gray);" "the ovuliferous scale in *Araucaria,* united with the bract, and resembling the ligule in *Isoëtes* (Potter);" "the envelope which protects the young leaf in palms, as *Chamaerops* and *Rhapis*;" "sealing growth in cones between the angles of the primary scales in *Dammara* Lam. (Church)."
II. Thiergart, 1938, p. 301: *Tr.* The main characteristic of these types [*Sequoia-pollenites polyformosus* Thiergart (1938)] is an exitus, which often appears surrounded by a swelling in a half or a complete circle; also it is furnished with a more or less long ligula. [see Fig. 212].

LIMB, n., pl. -s. (L. limbus, a border or hem). cf. amb, ambit, ambitus, circumscripto, contour, equatorial limb, extrema lineamenta, figura, Groups of Sporomorphae, limbus, morphography, optical section, outline, shape classes.

I. Jackson, 1928: "the border or expanded part of a gamopetalous corolla, as distinct from the tube or throat;" "the lamina of a leaf or of a petal;" "the margin of the leaf in Mosses when distinct in colour and cell structure."

II. a. Wodehouse, 1935, p. 543: "the visual boundary or edge of the apparent disk of a sphere. In pollen grains it is the same as the equator only when the grain is viewed with one of the poles exactly uppermost."

b. Harris, 1955, p. 26: "the apparent edge or outline of the spore—the periphery."

LIMBUS, n. cf. limb, velum.

a. Potonié and Kremp, 1955, p. 19: *Tr.* In the development of *sacci* . . . [Fig. 198–200] the bases of the columellae detach from the intexine, and the air space between intexine and exoexine expands. The exolamellae with columellae clinging to the innerside as infrastructure, then form the membrane of the saccus.

The sacci can either expand as a rounded vesicle . . . [Fig. 202] or (as in some Monosaccites) can spread out like an umbrella through an equatorial seam or *limbus* . . . [Fig. 198–200]. The limbus is a sharp crease at the edge of the saccus where distal and proximal saccus walls touch to form a narrow crease and probably more or less fuse together, forming a more solid flange which holds the umbrella rigid.

b. Merriam-Webster, 1956, p. 488: "Zoology and Botany. A border distinguished by color or structure."

LIMES ANNULI, n. Edge of the annulus.
Iversen and Troels-Smith, 1950, p. 33: *Tr.* Exterior demarcation line of the annulus. [*see* PORE].

LIMES COLPI, n. cf. colpate, colpus, colpus edge, furrow, furrow rim, margo.
Iversen and Troels-Smith, 1950, p. 32: *Tr.* Either the furrow or suture, or in the case of a thinned or absent exine, the exterior limit of the lighter area to which it gives rise. [*see* COLPUS].

LIMES EXTERIOR, n. cf. exterior boundary, figura, limes interior, LO-analysis, ornamentation.
Iversen and Troels-Smith, 1950, p. 42 and Fig. 636–637: *Tr.* In the analytical description of a pollen grain, one can use the following scheme:

A. Statement of the position of the pollen grain; that is what part outward—toward the observer (e.g., a pole, colpus, or intercolpium). [Iversen and Troels-Smith, 1950, p. 43: *Tr.*]

B. Statement of those parts of the pollen grain, which one wishes to describe (e.g., a wart or a pore).

C. Statement of the focusing in relation to the upper or lower boundary level of the pollen grain (compare pollen dimensions, . . . [*see* DIMENSION]). This can be given in the following way:

foc 0: focusing above the outer limes exterior of the pollen grain (compare foc 1). Foc 0 can be subdivided furthermore from above to below into a, b, c, etc., e.g., foc 0, a.

foc 1: focusing on the outer limes exterior of the pollen grain. Small distances above and below the boundary level can be designated with ÷ (above) and + (below).

foc 2: focus between foc 1 and 3. Foc 2 may be subdivided furthermore from above to below, e.g., into a, b, c, . . .

foc 3: focus on the interior boundary level of the exine of the pollen grain; small distances above and below are stated by ÷ or +.

foc 4: focus between foc 3 and 5; it may be subdivided furthermore into a, b, c, etc.

foc 5: focus on the center of the pollen grain; one sees the exine in a sharp profile. Small distances above and below are designated by ÷ or +.

foc ÷4 to ÷0: focus below the center of the pollen grain designated as above, but with a ÷ as a prefix.
sdb. Ingwersen, 1954, p. 39; [*see* DIMENSION].

LIMES INTERIOR, n., interior boundary, cf. figura, limes exterior, LO-analysis, ornamentation.

LIMES MARGINIS, n. cf. colpus, furrow.
Iversen and Troels-Smith, 1950, p. 33: *Tr.* Edge of the margo; exterior demarcation line of the margo. [*see* COLPUS].

LIMES PORI, n. cf. edge of the pore.
Iversen and Troels-Smith, 1950, p. 33: *Tr.* Edge of the pore: demarcation line of the pore; therefore, either the above mentioned furrow or suture or—in thinned or absent exine—the outer boundary of the lighter spots which result from it. [*see* PORE].

LINEA DEHISCENCE, n. Y-mark, cf. colpus, commissure, dehiscence, dehiscence ridge, furrow, germinal furrow, laesura, sulcus.

LINEAR TETRAD, n., pl. -s, [*see* TETRAD], cf. aggregation, pollen tetrad, spore.

LIP, n., pl. -s. cf. annulus, aspis, contact figure, costa pori, dissence, endannulus, fissura dehiscentis, formation of lips, halo, labra, labrum, laesura, margo, oculus, operculum, operculum pori, pore ring, protrudence, swelling, trilete mark, triradiate germination cleft.

I. Jackson, 1928: "one of the two divisions of a bilabiate corolla or calyx, that is, a gamopetalous or gamosepalous organ cleft into an upper (superior or posterior) and a lower (inferior or anterior) portion;" "the labellum of Orchids."

II. Potonié, 1934, p. 19–20: *Tr.* The margin of the exoexine immediately surrounding the pore is the *pore ring* or *annulus*. In *Corylus* sp. the pore ring is often slightly curved outward. In this case one speaks of the *lips* of the porus . . . [Fig. 391]. In *Alnus* this pattern is more pronounced. In the most extreme cases beak or chimney is spoken of. [*see* APERTURE].

sdb. Dijkstra-van V. Trip, 1946, p. 20.
Kosanke, 1950, p. 14; [*see* SPORE].
Potonié and Kremp, 1955, p. 11; [*see* Y-MARK].

LIRA, n., pl. -ae. (L. a ridge). cf. list, ridge, rugula, stripe, vallum.
Erdtman, 1952, p. 464: "the narrow ridges between the striae in a striate spore."

LIST, n., pl. -s. cf. crest line, lira, murus, ridge, rugula, stripe, vallum, vertex, wall.

LO-ANALYSIS, n. (L. from lux, light; O from obscuritas, darkness; analysis; Gr. releasing). cf. OL-analysis.
Erdtman, 1952, p. 21–22: "A sexine like that in . . . [Fig. 628-635] exhibits different patterns at different adjustments of the microscope. At high adjustment small white islands produced by the spinules are seen . . . [Fig. 629(1)]. When focusing at a slightly lower level the same islands turn ± dark . . . [Fig. 629 (2–3)]. At medium adjustment very small dark islands appear produced by the puncta in the tegillum . . . [Fig. 629(4)]. They later become bright . . . [Fig. 629(5)]. At low adjustment numerous small white islands, caused by the bacula, appear and later likewise turn dark . . . [Fig. 629(6–7)]. These patterns are referred to as S-, T-, and I-patterns respectively (S: suprategillar, T: tegillar, I: infrategillar).

"In pattern analyses made according to this method it would seem convenient to record the first shade of every system of islands (and not that of the channels) as they appear at successive adjustments of the microscope from high to low. Even if it is impossible to elucidate any sexine details in sporoderms seen in optical section it may sometimes be possible to establish—by 'LO-analysis' (L. lux, light; O obscuritas, darkness) of a sexine surface—whether the sporoderm in question is provided with LO-, OL-, or still more complicated patterns. Immersion objectives are, as a rule, necessary for observations of this kind. Nevertheless the method of penetrating the sexine and describing the patterns in the way just mentioned is often difficult and hazardous."
sdb. Beug, 1961, p. 11.

LOBATE, adj. (L. lobatus; Gr. lobos, the lower part of the ear).

I. Jackson, 1928: "divided into or bearing lobes."

II. Kosanke, 1950, p. 11: [*see* ORNAMENTATION and Fig. 640]. [A type of lobed structure as in *Densosporites.*]

LOLONGATE, adj. (L. longus, long; elongatus, extended). cf. aperture, furrow, shape classes.
Erdtman, 1952, p. 464: "lolongate ora are longitudinally elongated." [*see* Fig. 423].

LO-MUSTER, n., pl. -s. cf. LO-pattern.

LONGICOLPATE, adj. cf. colpate, furrow.
Pike, 1956, p. 51: "With long colpi. Colporate grains are longicolpate when the colpi are longer than the distance between their apices and the poles."

LONGITUDINAL, adj. (L. longus, long). cf. figura, longitudo.
Traverse, 1955, p. 91: "Referring either to an internal structure that is more or less parallel with the axis or to a superficial structure lying on a line perpendicular to the equator; that is, along a meridian."

LONGITUDO, n. (L. length). cf. dimension, figura, length.

I. Jackson, 1928: "means, botanically, in the direction of growth."

II. Iversen and Troels-Smith, 1950, p. 36: *Tr.* Length: a dimension parallel to the polar axis in bipolar grains. With this limitation, it is used as M measurement. [*see* DIMENSION].

LONGITUDO TRANSVERSA, n. cf. dimension, figura, transverse dimension.
Iversen and Troels-Smith, 1950, p. 36: *Tr.* Transverse dimension: a measurement perpendicular to the polar axis of bipolar grains. With this limitation, it is used as M measurement. [*see* DIMENSION].

LO-PATTERN, n., pl. -s. cf. LO-analysis, OL-pattern, OLO-pattern, S-pattern, ST-pattern, T-pattern.
Erdtman, 1952, p. 464: "any pattern which at high adjustment of the microscope appears as 'bright islands' (=L; from L. lux, light) separated by 'dark channels' and on the lower adjustment presents the reverse picture, viz., 'dark islands' (=O; from L. obscuritas, darkness) separated by 'bright channels.' . . ." [*see* LO-ANALYSIS].
sdb. Potonié and Kremp, 1955, p. 13; [*see* ORNAMENTATION and Fig. 638].
Pike, 1956, p. 51.

LOPHATE, LOPHATUS, adj. (Gr. lophos, crest). cf. echinolophate, lopho-reticulate, psilolophate, spinolophate, subechinolophate, sublophate.
a. Wodehouse, 1928, p. 933: "With the outer surface thrown into ridges, anastomosing or free. Examples: *Barnadesia spinosa, Vernonia jucunda.*"
b. Wodehouse, 1935, p. 543: "With the outer surface thrown into ridges, anastomosing or free, as for example in the grains of *Pacourina edulis* . . . [Fig. 750] or *Taraxacum officinale.*"

c. Erdtman, 1947, p. 106: "(Wodehouse, 1935, p. 543) . . . Ridges generally higher than the muri of a reticulum, sometimes vertically striate and (or) perforate."

d. Harris, 1955, p. 26: "ridged, with simple, well-defined, flange-like ridges, usually anastomosing. (See 'lopho-reticulate.') (Compare also 'costate,' 'cristate,' and 'rugulate.')" [*see* ORNAMENTATION].

sdb. Faegri and Iversen, 1950, p. 162.
Traverse, 1955, p. 94.
Couper, 1958, p. 102; [*see* ORNAMENTATION].

LOPHO-RETICULATE, adj. cf. lophate, reticulate.
Harris, 1955, p. 26: "a sculpture pattern comprising anastomosing ridges, the ridges being at least 1 micron in height." [*see* ORNAMENTATION].

sdb. Couper, 1958, p. 102; [*see* ORNAMENTATION].

LOXOCOLPATE, adj. cf. aperture, colpate, furrow.
Erdtman and Straka, 1961, p. 66: "colpi converging in pairs." [*see* GROUPS OF SPOROMORPHAE].

LUMEN, n., pl. -lumina. (L. light, opening). cf. brochus, lacuna, mesh.
I. Jackson, 1928: "the space which is bounded by the walls of an organ, as the central cavity of a cell."

II. Potonié, 1934, p. 11: *Tr.* The meshes of the network are to be called *lumina*. If the meshes or lumina of a network become larger and if moreover they are very regularly arranged, then Wodehouse (1928) speaks of *lacunae* . . . [*see* ORNAMENTATION].

sdb. Erdtman, 1943, p. 51.
Selling, 1947, p. 76.
Faegri and Iversen, 1950; [*see* ORNAMENTATION and Fig. 689].
Iversen and Troels-Smith, 1950, p. 36; [*see* ORNAMENTATION].
Pokrovskaya, et al., 1950, p. 2; [*see* LACUNAE].
Erdtman, 1952, p. 464.
Cranwell, 1953, p. 17.
Ingwersen, 1954, p. 40.
Harris, 1955, p. 19–21; [*see* ORNAMENTATION].
Potonié and Kremp, 1955, p. 15; [*see* ORNAMENTATION].
Traverse, 1955, p. 93.
Beug, 1961, p. 11.
Couper and Grebe, 1961, p. 8; [*see* ORNAMENTATION and Fig. 766–768].

LUNE, n. (L. luna, moon). cf. figura.
Wodehouse, 1935, p. 543: "an area on the surface of a sphere bounded by arcs of two great circles passing through the poles."

M

MACRO- (Gr. macros, long).
Jackson, 1928: "in Greek compounds = long;" "frequently but improperly used for mega-, or megalo-, large."

MACROSPORA, MACROSPORE, n., pl. -ae, -s. (Gr. spora, seed). cf. megaspore.
Jackson, 1928: "the larger kind of spore in vascular Cryptogams;" "the embryo-sac in Phanerogams."
sdb. Melchior and Werdermann, 1954, p. 16. [see Spore].

MACULAR, MACULATE, MACULOSE, adj. cf. flecked, maculatus.
Jackson, 1928: "blotched or spotted."

MACULATION, n. (L. macula, a spot).
Jackson, 1928: "the arrangement of spots on a plant."

MACULATUS, adj. (L. maculare, to spot). cf. blotched, flecked, infrapunctate, intrapunctate, maculate, maculose, maculosus, planitegillate, psilotegillate, punctate, punctitegillate, subpsilate.
a. Potonié, 1934, p. 10: Tr. In all such cases, but also when, in the cross-section, no rodlets are recognizable and where, in vertical view of the exine, punctate to maculate patterns become noticeable without the existence of a sculpture, one speaks of *punctation* or *flecks* (maculation).
  If possible, the exine is called punctate (in the formula p = punctatus) only if it is not possible to measure the size of the individual structure elements with an enlargement of some 400X; as soon as the elements can be measured with the aforesaid enlargement we speak of a spotted exine (in the formula m = maculatus). [see Punctate].
b. Faegri and Iversen, 1950, p. 162: "With structure, the elements of which can be measured by 400X magnification."

MACULOSUS, adj. cf. flecked, maculatus.

MAIN AXIS, n., pl. -es. cf. axis, polar axis, pole axis.
Potonié, 1934, p. 8: Tr. The line connecting the poles is termed the main axis . . . [see Equatorial Axis].

MAIN PORE, n., pl. -s. cf. endopore, exopore, pore.

MAMMILLA, n., pl. -ae. (L. a nipple or teat). cf. elevation, excrescence, process, projection, protuberance, protrusion, tuber, tubercle.
Jackson, 1928: "a nipple or projection; used for granular prominences on pollen-grains."

MAMMILLATE, adj. (L. mammillatus). cf. tuberculate, tuberose.
Jackson, 1928: "having teat-shaped processes."

MARGINAL, adj. (L. marginalis). cf. figura.
Jackson, 1928: "placed upon or attached to the edge."

MARGINAL FRILL, n., pl. -s. cf. aerostatic umbrella, velum.

MARGINAL RIDGE, n., pl. -s. cf. cap ridge, comb, crest, cristae, crista marginalis, crista proximalis, proximal crest.
Wodehouse, 1935, p. 543: "The slightly projecting rim of the cap or disk, e.g., in the grains of *Pinus* . . . [Fig. 205–206]."

MARGO, n., pl. margines. (L. an edge, border, margin). cf. annulus, aspis, colpus edge, costa pori, dissence, endannulus, furrow rim, halo, kyrtome labrum, limes colpi, lip, margo colpae, oculus, operculum, pore ring, protrudence, torus.
I. a. Faegri and Iversen, 1950, p. 22: "A similar area (*margo*) may often be found surrounding the furrow, and in this case the ektexine is nearly always reduced in thickness. This is the part of the exine, which carries out the harmomegathic movements. The sculpturing of the margo is different from that of the rest of the grain, the ektexine elements decrease. This development is accentuated in the furrows, where tectum and in some cases all granula disappear. Thus, the margo, represents the transition zone between the furrows and the remainder of the exine." [see Fig. 354, 356, 361].
b. Iversen and Troels-Smith, 1950, p. 32: Tr. Area which surrounds the colpus like a belt and which is differentiated from the remaining exine of the pollen grains by deviations within the ektexine.

84

sdb. Traverse, 1955, p. 93.
   Beug, 1961, p. 11.
II. Harris, 1955, p. 26: "a transition zone between the commissure(s) of the tetrad scar and the remainder of the exine. It is distinguishable by an increase in thickness or modification of the sculpture pattern, or both."
sdb. Couper, 1958, p. 113.

MARGO AEQUATORIALIS, n. cf. equatorial rim.

MARGO ARCUATA, n. cf. arcuate ridge, arcuate rim, contact figure, crista arcuata, curvatura.

MARGO COLPAE, n. cf. margo.
Pokrovskaya, et al., 1950: *Tr.* It is a thickened edge of the furrow in some microscopes.

MASSA, n., pl. -ae. (L. a lump). cf. contact figure, floating body, floats, swimming apparatus.
I. Jackson, 1928: "the mass or substance of a body."
II. a. Potonié and Kremp, 1955, p. 12: *Tr.* We call *massa* a granulated more or less shapeless sprout, which can be somewhat ball- to cone-shaped or perhaps also slightly three-lobed and which represents a transformation of the tecta of the Y-mark. We do not use the term *massula* which has been previously used because it designates pollen masses which are dispersed in clusters as in the case of the Mimosae.
   b. Potonié, 1956, p. 83 and Fig. 134–135: *Tr.* In *Cystosporites varius* (Wicher) the three sterile spores are connected with the fertile spore by a considerable complex of a foamy substance, a massa. It corresponds in constitution and in position to the foamy substance which is found at the apex of the megaspore of *Azolla.* [Potonié, 1956, p. 84: *Tr.*] On the contact area rest three or nine floating bodies. If only three are present, we have a picture as in *Cystosporites* where the contact area bears three rudimentary megaspores . . .
      In *Azolla,* there originates in the megasporangium, as well as in *Cystosporites,* only one fertile megaspore. The other megaspores become disorganized and form, together with the dissolved tapetum cells, the foam which becomes deposited as the exospore on the fertile megaspore. In this way there originates from the remainder of the three rudimentary spores of the tetrad of the fertile megaspore three, or three times three, floating bodies by redeposition of the substance. If one dissolves the floating body of *Azolla filiculoides* Lam. (a megaspore which has only three floating bodies) (cf. Maedler, 1954, pl. 5, Fig. 6, 7), there remains on the dissolved area the Y-mark, contact area, and curvaturae as in *Cystosporites* (cf. Potonié and Kremp, v. 1, pl. 10, Fig. 77).
   c. Couper and Grebe, 1961, p. 4: "A granulous

transformation of the tecta . . . [Fig. 130]. This term is used only in the description of megaspores."

MASSULA, n., pl. -ae. (L. a little lump). cf. aggregation, Groups of Sporomorphae, pollen mass, spore, tetrad.
I. Jackson, 1928: "the hardened frothy mucilage enclosing a group of microspores in Heterosporous Filicinae."
II. a. Jackson, 1928: "In Phanerogams, a group of cohering pollen-grains produced by one primary mother-cell, as in Orchideae; also styled pollen-mass." [*see* MASSA and CONTACT FIGURE].
   b. Potonié, 1934, p. 8: *Tr.* In some genera, the four daughter cells that come from the mother cell remain together as a *pollen tetrad* (Ericaceae). In others all descendants of a mother cell form a *massula* (pollen mass) of 8, 12, 16, 32, 64 pollen cells which are connected with one another (Mimosae). [*see* APERTURE, (MASSULAE . . .)].
   c. Sellings, 1947, p. 77: "a unit of more than four grains not = the entire product of one theca. Cf. *Tetrad* and *Pollinium.*"
sdb. Beug, 1961, p. 12.

MATRIX, n., pl. -ices. (L. the womb). cf. groundmass, ground substance.
I. Jackson, 1928: "the body on which a Fungus or Lichen grows."
II. Iversen and Troels-Smith, 1950, p. 34: "the homogeneous fundamental substance of the exine."
sdb. Cranwell, 1953, p. 17.
   Potonié and Kremp, 1955, p. 18; [*see* SPORODERMIS].

MATRIX POLLINIS, n.
Jackson, 1928: "The cell in which pollen-grains are developed; the pollen-mother-cell."

MAXIMAL, adj. (L. maximus, greatest). cf. dimension, figura.
Jackson, 1928: "employed to denote the utmost which an organism can endure as, the greatest degree of heat."

MEASUREMENT, n., pl. -s. cf. dimension, figura, mensura.

MEDIANUM COLPI, n. cf. colpus, colpus median, furrow.
Iversen and Troels-Smith, 1950, p. 32: *Tr.* A line which splits a colpus into two nearly symmetrical halves. One differentiates . . . medianum colpi longitudinalis (the longitudinal median of the colpus) and . . . medianum colpi transversalis (the transverse median of the colpus). [*see* COLPUS].

MEDIANUM COLPI LONGITUDINALIS, n. cf. medianum colpi.

MEDIANUM COLPI TRANSVERSALIS, n. cf. medianum colpi.

MEDIANUM INTERPORII, n.

Iversen and Troels-Smith, 1950, p. 34: *Tr.* Line which divides an interporium into two symmetrical halves. [*see* INTERPORIUM].

MEDINE, n. (L.) cf. intine, mesine.

Saad, 1963, p. 36: "the word medine (from medius) reflects the meaning of a layer intermediate in position between the exine and the intine." For further detail, *see* SPORODERMIS, Saad, 1963, and Fig. 570–577.

MEGASPORA, MEGASPORE, n., pl. -ae. -s. (Gr. megas, large; spora, seed). cf. macrospore.

a. Jackson, 1928: "the more correct form of MACROSPORE;" "the larger spores of vascular Cryptogams;" "used for OVULE;" "= embryo sac."

b. Dijkstra-van V. Trip, 1946, p. 21–22: *Tr.* In table 2 . . . some data on micro- and megaspores are assembled. From them, we may conclude that Zerndt's view is probably largely correct: microspores up to 100 μ, perhaps up to 200 μ, megaspores larger than 200 μ. Only rarely microspores occur above 200 μ, e.g., *Microsporites.* Megaspores below this limit are unknown.

In the table "micro spores" are put between quotation marks for the concept is *far from exact.* In our case it includes:

A. *Spores of spore plants—* a. True microspores, i.e., the spores of heterospore plants delivering the male prothalium (Lepidophytes, some Calamites and Noeggerrathiales).

b. Megaspores, namely various megaspores of the Calamites mentioned in a, e.g., (see Hartung, 1933) *Calamostachys Casheana* about 225 μ, *C. paniculata* 105–165 μ, *C. tuberculata* 90–100 μ, among others.

c. Isospores, i.e., either spores of heterosporic plants whose heterosporic character is not externally discernible (homospores) (probably various Calamites) or spores which deliver only one kind of prothallium, that means a prothallium with both male as well as female archegonia (monospores) (probably Calamites, Sphenophyllales, Leptosporangiate ferns).

B. *Pollen of seed plants—*a. Genuine pollen (Cordiates) (according to Florin, 1936, 110–120 μ).

b. Prepollen . . . (Pteridospermes). Among them there are some which measure much more than 100 μ (to mention only those actually being found in the coal in a dispersed condition): *Whittleseyinae*-prepollen, e.g., *Monoletes Schopf,* up to 490 μ (cf. pollen of *Dolerophyllum,* type 31 Zerndt); and the multicellular bodies described by Shimakura as "megaspore-like remains," some of which he illustrates. They largely agree with Florin's pictures (1937) of *Stephanospermum*-prepollen, 100–120 μ.

Dijkstra-van V. Trip, 1946, Table 2. Subdivision of the Spores According to Their Size.

Author	"Microspores"	Megaspores
Lange, 1927	smaller than 100 μ	artificial subdivision larger than 0.1 mm
Zerndt 1930, 1934	much smaller than 200 μ	intermediate zone dubious larger than 200 μ
Zerndt 1934, p. 27	*Microsporites karczewskii and M. gracilis* up to 300 μ, 420 μ, respectively	larger than 200 μ
Elovsky, 1930	30–60 μ	much larger, up to 2 mm
Ergolskaya, 1930	up to 120 μ	much larger
Sprunk and Thiessen, 1932	13–70 μ	500–3000 μ
Hartung, Calamariaceen, 1933	60–150 μ	100–350 μ
Raistrick, 1933, 1934	20–90 μ, up to 100 μ	
Wicher, 1934		Sieving fractions larger than 0.2 mm are investigated for megaspores
Shimakura, 1940		Opaque, oval to round bodies, 90–265 μ in size were described as megaspores; also translucent, oval bodies, multicellular between 100–200 μ
Hirmer, Noeggerathineae, 1940	90–130 μ	0.8–1.33 mm

sdb. Schopf, 1938, p. 10–13; [*see* SPORE].

MEIOSPORANGE, n., pl. -s. (Gr. meion, less; spora, a seed; angeion, a vessel).

Jackson, 1928: "Sauvageau's name for the smaller plurilocular sporangia enclosing zoospores of *Ectocarpus virescens,* Thuret."

MEIOSPORE, n., pl. -s.

Jackson, 1928: "the product of a MEIOSPORANGE arising through an ontogenetic reduction (Janet)."

MEMBRANA COLPI, n. colpus membrane, cf. colpus, furrow.

MEMBRANA PORI, n. pore membrane.

MEMBRANE, n. (L. membrana). cf. exine, furrow membrane, lamella, pore membrane, sublayer.

   I. Jackson, 1928: "a delicate pellicle of homogeneous tissue."

   II. Faegri and Iversen, 1950, p. 23: "Furrow or pore membrane." [*see* Fig. 349–350].

   III. Beug, 1961, p. 12: *Tr.* Is used here for exine . . .

MENSURA, n., pl. -ae. dimension, measurement, cf. figura.

MEOSPORE, n., pl. -s. meiospore.

MERIDIONAL, adj. (LL. meridionalis; Fr. meridies, midday, south). cf. figura.

Faegri and Iversen, 1950, p. 15: ". . . similarly surface features perpendicular to the equatorial plane are called *meridional.*"

MERIDIONAL AREA, n., pl. -s. cf. colpus, furrow, furrow-pleat, furrow-ruga.

Potonié, 1934, p. 15: *Tr.* We find a similar situation in variations of type 1 germinal apparatus in those pollen which show three large *primary meridional areas* . . . . These meridional areas can be roughly designated as the surfaces of sphere-sectors; however, their points do not reach the poles of the pollen grain, which is spherical in the ideal case. [Potonié, 1934, p. 23: *Tr.*] When the pollen grain swells, the edges of the sulcus separate, the furrow broadens and by expansion of the intexine finally becomes a broad secondary *meridional area* which is more or less pointed at both ends (polewards). Hence the floor of the meridional area is formed by expanded intexine; it is generally completely smooth while sculpture is situated only on the exoexine.

In the subfossil *Fagus,* as in the Tertiary pollen types which are morphologically or systematically related to *Fagus* pollen, the fossils never show meridional areas in an extended state.

However, there are also species which in the fossil state regularly exhibit only a broad, long meridional area, and in fact without a differentiated germinal apparatus. Thus we see, e.g., in pollen species morphologically related to the genus *Acer,* merely three simple meridional areas whose points approach the poles and which have a considerable breadth at the equator, i.e., where the differentiation of the germinal apparatus would be situated. Sometimes such pollen coats are completely compressed in the sagittal direction, whereby the intexine membrane filling the meridional areas, rupture and is in some instances completely lost. The pollen species of many of the Liliaceae for example are endowed with only one single simple meridional area. Therefore it may be expedient to differentiate the expanded meridional areas seen in fossils as *primary* (e.g., Liliaceae) from fossil, unexpanded, *secondary* meridional areas (e.g., Fagaceae). Primary meridional areas are already present as a broader surface in the recent freshly dried pollen grain. They here are folded inward completely, so that particularly this feature merits the name fold or ruga; more precisely, we are dealing with an *intexine ruga.*

The secondary meridional area attains greater breadth by swelling; and this by strong participation of expanding intexine. Here, where expansion is not too far advanced, the intexine again contracts by drying or by release of inner pressure.

It is probably a special development which leads to the primary meridional area.

For reasons of evolution, let us first consider the case of the pollen grain with three primary meridional areas. Of all known comparable pollen types, this case shows the most distinct homologies with the Carboniferous trilete spores provided with contact areas. The meridional areas can be readily perceived as contact areas which have secondarily lengthened beyond the equator. The structure or sculpture-furnished meridional stripes between the meridional areas would then, in their proximal part, be homologous to the Y-mark; they have lost, however, the dehiscence furrow, its function being taken over by the meridional area. Hence, dehiscence furrows are analogous to meridional areas.

We refer to such species as *Taxus, Larix, Sequoia, Taxodium, Cupressus,* and *Juniperus* as pollen without germinal apparatus. When the pollen of these genera are examined more closely, it appears that the exine through stronger swelling of the fresh pollen grain does not burst altogether irregularly. [Potonié, 1934, p. 24: *Tr.*] On the contrary, there often arises a quite sharp, straight fissure, a *fissura* . . . [Fig. 213]. A corresponding illustration of *Cupressus* was already given by Fritzsche (1832, pl. 1, fig. 16). This fissure is very characteristic for certain fossil pollen; we can refer to *Pollenites hiatus,* which is probably related to one of these genera. Weakening of the exoexine at a stripe corresponding to the fissure may also have led to the formation of the primary meridional area. [*see* APERTURA, Potonié, 1934; FISSURA, Potonié, 1934].

MERISPORE, n., pl. -s. (Gr. meros, a part).
Jackson, 1928: "the segment of a sporidesm."

MESEXINE, adj. (Gr. mesos, middle). cf. endosexine, insulation layer.
  a. Erdtman, 1943, p. 51: ". . . denotes, topographically, a layer between the ektexine and endexine though it may be formed by (and possibly should be referred to as) these layers or one of them."
  b. Selling, 1947, p. 77: "part of the exine between the ektexine and endexine . . . An arbitrary, merely topographical term."
  c. Faegri and Iversen, 1950, p. 162: "(Erdtman, 1943) = part of ektexine."
  d. Potonié and Kremp, 1955, p. 19: *Tr.* Other authors use the term mesexine which is supposed to lie between intexine and exoexine. Van Campo considers the mesexine as the interior part of our exoexine. The mesexine separates from the intexine during formation of the saccus *formation.* Accordingly, the mesexine is the insulation layer which consists of columellae, the endosexine of Erdtman.
  sdb. Pokrovskaya, et al., 1950.
      Harris, 1955, p. 26.

MESEXINIUM, n. cf. mesexine.

MESH, n., pl. -es. cf. brochus, lacuna, lumen.
  For Ornamentation of disaccate pollen grains [*see* SACCATE, Zaklinskaya, 1957, Fig. 224–225].

MESINE, n. (Gr. mesos, in the middle). cf. medine, sporodermis.
  Rowley, 1957: "Mesine is proposed as a new term for the laminated electron-dense layer between the intine and exine. This term is also used at present to refer to electron-dense laminated material found in a supra-exinous position." For further detail [*see* SPORODERMIS, Rowley, 1959, and Fig. 588a].

MESOBROCHATE, adj. cf. reticulate.
  Erdtman, 1952, p. 464: "with about 15–45 brochi per amb (in homobrochate grains)."

MESOCOLPIUM, n., pl. -ia. cf. intercolpium, mesorium.
  Erdtman, 1952, p. 464–465: "(Erdtman, 1951a; Faegri and Iversen, 1950): intercolpium: an area delimited by two adjacent colpi and by transverse lines drawn through the apices of the colpi. The lines form the equatorial limit of the apocolpia. A mesocolpium thus borders on two colpi and two apocolpia (in syncolpate pollen grains there are no apocolpia; in grains with unipolar syncolpatism there is one apocolpium only; in parasyncolpate grains the mesocolpia are delimited throughout by colpi)."
  sdb. Pike, 1956, p. 51.

MESONEXINE, n.
  a. Erdtman, 1948, p. 267: "a local occurrence form-

ing thickenings at apertures, etc." [*see* SPORODERMIS and Fig. 541].
  b. Faegri and Iversen, 1950, p. 162: "(Erdtman, 1948) v. nexine."

MESOPORATE, adj. (L.) cf. triporate.
  Norem, 1958, p. 670: "Equatorially placed pores . . . Note: the prefix "meso" may be dropped. The unmodified term "porate" is thereby restricted to equatorially placed pores . . . Tri(meso)porate."

MESOPORIUM, n., pl. -ia. cf. interporium.
  Erdtman, 1952, p. 465: "(Iversen and Troels-Smith, 1950, interporium): an area delimited by two adjacent pori and by their transverse common tangents. The tangents form the equatorial limit of the apoporia. A mesoporium thus borders on two pores and two apoporia."

MESORIUM, n., pl. -ia. cf. intercolpium, mesocolpium.
  Erdtman, 1952, p. 465: "an area delimited by two adjacent ora and by their transverse common tangents."

MESOSACCALE AREA, MESOSACCIUM, n., pl. -s, -ia.
  Erdtman, 1957, p. 3: "The surface of the corpus of a pollen grain with bladders can be divided into the following areas: n saccale areas, forming the floor of the sacci, n mesosaccale areas (mesosaccia), i.e., area between the sacci and in the same latitude as these . . ." [*see* SACCATE].

MESOSPORE, MESOSPORIUM, n., pl. -s. cf. endoexine, insulation layer, intexine, mesexine, mesosporoid.
  I. Jackson, 1928: "Dictel's term for an *Uredo*-spore which apparently will only germinate after a resting period;" "the middle portion of the spore of *Isoetes* (Fitting)."
  II. a. Dijkstra-van V. Trip, 1946, p. 22–23: *Tr.* In megaspores one distinguishes the following layers (from inside to outside): endospore, mesospore (sometimes), exospore, and sometimes perispore.
      The endospore is never preserved in the fossils: it is a thin cellulose membrane in recent spores.
      A mesospore, according to Fitting (1900), occurs in *Isoëtes* and *Selaginella.* Ergolskaya (1938) interprets the small dark band which he notices in the section within the exospore (exine) incorrectly as the endospore (intine). Wicher (1934) "for purely descriptive reasons" described in megaspores macerated from bituminous coal an "extremely thin, folded, hyaline membrane" as mesospore. This term was expanded by Zerndt (1934, p. 11) to an important palaeobotanic concept: the mesospore is supposed to be a completely closed and homogeneous spore membrane without a definite opening

place; it is fixed to the exospore at the inside of the contact areas; during germination it is supposed to tear along the triradiate ridges of the exospore. Although we do not know the real mesospore, it is not very likely that it is an artificial phenomenon caused by maceration as Sahabi (1936) thinks. Zerndt found it in *Triletes brasserti* (1934, p. 24). In *Triletes triangulatus and T. brasserti* we have recognized very distinctly the place where a possible mesospore is attached, a somewhat rough surface at the inside of the exospore at the place of the contact areas. Also Wicher mentioned these spore types among the species where a mesospore occurs. Nikitin (1934) believes that the limitation of the "Androtheca" of his Devonian megaspore *Kryshtofovichia africani* toward the true spore body is established by a thick granular mesospore.

The exospore is usually the carrier of the external sculpture of the spore. It follows from the countless coal petrographic pictures where megaspores are cut in section that the exospore is definitely not homogeneous. Slater and Eddy (1932) even described some of their types as "double layered."

We show in plate 8, fig. 78, a specimen of *Triletes mamillarius,* in which by the process of peeling off the exospore, an interior membrane becomes visible which probably is nothing more than an interior part of the exospore. A photograph with infra-red light by Leclerq (1933) indicates something similar. Also in the specimen of *Triletes hirsutus* shown in plate 8, fig. 79, it seems that an interior part of the exospore has become visible by over-maceration, it is shiny and partly transparent; we take *Sporites fumosus Schopf,* 1938, for something similar, probably it is also the same spore type.

b. Høeg, Bose, and Manum, 1955, p. 102–103 and Fig. 565–566: "In spores [of *Duosporites congoensis*] made translucent by means of alkali an internal body (= mesosporium) is observable, the diameter of which is about ½ to ⅔ of the diameter of the spore itself. It can easily be taken out by dissecting the spore under the binocular. It also sometimes becomes free if the spore, after maceration, is treated with ammonia, because the outer wall, the exosporium, may then break.

"The mesosporium is attached to the inner surface of the exosporium on the proximal side of the spore. The diameter of the attachment area is ½ to ⅔ of that of the contact area . . . The two areas are concentric. Very frequently the mesosporium is folded along the outline of the attachment area . . .

"In microtome sections of the attachment area (in macerated spores not treated with alkali) it is often difficult to observe any border between the meso- and exosporia. This may partly be due to the fact that our sections had to be cut rather thick (16–18 μ), because in thinner sections the spore wall mostly cracks or the mesosporium becomes detached from the exosporium. However, in some cases the border between them is observable.

The mesosporium wall has two layers. The outer layer is 3–7 μ thick and of a spongy structure, and has a fringed appearance within the attachment area . . . The inner layer is homogeneous and hyaline, c. 1 μ thick. If the chemical treatment of the spore is carried too far, the outer layer is completely lost and the inner hyaline one alone is left.

"If the mesosporium has become free, either by maceration or dissection, it is seen to have, on its proximal end, a triradiate mark which evidently is only an impression made by the triradiate mark of the exosporium. (In microtome sections no break in the mesosporium is seen here.) Along the sides of the branches of the triradiate mark on the mesosporium there is a row of dark spots, approximately 20 altogether . . . These are nipple-like projections, c. 5 μ high and 12–14 μ in diameter. The microtome sections show that they point inward, into the hollow of the spore . . ." [*see* Fig. 566].

c. Potonié and Kremp, 1955, p. 19: *Tr.* The *mesopore* is an interior part of the exospore and might correspond to the intexine. In Zerndt (1934, p. 10) the mesospore does not belong to the exospore. Other authors use the term *mesexine,* which is supposed to lie between intexine and exoexine. Van Campo's mesexine is the inner part of our exoexine. The mesexine separates from the intexine during formation of the saccus. Therefore, the mesexine is the insulation layer which consists of columellae; the endosexine of Erdtman. [*see* SPORODERMIS].

MESOSPOROID, n., pl. -s.

a. Potonié, 1958, p. 14–15: *Tr.* [In *Duosporites congoensis*] . . . the mesospore (the intexine) has separated from the exoexine and is only proximally attached to the exoexine as a mesosporoid. On the interior of the proximal side, the intexine contains small verrucae or coni which project toward the interior of the spore and which parallel the Y-rays.

According to Høeg, Bose, and Manum (1955, p. 105), the feature described by Potonié and Kremp, v. I, 1955, p. 53–55, pl. 1, fig. 8a, b, is also a mesosporoid. There, we deal with a spore assigned to *Laevigatisporites* . . .

Credit is due to Høeg, Bose, and Manum in

that, in a comparison with their own fossil material, they referred to the publication of Fitting (1900) [Bau and Entwicklungsgeschichte von Isoetes und Selaginella etc. — Bot. Z. v. 58, p. 107–164, pl. 5, 6, Leipzig]. In his study of extant species of *Isoetes* and *Selaginella,* this author found that in a certain stage of development of young megaspores, the intexine separates from the exoexine so that, except on the proximal side, an open space is formed which closes again at the maturity of the spore. The last fact is not mentioned by Høeg and associates.

In other words, in the extant *Isoetes* and *Selaginella,* the mesospore (the intexine) is separated from the exoexine only during a transitional stage. The intermediate space is filled with solution "which delivers material needed for the growing of the membranes." In this stage, the mesosporoid is attached only proximally to the Y-rays.

Fitting (1900, p. 115) says of *Isoetes* that the mesospore lies close to the exoexine in the mature spore; however, it is not "firmly connected to it and therefore, it always contracts from it when the spores are put into a dehydrating fluid."

In spite of many thin and polished section investigations of *Laevigatisporites* (that is in unmacerated material), we have seen nothing of a separated mesospore (-intexine); therefore, one would suspect that this separation occurs, if anywhere, in the maceration process. This does not mean that this separation might not have already occurred earlier in other species, for example in salt solutions during the fossilization stage.

Perhaps it also occurs in other species, more or less regularly, without the maceration procedure.

In the cases which were more closely investigated by us, namely *Laevigatisporites* (Potonié and Kremp, 1955, v. I, p. 54), we could not find any mesosporoid in macerated material, although it was observed in this genus several times.

In spite of the fact that dispersed spores, in general, are mature before shedding, one might sometimes find spores of an immature stage in macerated coals and this might possibly explain the difference . . .

Høeg, Bose, and Manum think that the presence of a mesosporoid might be of certain taxonomical value. They are probably right. The separation of the intexine from the exoexine might take place in certain species more easily than in others.

b. Potonié, 1958, p. 37: *Tr.* The Saccites are to be differentiated from Sporites with a mesosporoid. In the Saccites, a more or less partial inflation of the exoexine in the centrifugal direction occurs resulting in a change in the shape of the pollen grain so that only the intexine approximately preserves its original shape. On the other hand, the mesosporoid spores appear as if the intexine (that is, the mesospore) moved in the centripetal direction, after its more or less extended separation from the exoexine. In any case, the outer shape of the spore is preserved after the forming of the mesosporoid.

MEXINA, MEXINE, n.

Kupriyanova, 1956: *Tr.* . . . we have to conclude that it is impossible to apply the names ekt- and endexine to concrete layers, because under the name of ektexine is combined a whole group of layers (1) the uppermost granulated, easily soluble layer, first described by Afzelius in 1955, (2) one or several colummuar layers (sexina), (3) the lowest coarse-grained loose layer (mexina); the general exine of tetrads (synexine). [*see* SPORODERMIS and Fig. 540].

MICRO- (Gr. mikros, small). cf. scabrate.

Beug, 1961, p. 12: *Tr.* Microbaculate, -clavate, -gemmate, -echinate, -reticulate; also [used] as noun: microbacula, microclava, etc. In this way sculpture types are characterized where the elements are smaller than 1 μ in their largest dimension. Collective term (with exception of netlike sculpture types): scabrate, respectively scabra, -ae.

MICROCYST, n., pl. -s. (Gr. a bag).

Jackson, 1928: "an amoeboid cell which is surrounded by a membrane, the resting state of swarm cells of Myxogastres." [*see* Melchior and Werdermann, 1954, p. 14 under SPORE].

MICROFLORA, n., pl. -ae. (L. flora, goddess of flowers).

Jackson, 1928: "the alpine flora, especially when small and massed (Freshfield);" "the microscopic flora of a given locality."

MICROPUNCTUM, n., pl. -a. (L. punctum, small hole).

Erdtman, 1957, p. 3: "The outer wall (of the sacci) consists of thin ectosexine which is often perforated (shown in electromicrographs, not published, by Erdtman and Thorson in 1950). The small holes (micropuncta) are usually difficult to observe through an ordinary light microscope." [*see* SACCATE].

MICROSPORE, n., pl. -s. cf. megaspore, spore, sporomorph.

a. Jackson, 1928: "the smaller sized spore in heterosporous plants, as *Selaginella;*" "of late years applied to the pollen-grain."

b. Schopf, 1938, p. 11: "small male spores of heterosporous free-sporing plants are true microspores." [*see* SPORE].

sdb. Dijkstra-van V. Trip, 1946, p. 22; [*see* MEGA-

SPORE]. Melchior and Werdermann, 1954, p. 16, under SPORE.

MIOSPORE, n., pl. -s. (Gr. meion, less, smaller). cf. pollen, palynomorph, pollenospore, spore, sporomorph.

> Guennel, 1952, p. 9–10: "American workers have adopted the term "small spore" to denote spores of relatively small size, regardless of their functions. This term, however, seems rather vague and too general. In order to categorize spores of relatively small sizes and to differentiate them from macrospores or large spores, the term "miospore" is proposed. All fossil spores and spore-like bodies smaller than 0.20 mm, including homospores, true microspores, small megaspores, pollen grains, and pre-pollen, are arbitrarily called miospores. Large spores, on the other hand, are referred to as macrospores. The 0.20 mm measurement is used as the dividing line, because the standard screens which are used for sizing in the coal preparation process have openings of approximately 200 micra in size."

MONAD, n., pl. -s. (Gr. monos, one). cf. aggregation, Groups of Sporomorphae, spore, tetrad.
a. Jackson, 1928: "occasionally used for Zoospore."
b. Selling, 1947, p. 77: "pollen grain (spore) occurring singly."
c. Cranwell, 1953, p. 17: "Reichenbach, 1852, any grain occurring singly (cf. dyad, tetrad): to be distinguished from "pseudomonad" (Selling), signifying a grain whose development is known to have been compound, e.g., *Carex, Oreobolus*—typical of the pollen of the Cyperaceae, but not always recognizable as such."

MONO-,
> Merriam-Webster, 1960, p. 951: "a prefix meaning, single, alone."

MONOAPERTURATE, adj. (L.).
> Norem, 1958, p. 669: "With one aperture."

MONOCOLPATE, adj. (L. monocolpatus). cf. colpate, furrow, monocolpatus, monosulcate, sulcate, sulcatus.
a. Wodehouse, 1935, p. 543: "having a single germinal furrow or harmomegathus on one side of the grain. Example: *Ginkgo biloba* . . . [Fig. 245]. If the grain is encircled by a single furrow, it is regarded as dicolpate or zonate."
b. Selling, 1947, p. 9: "Grains colpate in a restricted sense. Grains monocolpate (= with one colpa without a pore or a transverse furrow)."
c. Erdtman, 1945, p. 190–191: [*see* SULCATE].
d. Traverse, 1955, p. 95: "with a single longitudinal furrow and no germ-pores."
sdb. Erdtman, 1943, p. 51.
> Faegri and Iversen, 1950, p. 128; [*see* GROUPS OF SPOROMORPHAE and Fig. 52].

Pokrovskaya, 1950.
Cranwell, 1953, p. 16; [*see* COLPATE].
Potonié, 1958.
Beug, 1961, p. 12.

MONOCOLPORATE, adj. cf. colpate, colporate, colpus, furrow, Groups of Sporomorphae, porate.

MONOLEPT, adj. cf. aperture.
> Erdtman and Straka, 1961, p. 65: ". . . (tenuitatiferous) spores: with one or several leptomata . . ."

MONOLETE, adj. (L. monoletus). cf. contact figure.
a. Erdtman, 1943, p. 51: "monolete; a spore with a single straight tetrad scar."
b. Selling, 1947, p. 77: "(spore) with one ± straight tetrad scar (on the proximal side of the spore)."
c. Traverse, 1955, p. 95: "Having a single scar in the coat(s), a result of the contact with other members of the tetrad."
sdb. Erdtman, 1947, p. 114.
> Pokrovskaya, et al., 1950.
> Erdtman, 1952, p. 465.
> Cranwell, 1953, p. 17.
> Thomson and Pflug, 1953, *see* Fig. 114.
> Beug, 1961, p. 12.

MONOLETE MARK, n. cf. contact figure, dehiscence fissure, laesura, monolete, monolete aperture.

MONOLETOID, adj. cf. contact figure.
> Erdtman, 1947, p. 114: "with one laesuroid-sulcoid aperture, not as distinct as a laesura." [*see* GROUPS OF SPOROMORPHAE].

MONOMORPHIC, adj. cf. figura, Groups of Sporomorphae, monomorphous, morphography, shape classes.
> Erdtman, 1952, p. 465: "in plants with monomorphic pollen all pollen grains are of the same (genetically fixed) type."

MONOMORPHOUS, adj. (Gr. morphe, shape). cf. monomorphic.
> Jackson, 1928: "of one form only, not polymorphic (Bailey)."

MONOPORATE, adj. cf. aperture, Groups of Sporomorphae, monopored, porate, pore.
a. Erdtman, 1947, p. 113: "with one pore." [*see* GROUPS OF SPOROMORPHAE].
b. Cranwell, 1953, p. 17: "with one pore, independent of a furrow, e.g., *Freycinetia,* all grasses."
sdb. Faegri and Iversen, 1950, p. 128; [*see* GROUPS OF SPOROMORPHAE].
> Norem, 1958, p. 669.
> Traverse, 1959, p. 95.
> Beug, 1961, p. 12.

MONOPORED, adj. cf. monoporate.
> Wodehouse, 1928, p. 933: "Possessing a single but

91

clearly defined germinal aperture. Example: *Phleum pratense.*"

MONOSACCATE, adj. cf. perisaccate.
Potonié and Kremp, 1954, p. 170. One-winged grains belonging to the subturma MONOSACCITES.

MONOSPORANGIATE, adj.
Jackson, 1928: "unisexual;" "applied to a flower with sporangia borne on separate axes, as the beech and oak;" "having one sporangium; further distinguished as macro- or micro-sporangiate, as they bear sporangia of the kind indicated."

MONOSPORE, n., pl. -s. cf. spore, sporomorph.
   I. Jackson, 1928: "a special spore in *Ectocarpus,* by Sauvageau considered to be a GEMMA."
   II. Dijkstra-van V. Trip, 1946, p. 22: *Tr.* . . . spore which delivers only one kind of prothallium, therefore a prothallium with both male as well as female archegonia . . . (probably Calamites, Sphenophyllales, leptosporangiate ferns). [*see* MEGASPORE].
   III. Melchior and Werdermann, 1954, p. 16: *Tr.* Formation of only one spore without reduction-division in the monosporangium; e.g. *Rhodophyta.* [*see* SPORE].

MONOSULCATE, adj. cf. colpate, furrow, monocolpate, monocolpatus, sulcate, sulcatus.
   a. Erdtman, 1947, p. 113: "with one sulcus; abberant forms." [*see* GROUPS OF SPOROMORPHAE].
   b. Selling, 1947, p. 8: "Grains monosulcate (= with one undivided sulcus). Cf. also monolete spores . . ."
   sbd. Kupriyanova, 1948; [*see* GROUPS OF SPOROMORPHAE].
      Norem, 1958, p. 669 and text—fig. 10.

MONOTREME, adj. cf. aperture.
Erdtman and Straka, 1961, p. 65: "(1-treme) spores: with one aperture." [*see* GROUPS OF SPOROMORPHAE].

MORPHOGRAPHY, n. (Gr. morphe, shape; graphe, a drawing). cf. anatomy, dimorphic, emphytic characters, eurypalynous, figura, fixiform, formula of spores and pollen, Groups of Sporomorphae, haptotypic character, harmomegathus, harmomegathy, heterotasythnic, isometric, isotasithynic, monomorphic, monomorphous, morphology, nonfixiform, organography, palynogram, pollen classes, pollen form index, shape, shape classes, size classes, size range, stenopalynous, tasithynic, taxonomy.
   a. Potonié, H., 1912, p. 3–5: *Tr.* . . . A sharp differentiation should be made between organography (one might also say morphography) and morphology. In the following it will be spoken of as morphologic characteristics only if dealing with theoretical discussions of the organisms, which derive from a comparative review of its shape relationships with equal regard given to the form relation of the organisms. Organography generally deals—at least in its original meaning—with shape relationships and their practical indexing without undertaking theoretical considerations . . .
      Organography deals only with construction and formation (the development, the organogeny). (Examples: leaf margins of the linden are serrate; young fern fronds are circinate).
   b. Jackson, 1928: "anatomy and descriptive histology (Vuillemin)."
   c. Potonié and Kremp, 1955, p. 2–3: *Tr.* The term morphology did not concern—in its original meaning by Goethe—the mere description of the shape, but the comparison of different shapes for theoretical purposes. Later the definition was expanded to include the pure descriptive work. Some scientists recognize this and now determine the mere descriptive work within morphology as the "purely morphologic treatment" (rein morphologische Behandlungsweise) (Niggli). Now the term morphology may be generally used in a more comprehensive way, meaning that which should be termed morphology in its original sense as well as that which is called the "purely morphologic" working method. We prefer for the "pure morphologic" work the term *morphography,* and use the term morphology in its original theoretical sense.
   sdb. Krutzsch, 1959, p. 41.

MORPHOLOGY, n. (Gr. morphe, shape; logos, discourse). cf. anatomy, Groups of Sporomorphae, morphography, taxonomy.
   a. Jackson, 1928: "the study of form and its development."
   b. Merriam-Webster, 1956: "1. the branch of biology that deals with the form and structure of animals and plants without regard to function; 2. the branch of linguistics that deals with the internal structure and forms of words; with syntax it forms a basic division of grammar; 3. any scientific study of form and structure, as in physical· geography, etc. 4. a. form and structure, as of an organism, regarded as a whole; b. morphological features collectively, as of a language."

MOTHER CELLS, n. (L. cellula, a small apartment). cf. maternal cell, spore.
Jackson, 1928: "those which divide to form other cells."

MULTIBACULATE, adj. (L. multus, many, much). cf. baculate.
Erdtman, 1952, p. 465: "muri supported by more than two rows of bacula are referred to as multibaculate."
sdb. Beug, 1961, p. 12.

MULTIHETEROCOLPATE, adj.
Norem, 1958, p. 672: "with colpi and pori on same grain."

MULTILATERAL, adj. (L. latus, a side). cf. polygonal, shape classes.
Jackson, 1928: "many-sided, having several flattened surfaces."

MURUS, n., pl. -i. (L. a wall). cf. list, ridge, wall.
a. Potonié, 1934, p. 11: *Tr.* Often the "reticula" consist only of smooth stripes or *lists* (muri). We then have the *reticulum* simplex. In both bases the exine is called *reticulate.* [*see* ORNAMENTATION (muri)].
b. Erdtman, 1943, p. 51: "low ridges separating the *lumina* of an ordinary *reticulum.*"
sdb. Faegri and Iversen, 1950, p. 21; [*see* ORNAMENTATION and Fig. 689–690].
Iversen and Troels-Smith, 1950, p. 36; [*see* ORNAMENTATION].
Pokrovskaya, et al., 1950.
Erdtman, 1952, p. 465.
Ingwersen, 1954, p. 40.
Potonié and Kremp, 1955, p. 15; [*see* ORNAMENTATION].
Traverse, 1955, p. 94.
Couper and Grebe, 1961, p. 8; [*see* ORNAMENTATION and Fig. 766–768].
Beug, 1961, p. 12.

MURUS SPORAE, n. cf. spore wall, sporodermis.

N

NEGATIVE RETICULUM, n., pl. -a. (L. negativus; Fr. negare, to deny; recticulum, a little net). cf. fossula, inverse reticulum, ornate, reticuloid, reticulum fossulare.

Kupriyanova, 1948, p. 70: *Tr.* If the tubercles have the form of *areas* (areolata) surrounded by narrow ditches (fossula) a *negative reticulate* structure is created (areolata, reticulum fossulare) . . . *Dracaena aurea . . . , Lysichiton kamtschatkense . . . ,* and *Chamaedorea concolor* . . . , as well as some others can serve as examples of the negative reticulate structure. [*see* ORNAMENTATION and Fig. 662].

  sdb. Erdtman, 1943, p. 52; [*see* Fig. 639F and Fig. 739].

    Potonié and Kremp, 1955, p. 14; [*see* ORNAMENTATION].

NETLIKE SCULPTURE, n., pl. -s. cf. reticulate.

Kupriyanova, 1948, p. 71, fig. 9–11: *Tr.* A term used with prefix words (simple, complex, etc.) to describe ornamentation types; for illustration see . . . [Fig. 662, 666–668]. [*see* ORNAMENTATION].

NETWORK, n., pl. -s. cf. fossula, reticuloid, reticulum.

NEXINE, NEXINIUM, n. cf. endexina, endexine, intexine, nexina, nexinium.

a. Erdtman, 1948, p. 387: "Nonsculptured exine . . ." Subdivisions: 1. endonexine, 2. mesonexine, and 3. ectonexine; [*see* SPORODERMIS].

b. Faegri and Iversen, 1950, p. 162: "(Erdtman, 1948) seems to correspond more or less to endexine."

c. Erdtman, 1952, p. 465: "the inner, non-sculptured part of the *exine.* Synonyms: Intexine(?) (Fritzsche, 1837, p. 28), 'Membran der Exine' (Fritzsche, 1. c., p. 31), Intexine (Potonié, 1934b), endexine (Erdtman, 1943a, p. 41)." [*See* SPORODERMIS, Erdtman, 1952, p. 19].

  sdb. Erdtman, 1952, p. 19; [*see* SPORODERMIS].

    Pike, 1956, p. 51.

d. Potonié and Kremp, 1955, p. 18: *Tr.* Nexine = intexine. [*see* SPORODERMIS].

e. Kupriyanova, 1956, fig 3: *Tr.* The term nexine must be retained for the most inner homogenic layer of exine, that is not stained with fuchsin . . . [*see* SPORODERMIS and Fig. 540].

  sdb. Pike, 1956, p. 51.

NON-APERTURATE, adj. (L. non, not). cf. acolpate, alete, aletus, asulcate, colpate, furrow, furrowless.

a. Erdtman, 1946, p. 71–72: "Non-aperturate grains (nap; *pollina nonaperturata*) may, eventually, be referred to as asulcate, acolpate, or arugate if derived from sulcate, colpate, or rugate pollen types."

b. Erdtman, 1947, p. 112: "Without apertures or aperturial areas." [*see* GROUPS OF SPOROMORPHAE].

c. Selling, 1947, p. 77: "Pollen without (preformed) apertures (i.e., pores or furrows of various kinds). Cf. *alete* spore."

d. Faegri and Iversen, 1950, p. 162: "nonaperturate (Erdtman, 1947) = inaperturate."

  sdb. Erdtman, 1952, p. 465.

    Norem, 1958, p. 668, text—fig. 6.

NONFIXIFORM, adj. cf. figura, Groups of Sporomorphae, morphography, shape classes.

Erdtman, 1947, p. 112: "without fixed shape." [*see* GROUPS OF SPOROMORPHAE].

  sdb. Erdtman, 1952, p. 465.

NOZZLE, n., pl. -s. bulla apicalis, dehiscence cone, gula, top vesicle.

N-STEPHANOPORATE, adj. cf. aperture, cribellate, porate, pore.

Traverse, 1955, p. 95: "Way of designating pollen with more than 3 pores." [*see* STEPHANOPORATE].

O

OB-,
  Jackson, 1928: "as a prefix; means inversely or oppositely; as obovate, inversely ovate; sometimes but incorrectly, used for sub-."

OBERVERMICULATE, adj. cf. rugose, rugulate.
  Kosanke, 1950, p. 11. [Rugulate, as in *Laevigatosporites pseudothiesseni*.] [*see* ORNAMENTATION and Fig. 640.]

OBINFUNDIBULIFORMIS, adj. (L. ob, reverse; infundibulu, a funnel; forma, shape). cf. aperture, infundibuliformis, porate, pore.
  Pokrovskaya, et al., 1950, p. 1: *Tr.* Reversed cone shape used to classify shape of certain pore canals. [*see* PORE CANAL].

OBLATE, adj. (L. oblatus; latus, broad). cf. shape classes.
  a. Jackson, 1928: "flattened at the poles, as an orange."
  b. Erdtman, 1952, p. 465: "distinctly flattened. This term is used exclusively in descriptions of radiosymmetric, isopolar spores where the ration between polar axis and equatorial diameter is 0.75–0.50 (6:8–4:8)."
  sdb. Erdtman, 1943, p. 44–45; see Fig. 521.
    Selling, 1947, p. 77.
    Pokrovskaya, et al., 1950.
    Faegri and Iversen, 1950, p. 162.
    Erdtman, 1952, p. 465.
    Ingwersen, 1954, p. 40.
    Pike, 1956, p. 51.
    Norem, 1958. [*see* SHAPE CLASSES and Fig. 167d].
    Beug, 1961, p. 12.

OBLATE SPHEROIDAL, adj. cf. shape classes.
  a. Erdtman, 1943, p. 45: "shape class oblate spheroidal has P:E ratio 1.00–0.88 (8:8–7:8)." [*see* GROUPS OF SPOROMORPHAE and Fig. 521].
  b. Selling, 1947, p. 77: "Ratio P:E (0.88–0.99)."
  c. Erdtman, 1952, p. 465: "this term is used exclusively in descriptions of radiosymmetric, isopolar spores, where the ratio between polar axis and equatorial diameter is 1.00–0.88 (8:8–7:8)."

OBLATOID, adj. cf. shape classes.
  Erdtman, 1952, p. 465: "subisopolar spores, which would be oblate if they were isopolar, are termed oblatoid. Cf. . . . [Fig. 519, 522, 527, 529]."

OCULUS, n., pl. -i. (L. an eye). cf. annulus, aspis, costa pori, dissence, endannulus, halo, labrum, lip, margo, operculum, pore ring, protrudence.
  I. Jackson, 1928: "the first appearance of a bud, especially on a tuber;" "the depression on the summit of some fruits as the apple."
  II. a. Thomson and Pflug, 1953, p. 31: *Tr.* Quite analogous to the anulus, the rodlets of the exolamella "b" can also be elongated into a ring-shaped circumpolar zone. This structure is called the "oculus." In contrast to the anulus, the oculus is smaller in cross-section, but larger in plan-view. It appears in the polar view as a dark, circular spot lying on the pore region or, as a dark, round disk or ring. [*see* APERTURE and Fig. 544].
    b. Thomson and Pflug, 1953, p. 20: *Tr.* An analogous swelling of the exolamella . . . [*see* DISSENCE].

OIDIUM, n., pl. -dia. (Gr. oion, an egg, + idion, a diminutive.)
  a. Jackson, 1928: "a term used to denote cancatenate conidia (Cooke): not to be confounded with the form genus *Oidium*, Link, the conidial stage of Erysipheae."
  b. Melchior and Werdermann, 1954, p. 14; [*see* SPORE].

OLIGOBACULATE, adj. (Gr. oligos, small, few). cf. baculate, simplibaculate.
  Erdtman, 1952, p. 465: "simplibaculate muri, where the number of bacula in the muri surrounding a lumen is less than twice the number of adjacent lumina, are referred to as oligobaculate."

OLIGOBROCHATE, adj. cf. reticulate.
  Erdtman, 1952, p. 465: "with comparatively few brochi (less than 15 brochi per amb)."

OLIGOFORATE, adj. cf. aperture, cribellate, foramen, foraminate, forate, Groups of Sporomorphae, perforate, porate, pore, polyforate.

Erdtman, 1952, p. 465: "with 12 foramina or less."

OLIGOPERICOLPATE, adj. cf. pericolpate, polypericolpate.

Norem, 1958, p. 671: ". . . 4–8 colpi . . . uniformly distributed over surface . . ."

OLIGOPERICOLPORATE, adj. cf. pericolporate, polypericolporate.

Norem, 1958, p. 671: ". . . 4–8 colpori . . . uniformly distributed over surface . . ."

OLIGOPERIPORATE, adj. cf. periporate, polyperiporate.

Norem, 1958, p. 670: ". . . 4–8 pores . . . uniformly distributed over surface . . ."

OLO-PATTERN, n. (L. from lux, light; O, from Obscuras, darkness). cf. LO-pattern.

Erdtman, 1952, p. 465–466: "a succession of three patterns (dark, bright, dark) at high, medium, and low adjustment of the microscope . . ." [see LO-ANALYSIS].

OL-PATTERN, n. cf. LO-pattern.

Erdtman, 1952, p. 465: "reverse to LO; LO-pattern (q.v.)." [see LO-ANALYSIS].

sdb. Potonié and Kremp, 1955, p. 13; [see ORNAMENTATION].

ONCUS, n., pl. -i. (Gr. onkos, bulk, mass). cf. intine, medine, sporodermis.

Hyde, 1954, p. 255 and Fig. 582–588. "The (Latin) pollen-morphological terms and definitions put forward by Erdtman (1943, 1952) and Iversen and Troels-Smith (1950), respectively, will doubtless be widely accepted, but, since they cover only grains which have been fossilized either by decay or by chemical treatment, and in which only the exine remains, it will probably be necessary in the interests of the study of pollen in the fresh condition to supplement these terms by some relating to the intine; one such is here proposed. [Hyde, 1954, p. 256.] "Strasburger (1882) and others have described conspicuous thickenings of the intine inside the apertures in different kinds of pollen grains. Zander (1935) called these thickenings Keimhöfe, "because they give the impression of empty hollows." He added that they greatly facilitate the recognition of many kinds of pollen and he listed nineteen taxa in which he had observed them. Erdtman (1943, p. 52) applied the term vestibulum (which had been used by Potonié, 1934, in the sense of a circular surface cavity lined with intexine) to a "small chamber under an aspidate pore." He went on: "Aspidate germ pores protrude as rounded domes. The protrusions are due to a thickening of the intine underlying the region of the pore." (The sentence here italicized was a quotation from Wodehouse, 1935, p. 363, whose studies were all made on non-fossilized pollen.) It might perhaps have been inferred therefore that Erdtman was using vestibulum as synonymous with Zander's Keimhöfe, though the context provides no evidence for or against this. In his later treatise (1952) Erdtman omits the term vestibulum altogether. Faegri and Iversen's definition (1950, p. 163) of vestibulum, viz., "a cavity forming the pore and separated from the interior of the grain by a low rim or by a separation between different layers of the exine," is clearly intended to exclude the intine, and the same applies to Iversen and Troels-Smith's (1950), viz., "a small anteroom 'Vorraum' within the pore which arises in such a way that the exine at the pore edge as seen in optical section has a split-in-two appearance.

"An English and international term is needed for Zander's Keimhöfe. Vestibullum (vestibule) having been precised in a different sense, the term oncus* (pl. onci) is here proposed instead. In pollen morphology an oncus may be defined therefore as a more or less conspicuous thickening of the intine occurring beneath the apertures of many kinds of pollen grains.

* Latinized from Greek onkos, bulk, mass, weight; whence in modern Greek also tumour, swelling. The combining form onco- has been employed in a few terms in medical science, but not previously in botany, and the use of oncus in the sense now proposed is unlikely to cause any confusion.

"Onci have been noted in porate grains belonging to the following families, as exemplified by the species mentioned, viz., among Dicotyledons: Betulaceae (Betula pubescens, B. verrucosa, Alnus glutinosa), Campanulaceae (Campanula rotundifolia), Corylaceae (Corylus avellana) . . . [Fig. 582–583], Juglandaceae (Juglans regia, Pterocarya taxifolia), Moraceae (Cannabis, sativa, Humulus lupulus, Morus alba), Ulmaceae (Ulmus glabra) . . . [Fig. 586–587], and Urticaceae (Parietaria diffusa, Urtica dioica); among Monocotyledons: Sparganiaceae (Sparganium ramosum) and Typhaceae (Typha latifolia). They have also been noted in the (colporate) grains of certain Rhamnaceae (viz., Frangula alnulus) . . . [Fig. 588] and (Rhamnus cathartica) and appear to be characteristic of the (colpate) grains of Oleaceae (e.g., Fraxinus excelsior, F. ornus) . . . [Fig. 584–585], (Jasminum nudiflorum, Syringa vulgaris, Forsythia sp.)."

OOGONE, OOGONIUM, n., pl. -a. (Gr. oion, an egg; gonos, offspring).

Jackson, 1928: "a female sexual organ, usually a spherical sac, containing one or more oospheres."

OOSPORE, n., pl. -s. (Gr.)

a. Jackson, 1928: "the immediate product of fertilization in an oophore."

b. Melchior and Werdermann, 1954, p. 14, [see SPORE].

OPEN PORE, n., pl. -s. cf. diaporus, exitus, germ pore, pore.

OPERCULATE, adj.
Beug, 1961, p. 12: *Tr.* Provided with an operculum.

OPERCULUM, n., pl. -a. (L. a lid). cf. annulus, aspis, costa pori, dissence, endannulus, halo, labrum, lip, margo, oculus, operculum pori, pontoperculate, pore lid, pore ring, protrudence.
  I. Jackson, 1928: "a lid or cover which separates by a transverse line of division, as in the pyxis and Moss capsules;" "also in some pollen grains;" "the cover of certain asci, which falls away at maturity (Traverso)."
  II. Wodehouse, 1935, p. 543: "a thickening of measurable bulk and clearly defined, of the pore membrane. Example: grass pollen . . . [Fig. 268–269], *Castalia* . . . [Fig. 270]."
  III. Erdtman, 1943, p. 51: "a thickening—clearly defined and of measurable bulk—of the pore membrane. On rare occasions the operculum may be represented by a number of more or less separate thickenings in the pore membrane."
  IV. Faegri and Iversen, 1950, p. 24: "In living grains both pores and furrows may be covered by an *operculum* (Wodehouse, 1928), an isolated part of the ektexine which is separated from the rest by a narrow zone in which the ektexine is missing or greatly reduced. As this zone forms an area of least resistance, the grain is generally ruptured along it when pressure is exerted from within, and especially in fossil material, pore opercula are frequently lost. If the furrow operculum is preserved, the effect may simulate that of a double furrow, separated in the middle by a boat-shaped part. Sometimes the two parts of the primary furrow are definitely separated from each other and may even coalesce with the. adjacent furrows (*Sanguisorba officinalis*)." [see Fig. 350].
  V. Erdtman, 1952, p. 466: "(Wodehouse, 1935, etc.): a thickening of measurable bulk and clearly defined of an aperture membrane (± circular in pori, elongate in colpi, etc.). Examples: . . . [Fig. 419–421]."
  VI. Traverse, 1955, p. 93: "An ektexinous modification of a pore membrane sitting on the membrane like a lid."
  VII. Pokrovskaya, et al., 1950: *Tr.* Small lid. Thickenings of the membrane of a pore which are expressed very distinctly and can be measured. In rare instances the cover is represented by a number of more or less isolated thickenings in

the pore membrane. [see GROUPS OF SPOROMORPHAE].

OPERCULUM COLPI, n. cf. colpus, furrow.
Iversen and Troels-Smith, 1950, p. 32: *Tr.* The thicker part of a colpus which is situated within a furrow or suture. The structure of the operculum is of similar construction as the remaining exine of the pollen grain. [see COLPUS].
sdb. Beug, 1961, p. 13.

OPERCULUM PORI, n. cf. operculum.
Iversen and Troels-Smith, 1950, p. 33: *Tr.* . . . the thicker part of a pore which is situated within a furrow or suture. The structure of the operculum is similar to that of the remaining exine of the pollen grain. [see PORE].
sdb. Beug, 1961, p. 13.

OPTICAL SECTION, n. (Gr. optikos, optic). cf. amb, ambit, ambitus, circumscriptio, contour, equatorial limb, extrema lineamenta, figura, focus 0–5, Groups of Sporomorphae, limb, limbus, morphography, outline.

ORATE, adj. colpate, colpoidorate, colpus, furrow, Groups of Sporomorphae, porate.
  a. Erdtman, 1947, p. 113: "with nonaspidate ora." [see GROUPS OF SPOROMORPHAE].
  b. Erdtman, 1952, p. 466: "with (os or) ora."

ORBICULAR, n. (L. orbiculus, dim. of orbis, orb). cf. circular, global, orbiculate, shape classes.

ORBICULATE, adj. (L. orbiculatus). cf. orbicular, shape classes.
Jackson, 1928: "disk-shaped."

ORDINATUS, adj. (L. in order).
Iversen and Troels-Smith, 1956, p. 46 and Fig. 621: *Tr.* . . . sculpture elements arranged in a design, antonym: inordinatus . . . [see ORNAMENTATION].

ORGANOGRAPHY, n. cf. morphography.

ORNAMENTATION, n. (L. ornamentum). cf. aboral lacuna, acantha, acanthon, acanthous, acidotus, acusporid, angustimurate, apsilate, areola, areolata, areolar, areolate, areolatus, bacularium, baculate, baculatus, baculum, bastioned, brochal, brochus, canaliculate, canaliculatus, capillate, capillus, cap ridge, caput, carinimurate, chagrenate, chagrenate-corrugate, cicatricose, cicatricosus, circumpolar lacuna, clava, clavate, club, collum, columella, columella conjuncta, columella digitata, columella simplex, columellate, comb, comb-like, conate, cone, cone-shaped, conus, convolute, corona, corrugate, corrugatus, costate, crest, crista, crista aequatorialis, crista marginalis, crista proximalis, cristate, cristo-reticulate, cristatus, crustate, curvimurate, density of pattern, duplibaculate, echinate, echinatus, echinolophate, echinulate, echinus, elementum elongatum, elementum punctualium, elevation, equatorial crest, equatorial fringe, equatorial lacuna, equatorial ridge, equidistant, excrescence, expansion fold, extrareticulate, extrareticulum, fence, fenestrate, fimbria, fimbria equatorialis, fimbriate, fimbriatus, flecked, foramen, forate, fossula, fossulate, fossulatus, fovea, foveate, foveola, foveolate, foveo-reticulate, fragmentimurate, gemma, gemmate, gemmatus, germinal lacuna, granifer, granula, granulate, granulatus, granule, granulose, granulum, granum, hair, hamulate, heterobrochate, heterorugate, homobrochate, hyaline smooth, infra-granulate, infra-punctate, infra-reticulate, infra-reticulum, infra-sculpture, infractectal, infrategillar, inordinatus, intectate, intectatus, intectate-reticulate, interlacunar crest, interlacunar crista, interlacunar margina, interlacunar muri, interlacunar ridge, interlacunaris, interporal lacuna, intra-baculate, intra-baculatus, intra-granulate, intra-punctate, intra-reticulate, intra-reticulatus, intra-rugulate, intra-rugulatus, intra-sculpture, intra-striate, intra-striatus, I-pattern, lacuna, lacuna aequatorialis, lacunar, lacune, laevigate, laevigatus, laevis, latimurate, levigate, levis, lira, list, LO-analysis, lobate, LO-muster, lophate, lophatus, lopho-reticulate, lumen, maculate, maculation, maculatus, macular, maculose, maculosus, mammilla, mammillate, marginal ridge, mesh, mesobrochate, micropunctum, multibaculate, murus, negative reticulum, netlike sculpture, network, obvermiculate, oligobaculate, oligobrochate, oligoforate, OLO-pattern, OL-pattern, ordinatus, ornamentation, ornamentatum, ornate, papilla, papillate, paraporal crest, paraporal lacuna, paraporal ridge, pattern, paxilla, paxillate, perforate, perforate tectum, pila, pilarium, pilate, pilatus, piliferous exine, piliferous, piliformis, pilum, pit, pitted, planitegillate, polar lacuna, polybrochate, polyforate, poral lacuna, porphyritic structure, prismatic structure, process, process-projections, projection, protuberance, protrusion, proximal crest, psilate, psilatus, psilolophate, psilolophatus, psiloluminate, psilotegillate, punctate, punctate-reticulate, punctitegillate, punctum, pustulate, ramibaculate, rectimurate, reticula cristata, reticula composita sepibus evallis, reticula sepibus densis, reticula sepibus evallis, reticulate, reticulatus, reticuloid, reticulum, reticulum cristatum, reticulum fossulare, reticulum simplex, retipilariate, retipilate, ribbed, ridge, ridged, rod, rodlet, rodlet-layer, rugate, rugatus, rugoporate, rugose, rugosus, rugous, rugulate, rugulatus, rugulose, rugulostriate, scabrate, scabratus, scrobicula, scrobiculate, scrobiculatus, scrobiculus, sculptine, sculptinium, sculptured, sculpturing element, sepes a vallis, sepes densi, setaceous, setose, setosus, setula, setulate, simplibaculate, smooth, sparse, S-pattern, spina, spine, spinolophate, spinolophatus, spinose, spinosus, spinula, spinule, spongy, sporodermis, ST-patterns, streak, stria, striate, striation, striato-reticulate, striatus, strio-rugulate, structure, subechinate, subechinolophate, subpilate, subpsilate, subreticulate, suprareticulate, suprategillar, sympilate, tectate, tectate-reticulate, tectatus, tectoid, tectum perforatum, tegillate, textura, texture, thorn-like sculpture element, T-pattern, translucent, triacontarugate, tuber, tubercle, tubercled, tubercular, tuberculata, tuberculate, tuberculose, tuberculous, tuberculum, tuberiferous, tuberose, tuberosus, tubulosus, tubulus, unadorned, unditegillate, undulate, valla, vallum, vermicular, vermiculate, vermiculus, verruca, verrucate, verrucatus, verrucose, verrucosus, wall, wart, warty sculpture, waved, wedge, winged, wrinkled; for ornamentation of disaccate grains *see* SACCATE.

a. Potonié, 1934, p. 10: *Tr.*

Punctation, maculation—In all such cases when in the cross section no rodlets are recognizable and where in the vertical view of the exine, punctate to maculate patterns become noticeable without the existence of a sculpture, one speaks of *punctation* (punctatus, maculatus) or *flecking* (maculation).

If possible the exine is called punctate (in the formula p = punctatus) only if it is not possible to measure the size of the individual structure elements with an enlargement of some 400X; as soon as the elements can be measured with the aforesaid enlargement, we speak of a spotted exine (in the formula m = maculatus) . . . [Potonié, 1934, p. 11: *Tr.*]

Sculpturing elements, grana, verrucae, pila—The sculpture elements very often consist of regularly arranged, more or less broadly projecting, irregularities of the exoexine. There are granules (grana) . . . [Fig. 649], little warts (verrucae) . . . [Fig. 655], spines (spinae) . . . [Fig. 652 and 656], rodlets (bacula) . . . [Fig. 651], clubs (pila) . . . [Fig. 653] which stand more or less closely together or in a certain way and in part in quite different shapes and sizes. There occur also combinations of different sculpture elements, e.g., a granulation or maculation between spines. The grana are generally more or less round to globular, and the verrucae have an irregular outline. *Pila* are very remarkable. In *Pollenites margaritatus* for example, they are so close together that the heads of the clubs almost touch . . . [Fig. 654] giving an impression of a network in vertical view. By means of this confluent canopy, there arises below the heads an insulating air layer in which the stems of the heads stand. H. Fischer uses some examples to discuss the fact that the heads or the upper ends of the sculpture elements could be connected (e.g., by cross beams), in some cases they even coalesce to an almost uniform membrane; this is our exolamella. By this, it is recognizable in what way the rodlet-impregnated insulation layers of the exoexine are morphologically related to sculpture.

Fischer further shows, that sometimes even two insulation layers can lie over each other.

Cristae, reticulum cristatum—A frequent type of arrangement of the above named sculpture elements is that of the cristae. These are fences or combs that are built up from pila, rodlets, etc., which sometimes may even be laterally connected and which can join to form networks (reticula cristata) . . . [Fig. 657].

Muri, reticulum simplex, reticulate, bastionate, cicatricose, canaliculate, lumen, lacunae, germinal lacuna—Often the "reticula" consist only of smooth stripes or *lists* (muri). We then have the *reticulum simplex*. In both cases the exine is called *reticulate*. The outline of the grain appears in this case bastionate . . . [Fig. 650], whereby (with not too large a lumen) the bastion prongs appear to be connected by "skins" . . . [Fig. 650].

If the stripes or lists are not arranged like a net, the exine is *cicatricose;* however, this is only the case if the grooves between the individual muri are just as broad as or broader than the muri. In the opposite case the exine is *canaliculate*. The meshes of the network are called *lumina*. If the meshes or lumina of a network become larger and if moreover, they are very regularly arranged, then Wodehouse (1928) speaks of *lacunae;* according to their position he gives the lacunae special names. The *polar lacuna* is located at the pole; the *circumpolar lacunae* surround the polar lacuna; in the *poral lacuna* is the germinal; we, therefore, prefer to use the general [Potonié, 1934, p. 12: *Tr.*] term *germinal lacuna* (for more details one should compare the excellent explanations by Wodehouse, 1928, e.g., p. 933).

Interlacunar lists—The lists which separate the lacuna are called by Wodehouse *interlacunar lists* (interlacunar cristae are to be distinguished from interlacunar margina).

Scrobiculi—If the lumina become very small and the stripes between them broader than the lumina, then the lumina become *scrobiculi*. Frequently it takes some effort in order to discover whether it is a true net sculpture or a punctation, maculation, granulation, or verrucation. The space between the puncta, macules, granules, and warts often gives the impression of stripes or lists by which more or less big net lumina seem to be separated (especially if these elements are arranged very regularly and if one does not look too closely). One gets the true picture if one watches to see if the outline of the microfossil is smooth or differentiated. Furthermore, one's observations are better if one uses a binocular microscope in the research work.

b. Erdtman, 1943, p. 51–52: "ornamentation, sculpturing; form elements appearing as a relief on the surface of the exine. The exine may be provided with spines, spinules, warts, granules, pila (small rods with rounded, swollen top-end), pits, streaks, reticulations, etc., and the pollen grains are accordingly described as echinate, subechinate, verrucate, granulate, piliferous, scrobiculate, striate, reticulate, etc. . . . [Fig. 639] is a diagram illustrating some of these types of ornamentation. In the upper row (1), they are seen from the side; in the middle row (2), from above at a high adjustment; and in the lower row (3), from above, but at a lower adjustment . . . The distribution of dark and bright areas in the lower rows is due to the character of the ornamentation, but similar patterns may also be produced by the texture of the exine. A network, or *reticulum,* is formed by anastomosing ridges (*muri*) on the surface of the exine, enclosing small, frequently more or less irregular, spaces (*Lumina*). Anastomosing grooves in the surface of the exine, enclosing small elevated exine surfaces, would constitute a negative, or inverse, *reticulum* . . . [Fig. 639:F]."

c. Kupriyanova, 1948, p. 70–71 and Fig. 658–669: *Tr.* The sculpture (ornamentation) of the exine must be differentiated from the texture (textura); the latter is created by partial separation of the exine layers, in this way forming a slight roughness on the surface of the exine; for example, we can point out the pollen grains of the grasses. The sculpture of the exine of monocotyledons is quite diverse.

The reticulate sculpture (reticulata) is most characteristic, the thornlike sculpture (echinata, spinosa) is somewhat rare. Both the first and the second sculpture are characteristic of most all groups of the monocotyledons and they are interrelated by means of some transitional stages. Spines (spinae) can be different in size and construction, in staining the preparations in some cases only the spines are tinted, while the membrane-like exine that covers the entire surface of the pollen grain remains colorless, as happens in staining grains having a pore or furrow membrane. Such staining indicates that the spines are composed of the ektexine element of the membrane, while the surface of the pollen is wanting this layer and is covered only with a layer of the endexine. As a good example there is pollen of *Abalboda* . . . The spines of *Eminium alberti* and large tuberculate sculpture of *Arum korolkovii* are not stained with fuchsin as is the other part of the exine. When treated with alkali, all such thorns and tubercules disintegrate.

Sometimes the shape and size of spines are different on the same pollen grains; for instance the pollen grains of *Aphylanthes monospeliensis* . . . possess both large conical and small somewhat flattened spines.

Along with the usual spines, the exine of pollen

grains of *Lamandra effusa* has thorns with *club-like thickenings* (clavata, piliformis) suggesting external glandules.

The flattened thorn appearance can be called *tubercels* (tuberculata).

If the tubercels have the form of *areas* (areolata) surrounded by narrow ditches (fossula) a *negative reticulate* structure is created (areolata, reticulum fossulare).

Dracaena aurea . . . Lysichiton kamtschatkense . . . and *Chamaedorea concolor . . .* as well as some others can serve as examples of the negative reticulate structure.

The most diminued sculpture should be considered as granulate sculpture (granulata). Erdtman defines it as follows: having small grains, the basal diameter of which does not exceed their height. This sculpture differs from the tuberculate structure by being much smaller.

Warty sculpture (verrucosa) that is, exine with warts, the basal diameter of which is greater than their height. The verrucate sculpture can be smaller or larger but is always with more or less flat protuberances.

Pitted (foveolata) sculpture in which the pits are represented by narrow indentations in the upper layer of the exine. Such sculpture is found in the representatives of the family *Restionaceae . . .*

An example of the coarse pitted sculpture is the pollen of some genera of the family *Centrolepidaceae . . .* [Fig. 665].

The most widespread sculpture element (among pollen of monocotyledonous plants) is represented by the netlike sculpture (reticulata). The netlike state can be of two types: *a little net of rodlike forms* (sepes a vallis)—when the walls of the net consist of joined rods providing the upper surface of the walls with a beadlike character. This sculpture is characteristic of pollen of the Lily, Amaryllis, Iris, and Orchid families and to a small degree of the Palm family. Walls of the net can be simple (reticulum simplex) and complex (reticulum compositum). In the first case the walls of the net consist of one row of rods . . . while in the second, it consists of several rows, usually two, jointed together; sometimes the rod rows diverge, in turn forming cell-like walls . . .

When the walls of a net do not consist of rods and remain unbroken (sepes densi), the upper surface does not show a beadlike pattern. An example of such sculpture are the families *Aponogetonaceae* and *Araceae.*

Discontinuous walls of the net or pollen of representatives of the family *Potoamogetonaceae* must be considered as the result of the destruction of the walls related to the general process of the reduction of exine. Pollen grains having wrinkled sculpture have been also found, though very seldom. They were discolored in *Tacca cristata.*

Frequently it is very difficult to distinguish the fine pitted sculpture from the fine tuberculous sculpture. In such cases a graphic scheme referred to by Erdtman (1943, p. 51) and A. L. Takhtadzhan and A. A. Yatsenko-Khmelevsky (1945, p. 41) must be used. The following must be taken into consideration: with the objective brought down, the pits appear as light spots and tubercels appear dark; while with the objective raised the picture is reversed. Also, it should be noted that it is convenient to study the sculpture on stained preparations.

d. Faegri and Iversen, 1950, p. 25–27 and Fig. 623–626, 693–697: "The sculpturing of the pollen grain is generally a fairly constant character and is in most cases an excellent means of recognition of pollen. It is, however, frequently difficult to describe in exact terms, and consequently somewhat unsatisfactory for analytical description. This is also the reason why we use as few terms as possible, and only distinguish between the following types, which are based upon the form and arrangement of *sculpturing elements,* i.e., those elements which project beyond an imaginary even surface, either the endexine in intectate pollen or an imaginary surface touching the lowermost parts of the tectum.*

* These terms refer to sculpturing only. Corresponding structure types are sometimes designated as *intra*-reticulate, *intra*-striate, etc.

"Table of Sculpturing Types:

"A. Sculpturing elements s.s. absent

B. Surface even or diameter of pits $<1\mu$ . . . . psilate (*Betula*)

BB. Surface pitted, diameter of pits $\leqq 1\mu$ . . . . foveolate (*Lycopodium selago*)

BBB. Surface with grooves . . . . fossulate (*Ledum palustre*)

AA. Radial projection of sculpturing elements ± isodiametric

B. No dimension $\leqq 1\mu$ . . . . scabrate (*Artemisia campestris*)

BB. At least one dimension $\leqq 1\mu$ . .

C. Sculpturing elements not pointed

D. Greatest diameter of radial projection greater than height of element

E. Lower part of element not constricted . . . . verrucate (*Plantago major*)

EE. Lower part of element constricted . . . . gemmate (*Juniperus*)

DD. Height of element greater than greatest diameter of projection

  E. Upper end of element not thicker than base . . . . baculate (*Nympheae*)*

* Mixed with clavae and echinae . . .

  EE. Upper end of element thicker than base . . . . clavate (*Ilex*)

CC. Sculpturing elements pointed . echinate (*Malva*)

AAA. Radial projections elongated (length at least twice the breadth) . . . .

"B. Elements irregularly distributed . . . . rugulate (*Nymphoides peltatum*)

BB. Elements ± parallel . . . . striate (*Menyanthes*)

BBB. Elements forming a reticulum . . . . reticulate (*Fritillaria*)"

Cf. Fig. 624–626

e. Iversen and Troels-Smith, 1950, p. 35 and Fig. 620–627, 688–709: *Tr.*

Sculpture of the exine—Elementa punctualia = dot-like sculpture elements: the greatest diameter* is less than twice as large as the smallest one.

  * In the case of sculpture and structure elements, the measurement parallel to the surface is designated as diameter, and the measurement perpendicular to the surface is designated as height.

ver = verrucae = warts: the largest diameter is as large as the height or larger. cf. footnote above; sculpture elements are neither pointed nor constricted.

gem = gemmae = grains: like verrucae, but with proximal constriction.

bac = bacula = rodlets: the largest diameter is less than the height; sculpture elements are neither pointed nor wedge-shaped.

cla = clavae = clubs: like bacula, but club-shaped.

ech = echini = thorns: pointed sculpture elements. The largest diameter can be larger or smaller than the height.

Elementa elongata = elongated sculpture elements: the largest diameter is more than twice as large as the smallest. Possibly the sculpture elements can be formed by a close union of dotlike elements.

[Iversen and Troels-Smith, 1950, p. 36: *Tr.*]

val = vallae = ridges, combs, etc. of different shape: oblong, vermiculately sinuous winding or of irregular outline. [The correct plural ending seems to be valli. From L. vallum, pl. -i, a wall.]

ret = reticulum = network: formed by the regular arrangement of oblong sculpture elements.

mur = muri: valla of a reticulum.

pl, lum = plateae luminosae: exine surface between the sculpture elements.

lum = lumina (sing.: lumen): the meshes of a reticulum surrounded by muri.

[Iversen and Troels-Smith, 1950, p. 46: *Tr.*]

Pollen sculpture types—Abbreviations of the terms are taken from the first 3 letters; e.g., psi = psilatus:

A. True sculpture elements absent

  B. Indentations lacking or less than 1μ . psilatus

  BB. With holes or pits $\geqq 1\mu$ . . . . foveolatus*

  * The bounds between foveolatus and reticulatus are determined in the following way: the diameter of the lumina equals, or is larger than the breadth of the bordering muri, (reticulate). The diameter of the fovea is shorter than the shortest distance to the neighboring fovea (foveolate).

  BBB. With scattered, elongated indentations . . . . fossulatus†

  † Fossulate-type presupposes that the fossulae do not anastomose to form sculpture elements (verrucae or valla).

AA. Sculpture elements present, all dot-like

  B. All dimensions $<1\mu$ . . . . scabratus

  BB. At least one of the dimensions = or is more than 1μ

    C. Sculpture elements not pointed

    D. Largest diameter > than the height

      E. Without proximal constriction . . verrucatus

      EE. With proximal constriction . . . . gemmatus

    DD. Largest diameter smaller than the height

      E. Without distal thickening . . . . . baculatus

      EE. With distal thickening . . . . clavatus

    CC. Sculpture elements pointed . . . . . echinatus

AAA. Sculpture elements present, all or at least some elongated

  B. Without order or without one particular arrangement . . . . rugulatus

  BB. With a predominantly parallel arrangement . . . . striatus

  BBB. Network arrangement . . . . reticulatus

One can differentiate two types according to the distribution of the dotlike sculpture elements:

inord = inordinatus: sculpture elements more or less incidentally dispersed.

ord = ordinatus: sculpture elements arranged in a design.

Pollen structure types—

A. Tectum not present . . . . intec = intectatus

AA. Tectum present .... tec =
tectatus

In tectate pollen grains one can differentiate in analogy to pollen sculpture types (see above) the following structure types according to distribution of granulae beneath the tectum.

intra-bac = intra-baculatus
intra-rug = intra-rugulatus
intra-str = intra-striatus
intra-ret = intra-reticulatus

[Iversen and Troels-Smith, 1950, p. 47: *Tr.*] Within the intrabaculate pollen grains one may differentiate two types (cf. p. 46):

inord = inordinatus
ord = ordinatus

. . .

f. Kosanke, 1950, p. 11: [*see* Fig. 640 showing ORNAMENTATION TYPES].

g. Pokrovskaya, et al., 1950: *Tr.* Sculpture, ornamented. Various elements which make a relief on the surface of exine. The following sculpture elements are recognized; grains, protrusions, warts, bands, combs, spines, club-shaped growth, reticulate thickenings, and pits.

The exine of a microspore is accordingly described as: granulata-granulous or granulate, tuberculata-tuberculate; verrucata-wartlike; striata-striate; cristata-comblike; echinata-thornlike; pilifera-club-carrying; reticulata-reticulate; scrobiculata-pitted. [For ornamentation of saccate pollen grains *see* Fig. 222–225].

h. Thomson and Pflug, 1953, p. 21 and Fig. 670–677: *Tr.* "Gefuege" cf. texture or pattern is the overall term for "structure" and "sculpture" differentiations. The structure of a lamella is defined according to its shape, size, number and arrangement of the building elements which compose the lamella. Accordingly, the sculpture is formed by the relief-building parts of the exine surface. Consequently, it is comprised of externally exposed structural elements of the exolamellae.

In some cases, it is difficult to decide whether structure or sculpture is at hand; then it must be sometimes assumed that the sculpture-forming exolamella is covered with a delicate structureless layer. Sculptures often lie over structures or lamellae of different structure may follow one another. In addition, the structure or sculpture of fossil exines is exposed to various modifying, changing secondary influences. Thus, "secondary structures" can often develop; for example, smooth surfaces may become roughened, sculptures may be ground off by corrosion, etc. *The authors assign a much higher systematic value to the structure characteristics of the germinal apparatus than to the structure or sculptural features.* Examples that have been studied show variations of structure and sculpture in species of the same genus, but on the other hand, convergence in species of different orders, etc. In our present [Thomson and Pflug, 1953, p. 22: *Tr.*] palynological work in the Tertiary, which applies especially to the separation of units corresponding to the botanical genera, sculpture and structure characteristics have only a subordinate value.

For descriptions, the following notations are used (partly according to R. Potonié, 1934, G. Erdtman, Faegri and Iversen, 1950) . . .

A. Homogeneous exine: appears not to be composed of particles (aperture 1.32).
  1. Hyaline smooth: completely transparent, strong-reflective.
  2. Smooth: completely transparent, weakly re-ly reflective.
  3. Chagrenate: translucent, individual structure particles are not recognizable.
B. Inhomogenous exine: appears to be composed of particles.
  I. Structure: structure particles below the surface of the exine.
    1. Intrapunctate: inhomogeneities in the form of dotlike bodies.
    2. Intrabaculate: distinct layers of rodlets present.
    3. Intragranulate: round grains with a diameter of less than 1 micron.
    4. Intrarugulate: irregularly arranged with elongated structure elements.
    5. Intrastriate: elongated structure elements arranged parallel, radially, etc.
    6. Intrareticulate: elongated structure elements arranged like nets.
  II. Sculpture: relief building structure particles of the exine surface.
    1. Baculate: rodlet sculpture.
    2. Verrucate: wart sculpture.
[Thomson and Pflug, 1953, p. 23: *Tr.*]
    3. Gemmate: small pearl-like sculpture.
    4. Clavate: small clublike sculpture.
    5. Echinate: spine sculpture.
    6. Rugulate: elongated sculpture elements irregularly arranged.
    7. Striate: elongated sculpture elements which are dominant in one direction.
    8. Cicatricose: parallel ridge sculpture. The ridges (muri) are as wide as or not as wide as the space between them. The ridges have the same overall crest height.
    9. Reticulate: muri as in 8, arranged like nets.
    10. Corrugate: rows of laterally fused warts with a varying crest height.
    11. Foveolate: circular pits.
    12. Fossulate: irregular depressions.
    13. Canaliculate: depressions arranged into par-

allel channels that are narrower than the bastiones between them.

Between these typical structure and sculptural forms are many transitions.

i. Harris, 1955, p. 18–21:

"(*I*) *Patternless*—The outermost layer may be fissured, fragmented, or ill-defined (which in certain cases is a characteristic feature and of diagnostic value), but lacking in uniform sculpture elements to which the terminology of groups (III) or (IV) would be applicable.

"(*II*) *Smooth or minutely sculptured*—Elevations or indentations, if present, are less than 1 μ in diameter.

Term	Definition	Example
Levigate	Smooth	*Sticherus cunninghamii*
Scrabrate	Flecked: with minute pits or elevations the psilate in part and scabrate of Faegri and Iversen (1950)	*Paesia scaberula*
Striate	Linear or 'fingerprint' patterns. The sculpture elements, though minute, exhibit parallelism	

"(*III*) *With cavities or depressions*—The pattern is regarded as comprising pits, indentations, or cavities in the general surface, not less than 1 μ in diameter.

Term	Definition	Example
Foveolate	The cavities or lumina may be up to 2 μ in diameter or, if larger, are too widely separated to form a reticulum	*Lycopodium varium*
Fossulate	Cavities elongate, regular to irregular, but not anastomosing	
Foveo-reticulate	The pits are large enough and close enough together to form a reticulum comprising the cavities (lumina or lacunae) and the intervening walls (muri) which separate them	*Phylloglossum drummondii*
Vermiculate	With grooves like worm tracks. The example is from Kosanke (1950), where the term is used (cf. fossulate)	*Punctati-sporites vermiculatus*

Areolate	With grooves which anastomose to form a reticulum. (A reticulate pattern may be seen in some verrucate exines, but the raised portions would provide the diagnostic character.)	*Anogramma leptophylla*

"(*IV*) *With projections or elevations*—The pattern is regarded as comprising elevations of, or projections from, the general surface, these being not less than 1 μ in height. The distinction here made between projections and elevations is a matter of convenience and is not necessarily of structural significance. Similarly, the term *papillate* as here used is synonymous with *baculate* as defined by Faegri and Iversen (1950).

"A. *Projections* irrespective of height, which are less than twice as long as wide. These may be grouped according to the characters of the apex, trunk, and base. The trunk or base is regarded as broad when a horizontal axis is equal to or greater than the vertical axis and narrow when the vertical axis is the greater.

Term	Definition	Example
Echinate	Apex more or less sharp. Trunk tapering with broad base.	*Pellaea* spp.
Echinulate	Narrow base.	
Setulate	Trunk not tapering. Apex not sharp. Trunk constricted.	*Polystichum sylvaticum*
Gemmate	Broad (greatest diameter).	
Clavate	Narrow (greatest diameter). Trunk not constricted.	
Papillate	Narrow.	*Hymenophyllum sanguinolentum*
Granulate	More or less isodiametric, not less than 1 μ or more than 1/20 equatorial diameter of spore if the latter is over 20 μ.	*Hymenophyllum* spp.
Verrucate	Broad, or, if more or less isodiametric larger than granulate.	*Lygodium articulatum*
Saccate	Projections which appear to be flaccid and hollow—classified in next section.	

"B. *Elevations*. These are long in surface view (length more than twice breadth) except where otherwise specified.

Term	Definition	Example
Rugulate	Wrinkled elements irregularly distributed.	*Botrychium australe*

NB.—Rugulae are long and narrow in surface view, whereas verrucae, if elongate are relatively broad. Intermediates can only be indicated by hyphenating the two terms.

Costate	Ribbed. Regular, well-defined elevations or corrugations more or less encircling the spore.	
Undulate	Waved, more or less isodiametric.	*Gleichenia circinata*
Saccate	Pouched, cavate, elevations approximately isodiametric.	

NB.—The walls of saccate elevations tend to be flaccid and hollow looking, whereas verrucate projections are firmer and appear to be solid.

Rugulo-saccate	Pouched, cavate, elevations, "long."	*Asplenium falcatum*
Lophate	Ridged, with simple flangelike ridges, either anastomosing or, if free, seldom much shorter than the shortest diameter of the spore.	
Lopho-reticulate	Ridges anastomosing.	*Ophioglossum vulgatum*
Cristate	Ridges suturelike, as if formed by portions of the outer wall uniting in a seam or crest.	*Asplenium flaccidum*
Cristo-reticulate	Crests anastomosing to form a network.	*Asplenium richardi*

"Note: Striate, if excluded from group (II) above because of a vertical axis not less than 1 µ in diameter, would have to be considered with this group, with costate, rugulo-striate, or strio-rugulate (*Thelypteris uliginosa*) as available alternatives.

"*The use of per- and sub- as prefixes*—For the purpose of defining the descriptive terms used in the more detailed categories, a threshold size limit for the sculpture elements was arbitrarily fixed at one micron. (Compare Faegri and Iversen, 1950.) The pattern may be readily classifiable at magnifications such as X400 even though the sculpture elements are smaller than one micron, but the necessity for critical measurements at this size level should be avoided. Border-line cases might be difficult to determine because of a slight variability. The use of the prefixes *per-* and *sub-* was suggested by Erdtman (1947), though he gave no rules. The convention is here adopted that sculpture patterns may be classified as in the more detailed categories ((III) and (IV) above) without accurate measurement if the term *sub-* is prefixed,

this prefix implying that the size of the sculpture elements is at or below the minimum.

"When the greatest height of a projection or the greatest transverse diameter of a cavity (lumen or lacuna) equals or exceeds one-tenth the equatorial diameter of the spore, this may be denoted by adding the prefix *per-* to the descriptive term.

"*Hyphenated terms*—If terms are used without adequate definition, a situation may arise where, for example, the exine of one species is described as *spinose* and that of a nearly related species as *echinate* either because the author regards the terms as synonymous or because there is a real, though unspecified, difference. Two factors which tend to discourage the adequate and strict definition of terms are the occurrence of border-line cases, and a reluctance to multiply terms. These two difficulties can be met to some extent by hyphenating terms, and by relying on the context to show whether distinct elements of both kinds of sculpture are present in a mixture or whether the pattern is so intermediate in character that within the limits of a small range of variability it could fall in either of the two descriptive categories.

"*Density of pattern*—the appearance of the sculpture pattern depends not only on the relative size of the constituent elements, but also on the distance between them. The practice is to describe the sculpture pattern as *dense* when the space separating the projections is less than one-twentieth the equatorial diameter of the spore and is *sparse* when in the main it is greater than one-tenth the equatorial diameter. Pits or cavities might be similarly treated, except when, together with the intervening walls, they form a compound pattern such as foveo-reticulate or lopho-reticulate. In other cases it is either stated or implied that the pattern is of medium density."

j. Potonié and Kremp, 1955, p. 13 and Fig. 157–159, 198–199, 638, 732, 738–746, 773–774, 802: *Tr.* In the description of the spore exine, it is necessary to distinguish between its structure, infrasculpture (= intersculpture), and sculpture . . .

Infrasculpture—We speak of infrasculpture (also intrasculpture) if, as in the case of saccus formation, the exine lamellae detach from each other and a relief remains on the inner saccus wall. This relief is formed by the elements which stand as "structure" of the insulation layer between the exolamella and the intexine and originally joined them together . . . [Fig. 198–199 and diagram placed in this text under Sporodermis].

Infrareticulate—Definitions such as intra- or infra-punctate, -granulate, -reticulate have been used (the last term in Iversen and Troels-Smith, 1950) to describe structural elements that are between unseparated lamellae of the exine and they were often mistakenly called granulation, reticulation, etc.

It must be said that structure and sculpture can grade into each other. One reason for this, among others, is that structural elements (that is, elements of the insulation layer) can project, even though smooth lamellae are draped over them . . .

OL- and LO-patterns—German words for structure and sculpture are *Verzierung* and *Muster*. Erdtman (1952, p. 22) distinguishes between LO-pattern and OL-pattern (L = Lux, O = Obscuritas). Under high focus of the objective, LO-patterns appear as light islands separated by darker zones and, in lower focusing, show the reversed picture. Under high focus OL-patterns appear as dark islands, separated by lighter zones and in lower focusing, show the reverse . . . [Fig. 638].

Infrapunctate—An exine with structure can have a ± smooth surface, that is a smooth outline . . . [*Leiotriletes sphaeotriangulus* (Loose)]; it does not have to show a sculpture. However, it can show a distinct design in plan view, for instance a punctation, maculation, or netlike design . . . [*Calamospora pedata* (Kosanke)], or a streaky, cloudy, flowlike structure or chagrenate appearance, etc. These apply only for the inner structure of the exine and also sometimes for its infrasculpture. Therefore, terms like punctation, etc., have never been used by us for surface sculptures; however, as already mentioned, they have previously been confused with granulation, etc. Therefore, the terms *infrapunctate, infragranulate, infrareticulate* (Iversen and Troels-Smith) appear to be practical to us.

One would speak of infrapunctation, -maculation, only in those cases where one can see punctated, maculated, etc. designs in the plan view of the exine without the existence of an outer sculpture. Therefore, these constructions are called structures of the exine. Those elements which are on top of the exolamella of the exine form the outermost "layer."

Sculpture—Only those form elements which stand out in relief on the surface of the exine can be called sculpture. [Potonié and Kremp, 1955, p. 14: *Tr.*] The sculpture elements consist of often more or less regularly arranged ± highly raised irregularities of the exoexine. There are little grains (grana, granula), little warts (verrucae), cones (coni), spines (spinae), rodlets (bacula), clubs (pila), hairs (capilli) of quite different shape and size and which stand more or less closely together or are grouped in a certain way. Combinations of different structural and sculptural elements also occur.

Granulae, grana—The granulae or grana are more or less circular in plan view; as seen from the side, they are more or less circular, semi-circular, or semi-oval. We speak of grana only if their size does not change too much on the individual spore which they decorate or if it only changes gradually from one area of the spore surface to the other . . . [Fig. 732, 738].

Verrucae—The *verrucae* have a more irregular outline, and, more important, on one and the same spore they vary somewhat in shape and size; in plan view one can notice smaller verrucae next to somewhat larger ones, round next to less round, lobulate next to bean-shaped, etc. Consequently, when the elements stand closely together, they form a more irregular mosaic than the grana . . . [Fig. 741]. Furthermore, the verrucae are usually not as high or not much higher than their width. Usually their greatest width is no more significant than their least; however, a ratio of 2:1 may occur. The surface of the verrucae can be rounded, flat, or irregular.

Reticuloid, extra reticulum—If the verrucae are close together, they become polygonal in plan view. In this case, their mosaic consists of elements which are separated by more or less straight clefts . . . [Fig. 739]. The cleft system forms a *negative reticulum* or *reticuloid*. (We distinguish the reticulum = *extra reticulum*, see below, from the infrareticulum and from the reticuloid or negative reticulum.)

Pila—The *pila* show a terminal thickening on a slender neck.

Coni—As seen from the side, the *cones* or *coni* . . . [Fig. 743] are pointed, blunt, or rounded cone-shaped. Their height does not exceed twice the diameter of their base; the latter can be extended.

Spinae—The spines or spinae . . . [Fig. 740] narrow more slowly than the coni, also they often show a thickened base. In this case, the length of that part of the spinae which is above the basal thickening is more than twice the dimension of the diameter of the spinae as measured above the thickening. Sometimes the spinae are distinctly pointed; occasionally the points are broken off. The spinae are more or less straight.

Bacula—The rodlets or bacula . . . [Fig. 745–746] are cylindrical. The narrowing is only very slight, if present; however, the base can sometimes be a little thicker. In addition the bacula are straight and often end abruptly.

Capilli—The hairs, capilli, or fimbriae . . . [Fig. 159, 744] are cylindrical or almost ribbonlike, especially they are not straight but more or less winding or coiled, knobbed, forked, branched, or anastomozed . . . [Fig. 157, 158] and perhaps also somewhat laterally expanded. Even the simple, more or less cylindrical, flexuose [Potonié and Kremp, 1955, p. 15: *Tr.*] hairs are probably almost never completely cylindrical. Their base may be thickened and the narrowing in the main part of the hair, if it occurs, is very small. Sometimes they are rather abruptly pointed.

Cristae—A special kind of arrangement of sculptural elements is that of cristae . . . [Fig. 742 and *Cristatisporites indignabundus* (Loose)]. These are fences or combs which may also join to form networks. They have been built by occasionally laterally fused cones, rodlets, etc.

Reticulum, muri, extra reticulate, curvimurate—The "reticula" . . . [Fig. 773–774] usually consist of smooth low *stripes* or higher *ridges* (muri) which join to form a net. In both cases, the exine is designated as *extrareticulate*. In the second case, the outline of the grain appears bastionical . . . [Fig. 802]. If there are especially high muri, the bastion-prongs appear connected by "skin," that is by muri as seen from the side . . . [Fig. 802]. The meshes of the networks are called *lumen*. Muri whose crests more or less undulate are called *curvimurate*, whereas muri which are more or less interrupted are called *fragmentimurate*.

Cicatricose, canaliculate—If the stripes or ridges are not, or nearly not, arranged into a net and if they run, for example parallel to one another for a long distance, then the exine is *cicatricose*. However, this is true only when the raised stripes are not too wide and if the channels which separate them are not too narrow. In this case, the exine is to be called *canaliculate*.

Scrobiculae, foveolae—If the lumina of the reticula become very small and the spaces between the lumina larger than the lumina, then the lumina are called *scrobiculae;* larger scobiculae are called *foveolae*.

Extrareticulum—Often it is difficult to determine if we are dealing with a reticulum = *extrareticulum*, with an infrareticulum, or a reticuloid (= negative reticulum). In the beginning, the spaces between punctae, maculations, granulae, and verrucae often give the impression of stripes or ridges (muri) by which more or less large net lumina appear to be separated from one another. One sometimes can find the true situation by considering the outline (extrema lineamenta) of the microfossil.

k. Zaklinskaja, 1957: For ornamentations of disaccate pollen grains *see* Figs. 222–225.

l. Couper, 1958, p. 102: "The definitions are in most cases summaries of more detailed definitions given by W. F. Harris (1955) who, however, pointed out that they were not necessarily those of the authors who first introduced the terms.

"In defining the terms used in the description of sculpture, Harris arbitrarily fixed the threshold size limit for the sculpture elements at 1 μ. However, with sufficiently high magnification, the sculpture pattern may be readily classifiable even though the

sculpture elements are less than 1 μ. Thus, he adopted the convention that such sculpture patterns may be classified as in the detailed categories II and III (see below) if the term 'sub' is prefixed, this prefix implying that the size of the sculpture elements is at, or below the minimum. Harris also added the prefix 'per' to the descriptive term when the greatest height of a projection or the greatest diameter of a cavity equals or exceeds one-tenth of the equatorial diameter of the spore.

"Terms Used in the Description of Sculpture

I—Minutely Sculptured

Scabrate—flecked; with minute pits or elevations less than 1 μ in size.

II—With Cavities or Depressions

(The pattern is regarded as comprising pits, indentations, or cavities in the general surface, not less than 1 μ in size.)

Foveolate—with cavities (pits) up to 2 μ in diameter, or if larger, too widely separated to form a reticulum.

Foveo-reticulate—with pits large enough and close enough together to form a reticulum, comprised of the pits (lumina or lacunae) and the intervening walls (muri) which separate them.

III—With Projections or Elevations

(The pattern is regarded as comprising elevations of, or projections from, the general surface. Projections or elevations not less than 1 μ in height.)

(a) Projections

Echinate—apex more or less sharp, trunk tapering with broad base.

Papillate—apex more or less rounded to truncate, trunk not markedly tapering.

Granulate—more or less isodiametric, not more than 1/20th of the equatorial diameter of the spore if the latter is over 20 μ.

Verrucate—more or less isodiametric, larger than granules.

(b) Elevations

Costate—ribbed; regular, well defined elevations or corrugations more or less encircling the spore.

Lophate—ridged; with simple flange-like ridges (usually anastomosing).

Lophate-reticulate—ridges anastomosing.

Rugulate—wrinkled; elements irregularly distributed. (Rugulae are long and narrow in surface view, whereas, verrucae if elongate, are relatively broad.)"

m. Krutzsch, 1959, p. 38–40 and Fig. 563, 735, 769–772: *Tr.* Sculpture is the structure of the spore wall situated on the surface. It is therefore a special case of structure which is limited to the interior of the spore wall or on its inner side (= surface). Between structure and sculpture naturally transitions occur, especially in more delicate ornamentations. This is because the structure elements may be present also in the different parts of the wall, at the same time, and in a different stage of development. Therefore, structure and sculpture can occur as a combination in one and the same spore-specimen.

For the names of the structure and sculpture types we usually follow Thomson and Pflug (1953, p. 22–23).

The concept *structure* implies "inner structure," *sculpture* defines the "outer appearance" of the wall.

In use of concepts such as medio- or intrapunctate it is quite unnecessary to add the word "structure;" similarly in the concept punctate the word "sculpture" is not needed. But if one wants to indicate a punctate structure, then one must append the word structure, since otherwise it refers only to sculpture.*

　* Moreover we further put it in quotation marks " " .

Proposal: It is therefore useful to always add the "necessary" mentioned prefix to the terms structure and sculpture. Whereby one must apply the prefix "extra-"—which of course has not yet come into use here—to sculpture-type terms without connoting sculpture in order to avoid misunderstanding and errors (cf. R. Potonié and Kremp, 1954, p. 166 among other publications). Therein the term punctate would only be a neutral general expression. In connection with this compare the following scheme:

Scheme:

Structure ⎫　　　　(extra-) punctate (etc.) -
　or　　　⎬ punctate　　　　　sculpture
Sculpture ⎭　　　 mediopunctate (etc.) structure-
　　　　　　　　　 intrapunctate (etc.) structure-
　　　　　　　　　　　　(= infrapunctate)

Conceptional framework of some weaker (less distinct) ornamentation types (at an aperture of about 1.3):

a). chagreen: single elements only more or less indistinctly resolvable; in part secondary and preservation conditions can be present. (Transitions to b). exist.) Structure and sculpture not differentiated.

b). punctate: single elements easily solvable, have dotlike appearance. (Transition to c). exists.) According to prefix, structure or sculpture.

c). granulate: single elements (1–$2\,\mu$) no longer dotlike, but look plain. (Transitions to sculpture- and structure-types with little knobs and warts, etc.)

d). baculate: rodlike, longish, often very dainty single elements, which have a dotlike appearance in plan view. In spores predominately present as sculpture, less so as structure. (Transitions to b). and c). exist.)

Some new structure-types—a). hamulate: hitherto found only as sculpture form; little hook elements in irregular, disorderly arrangement; no distinct re-

lief of an organized netlike character . . . [Fig. 735] (e.g., occurring in *Lycopodiales*).

b). foveate: in contrast to foveolate sculpture, where small, lacunae of different sizes are distributed more or less irregularly, here exist more or less regularly big, bigger, or small roundish lacunae between elevated parts of the wall. In the habitual end result neither the lacunae nor the netlike positive reticulate parts of the wall predominate, but they counterbalance each other. A special linear arrangement does not exist. Transitions to foveolate, reticulate, corrugate, and other structures are possible . . . [Fig. 771].

c). spongy: the wall is stuffed with altering meshlike little lacunae. They are only occasionally resolvable with the light microscope. According to Erdtman structure discerned with an electron microscope in the preceding example shows only the mode of texture, structure, and ornamentation . . . [Fig. 769] and fig. 6 of Erdtman (1954).

d). foraminate: the wall shows small roundish apertures, certainly no "pores" in germinal sense. Noticed in combination with structure(s) . . . [Fig. 772].

e). pustulate: the wall is interspersed with roundish to oval lacunae of different sizes, which give the wall a vesicular (pustulatelike) structure (character) . . . [Fig. 563].

f). acusporide: the wall is more or less interspersed by needle-stitchlike breaches. In case of concentration an intrabaculate structure may occur . . . [Fig. 770].

n. Couper and Grebe, 1961, p. 6 and Fig. 678–687, 766–768: *Tr.* Sculptural elements and patterns are considered under two main headings: those based on projections *from* the general surface (apiculate elements and patterns) and those based on elevations *of* the general surface (*muronate* elements and patterns).

In describing sculpture, the *size range* of sculptural elements (for example, height and diameter of *verrucae, spinae,* or diameter of *lumina,* width of *muri,* etc.) should be recorded. Also, the distance apart of elements and their distribution (i.e., whether confined to or better developed in certain areas) should be noted.

[Couper and Grebe, 1961, p. 7: *Tr.*]

When the *greatest* dimension (for example, diameter, height of spina, width of lumen, etc.) of a sculptural element is less than $1\,\mu$, the prefix *micro* can be used. Thus, for example, *micro-verrucate, micro-spinose.*

a). *APICULATE ELEMENTS AND PATTERNS*
　1. *Those with unbranched, non-pointed terminations*

　　Verrucae: projections whose greatest dimension is more than $1\,\mu$: base of element more or less isodiametric and diameter equal to or greater than height; top of element rounded to slightly flattened, never pointed . . . [Fig. 678]. A slight constriction of the base is permissible, as in . . . [Fig. 678] but if the element is carried on a distinct neck or stalk it is a *pilum* (see below).

Verrucate sculpture: a sculptural pattern formed by verrucae.

Grana: the same as verrucae but greatest diameter of elements less than 1 μ.

Granulate sculpture: a sculptural pattern formed by *grana.*

Pila: projections consisting essentially of a more or less spherical head carried on a neck or stalk . . . [Fig. 679].

Pilate sculpture: a sculptural pattern formed by *pila.*

Bacula: projections in which the height is greater than the basal diameter; bases more or less rounded in surface view; sides of projection when seen in optical section converging upward to more or less parallel; tops flat.

Baculate sculpture: a sculptural pattern formed of *bacula.*

2. *Those with unbranched, pointed terminations*

Spinae: projections in which the height is at least 2 times the basal diameter; bases more or less rounded in surface view; tops pointed . . . [Fig. 681].

Spinose sculpture: a sculptural pattern formed by *spinae.*

[Couper and Grebe, 1961, p. 8: *Tr.*]

Coni: projections in which the height is less than 2 times the basal diameter; bases more or less rounded in surface view; tops pointed . . . [Fig. 682].

Conate sculpture: a sculptural pattern formed by *coni.*

3. *Those with branched terminations*

Capilli: projections in which the terminations are branched or serrated; tops of branches can be rounded, pointed, or flat; the shaft can be either rounded or ribbonlike (i.e., flattened in one dimension); height greater than basal diameter . . . [Fig. 683].

Capillate sculpture: a sculptural pattern formed by *capilli.*

b). *MURONATE SCULPTURE ELEMENTS AND PATTERNS*

1. Those in which the *elevations* of the general surface leave no definite regular, reticulate pattern of lumina between them.

Cristae: elevations of the surface with elongate (at least 2 times long as broad) curved bases; occasionally fusing together to form irregular *reticulum;* tops of elements more or less pointed

or serrated to slightly rounded . . . [Fig. 684].

Cristate sculpture: a sculptural pattern formed by *cristae.*

Rugulae: elevations with elongated bases (at least 2 times long as broad), bases curved to irregular in shape in surface view; greatest dimension of bases at least 2 times the height; tops flat to slightly rounded . . . [Fig. 685].

Rugulate sculpture: a sculptural pattern formed of *rugulae.*

2. Those in which the lumina in the general surface of the spore do not form a definite, regular reticulate pattern.

[Couper and Grebe, 1961, p. 9: *Tr.*]

Foveolae: rounded lumina, 1–2 μ in diameter, or if larger are too widely spaced to form a reticulum . . . [Fig. 687].

Foveolate sculpture: a sculptural pattern formed by *foveolae.* Sculpture should still be described as foveolate (even if the diameter of the lumina is greater than 2 μ) when the diameter of the lumina is equal to or less than the distance between adjoining lumina . . . [Fig. 687].

Vermiculi: elongated, narrow, undulating, irregularly spaced lumina, not forming a definite pattern . . . [Fig. 686].

Vermiculate sculpture: a sculptural pattern formed by *vermiculi.*

3. Those in which the sculptural elements (muri, lumina, striae) form a more or less definite reticulate or striate pattern.

There are 3 terms which can be used for the sculptural elements of reticulate and striate sculptural patterns:

Muri: the elevations bounding the *lumina* of *reticulate* sculptural patterns and the *striae* of *striate* sculpture patterns . . . [Fig. 766–767]. The sides of the elevations when seen in optical section can be parallel, converging or diverging and the tops rounded, flat, or pointed or serrated . . . [Fig. 768].

Lumina: the pits (depressions) between the elevations of a reticulate sculptural pattern. The lumina of a reticulum can be either more or less rounded or polygonal in surface view . . . [Fig. 766].

Striae: the elongated depressions between the muri of a striate sculpture pattern . . . [Fig. 767].

Reticulate-sculpture: the sculptural pattern formed by the *muri* and adjoining *lumina.* When describing a reticulate sculptural pattern, the shape (rounded or polygonal) and relative size of lumina compared with the intervening elevations (muri) together with the nature of the sides (converging, diverging, parallel) and tops of the muri (rounded, flat, pointed, or serrated)

should be noted if possible.
[Couper and Grebe, 1961, p. 10: *Tr.*]

Striate sculpture: the sculptural pattern formed by *muri* and *striae.* When describing striate sculpture patterns, the relative width of the *striae* compared with the *muri* and the nature of the muri should be noted.

ORNAMENTUM, n., pl. -a. (L. ornament).

Faegri and Iversen, 1950, p. 162 "(Potonié, 1934) = structure + sculpturing. (Erdtman, 1943) = sculpturing."

ORNATE, adj. (L. ornatus, past part. of ornare, to adorn). cf. curvimurate, negative reticulate.

Erdtman, 1952, p. 466: "with ± reticuloid sculpturing: 'reticulum' usually ± 'curvimurate;' luminal areas often anastomosing, sometimes with isolated 'mural' islands: 'muri' usually comparatively wide." [*see* Fig. 790–792].

OROID, adj. cf. colpate, colporate, furrow, Groups of Sporomorphae, porate.

Erdtman, 1952, p. 466: "inner part, more or less similar to an os, of a composite aperture."

OROIDIFEROUS, adj. cf. colpate, colporate, furrow, Groups of Sporomorphae, porate.

Erdtman, 1952, p. 461: "colporoidate: with oroidiferous colpi." cf. COLPOROIDATE.

ORTHOCOLPATE, adj. cf. aperture, colpate, furrow, Groups of Sporomorphae.

Erdtman and Straka, 1961, p. 66: "Central line of the zone cutting the colpi at right angles." [*see* GROUPS OF SPOROMORPHAE].

OS, n., pl. ora. (L. the mouth). cf. colpate, equatorial pore, furrow, pore.

a. Jackson, 1928: "a mouth or orifice."

b. Erdtman, 1947, p. 105: "Equatorially arranged pores are termed 'ora' (L. os, oris; plur. ora). A pollen grain of *Thelygonum cynocrambe* with six equatorial pores ('ora') is thus hexorate; a typical *Betula* pollen is triorate. In colporate grains every colpa is provided with an 'os' not with a 'porus.' The term colporate remains unchanged, although the term 'os' was not thought of at the time it was proposed."

c. Erdtman, 1948, p. 268–269: "Tricolporate grains are provided with three oriferous (lat. os, gen. oris) colpae; 'os' denotes an equatorial pore or aperturoid spot; in pollen grains of *Eucommia* it is somewhat difficult to decide readily whether a tricolpate or a tricolporate status prevails."

d. Faegri and Iversen, 1950, p. 162: "(Erdtman, 1943) = equatorial pore."

e. Erdtman, 1952, p. 466: "the inner part of a composite aperture."

f. Erdtman, 1952, p. 15: "In structure certain apertures (laesurae, hila, sulci, and sulculi) are simple, whereas others (colpi, rugae, foramina, and pori) are either simple or composite. Simple apertures are formed so to speak by a succession of ± congruent holes in the different spore wall layers. In composite apertures the successive openings (they are usually but two) are not congruent: the superficial parts of the apertures have the shape of colpi, rupi, rugae, foramina, or pori, whilst the underlying parts (deviating in shape, size, or both) are either circular, lolongate (longitudinally elongate; L. longus, long, and elongatus, extended), or lalongate (transversely extended; L. latus, broad). They thus exhibit the various shapes of a mouth and are accordingly termed ora (mouths; L. os, plur. ora). Grains with lalongate ora meeting laterally to form a continuous oral zone are zonorate."

sdb. Pike, 1956, p. 51.

OUTLINE, n., pl. -s. cf. amb, ambit, ambitous, circumscriptio, contour, equatorial limb, extrema lineamenta, figura, Groups of Sporomorphae, limb, limbus, morphography, optical section, shape classes.

I. Jackson, 1928: "the continuous boundary-line of an organ, as of a leaf."

II. Harris, 1955, p. 26: "for the sake of brevity, when describing the contours of spores the convention has been adopted in this manual of using the word *outline* for the polar view, *profile* for the lateral view, and *cross section* when the shorter equatorial axis of monolete spores is described."

OVAL, adj. (L. ovum, egg). cf. ovatus, shape, shape classes.

a. Jackson, 1928: "broadly elliptic."

b. Couper and Grebe, 1961: [*see* SHAPE].

OVATE, OVATUS, adj. cf. oval, shape classes.

Jackson, 1928: "shaped like a longitudinal section of a hen's egg, the broader end basal;" "used for ovoid."

OVULE, n., pl. -s.

Jackson, 1928: "*Ovulum,* the young seed in the *ovary,* the organ which after fertilization develops into a seed."

P

PALYNOGRAM, n. (Gr. paluno, to strew; *see also* under PALYNOLOGY; gram, a combining form denoting something drawn; gramma, letter). cf. figura, formula of spores and pollen symbols, Groups of Sporomorphae, morphography, pollen and spore formula, shape classes.

a. Erdtman, 1952, p. 466: "(Klaus and Erdtman in Erdtman, 1951 . . .): a visual representation of a spore providing the main palynological data (on polarity, symmetry, apertures, shape, size, sporoderm stratification, sculptine patterns, amb, etc.)."

b. Erdtman, 1952, p. 24 and Fig. 76–99: "The chief results of a morphological analysis of a pollen grain or spore can be exhibited by a palynogram. A palynogram is a ± diagrammatic representation of a pollen grain or spore, its symmetry, apertures, shape, size, sporoderm stratification, sexine patterns, etc. Pollen grains and spores with distinct polarity are shown either in equatorial or in polar view. In the former case—and provided that the palynogram exhibits an isopolar, radiosymmetric pollen grain with equatorial apertures —the centre is occupied by an aperture (. . . [Fig. 76–81]; a horizontal line through the centre of these figures represents the equator). In the latter case one of the poles is exactly at the centre of the palynogram with one of the apertures vertically below it [Fig. 82–87]. A tetrahedral tetrad is shown in [Fig. 94]. Palynograms of heteropolar, bilateral (1-sulcate) pollen grains are shown in [Fig. 95] (distalipolar view of a spinulose grain) and [Fig. 96] (equatorial view of a reticulate grain, distal pole facing upwards). A 1-lete spore is given in [Fig. 97] (proximalipolar view) and [Fig. 98] (equatorial view; distal pole facing upwards). [Fig. 99] shows the proximal face of a 3-lete spore (one laesura facing vertically upwards).

Part of the palynograms shows the surface; the rest is an optical section through the pollen or spore wall in question. Detail figures [in Fig. 76–93, 95–99] outlines merely showing the size of the detail figures) supplement the main figure; if the latter shows a pollen grain in polar view, a detail figure in the bottom left hand corner [cf. Fig. 85–87] shows the same grain in equatorial view (x250); another—above the main figure—exhibits the spore wall stratification (x2000), whereas square detail figures on the right provide the results of an LO-analysis of the sexine patterns [the upper square shows the distribution of bright and dark areas at high adjustment of the microscope, the lower square(s) the same at (medium and) low focus]. For the sake of uniformity it is suggested that the sides of the squares should be 5, 10, or 15 mm."

PALYNOLOGY, n. (Gr. παλυνω (paluno), to strew or sprinkle. cf. (pale), fine meal; cognate with Latin *pollen,* flour, dust; Gr. logia, a combining form denoting doctrine, theory, or science).

I. a. Hyde, 1944, p. 146. "The English expression *pollen analysis* seems to have been first used in 1924: it has been employed to cover the whole field of recent palaeobotanical research centering on plant spores,* and is used in this sense in the title of Dr. Erdtman's book [1943] . . . ."

* In view of the admitted inadequacy of the expression *pollen analysis* it has recently been proposed to substitute for it the new word *palynology* (Gr. . . . the study of pollen and other plant spores and their dispersal, and applications thereof).

b. "Anonymous," 1945, p. 264: "The need for a better name [than pollen analysis] has been expressed in *Pollen Analysis Circular* [No. 8, 1944], a cyclostyled research bulletin edited by Prof. Paul B. Sears, of Oberlin College, Ohio. Messrs, H. A. Hyde and D. A. Williams, of the National Museum of Wales and Llandough Hospital, Cardiff, respectively, in the October issue of that *Circular* suggest the term *palynology* (Gr. παλυνω (paluno), to strew or sprinkle. cf. παλη (pale,) fine meal; cognate with Latin *pollen,* flour, dust) for the study of pollen and other spores and their dispersal, and applications thereof. It is hoped that the sequence of consonants p-l-n (suggesting pollen, but with a difference) and the general

110

euphony of the new word will commend it."

   c. Cranwell, 1953, p. 17: "a term introduced in Wales, in 1944, by Hyde and Williams, to embrace all pollen and spore studies."

II. Funkhauser, 1959, p. 126: "As originally proposed, the word palynology embraced only the study of pollen and spores. The paleontologists have largely been responsible for extending its coverage to all organisms or parts of organisms that fall in the general spore-pollen size range. Many fossils of non-spore-pollen palynology are recovered along with spores and pollen, using the same extraction techniques. A large proportion of these are algal, including the coccolithophorids, dinoflagellates, diatoms, desmids, *Pediastrum,* and many of the hystrichosphaerids. There are also fungal elements, as well as fragments of larger plants, such as tissues and cells. In zoological palynology are microforaminifera, tintinnids, radiolaria, and the like. The rapidly expanding interest in these groups has been greatly stimulated by their proved or potential value in stratigraphic and environmental studies . . ."

III. Sladkov, 1962, p. 13: "The term 'palynology' introduced abroad . . . has been accepted without its precise definition and has been widely popularized during the last decade. That is why many investigators accept the existence of 'palynology' as a special science on pollen, as well as the existence of one 'spore and pollen method'. There is also a tendency to replace the term 'spore and pollen analysis' by the term 'palynological analysis.'

   ". . . If we review the actual sum of our knowledge and of methods used, which are included in the designations 'palynology' and 'spore and pollen method', we will come to the conclusion that these terms are applied to a mosaic complex of data relating to various branches of botany and interrelated only by the fact that all the studies involved are to a certain extent associated with researches on spores and pollen.

   ". . . Two different methods of research should be distinguished—the method of spore and pollen analysis and the method of comparative spore and pollen morphology (palynomorphological method) but not one 'spore and pollen method' and 'palynology' as a separate science. The use of these terms and their derivatives ('palynological analysis', for instance) in the meaning of methods is indesirable.

   ". . . The spore and pollen analysis is a special branch of paleobotany, it uses statistical data on fossil spores and pollen. Palynomorphology is a branch in the morphology of plants, it studies the structure of spores and pollen in re-cent plants. The branch of paleobotany devoted to the morphological study of spores and pollen from the fructification of fossil plants can be called paleopalynomorphology.

   ". . . Researches on spore and pollen morphology carried out for the special purpose of spore and pollen analysis suggest the same treatment of standard material to which rock samples are subjected in the process of their preparation for analysis. As a result of such a treatment not only the intine and pollen grain content are destroyed but sometimes even some elements of the exine. For studies of the external envelopes of spores and pollen only it might be advisable to introduce a special determination, like, for instance, 'tectomorphologic'."

PALYNOMORPH, n., pl. -s. (Gr. paluno + morphe, form). cf. miospore, pollen, pollenospore, polospore, spore, sporomorph.

Tschudy, 1961, p. 53: "A term suggested by Dr. Richard A. Scott of the United States Geological Survey—was proposed to include the forms, such as pollen and spores, hystrichospheres, dinoflagellates, microforaminifers, and other kinds of organisms, encountered on slides examined by palynologists. Since the term palynology has gained rather wide acceptance, the companion term palynomorph is proposed as an inclusive appellative encompassing the types of microfossils found in palynological preparations. Dr. Scott's suggestion fills a need, and is acceptable because of its simplicity and inclusiveness."

PANAPERTURATE, adj.

Erdtman, 1958, p. 4: "Apertures ± uniformly distributed over the surface." [*see* Groups of Sporomorphae].

PANCOLPATE, adj. cf. colpate, furrow.

Erdtman, 1958, p. 137 and Fig. 105: [Apertures ± uniformly distributed over the surface. *see* Groups of Sporomorphae].

PANPORATE, adj. cf. aperture, cribellate, forate, Groups of Sporomorphae, porate, pore.

Erdtman, 1958, p. 137 and Fig. 105: [Apertures ± uniformly distributed over the surface. *see* Groups of Sporomorphae].

PANTOTREME, adj. cf. aperture.

Erdtman and Straka, 1961, p. 66: "spores with the aperture more or less uniformly distributed over the spore surface." [*see* Groups of Sporomorphae].

PAPILLA, n., pl. -ae. (L. a nipple). cf. germinal papilla, ligula.

   I. Jackson, 1928: "soft superficial glands or protuberances;" " 'also the aciculae of certain Fungals' (Lindley)."

II. Beug, 1961, p. 12: *Tr.* In plan view roundish, protuberant area of the exine provided for the exit of the pollen tube.

PAPILLATE, adj. (L. papillatus, bud-shaped). cf. baculate, flecked, maculate, punctate, scabrate, subpsilate.
a. Jackson, 1928: "having papillae."
b. Kosanke, 1950, p. 11: [*see* ORNAMENTATION and Fig. 640:C].
c. Harris, 1955, p. 19: "Trunk (of the projection) not constricted. Narrow . . . the vertical axis is the greater. Example: *Hymenophyllum sanguinolentum.*" [*see* ORNAMENTATION].
d. Traverse, 1955, p. 94: "Sculpturing of papillae. The term was used whenever the protuberances were distinctly nipple-like and well spaced. The size range includes elements that would otherwise be included either under scabrate (no dimension more than 1μ) or verrucate (at least one dimension more than 1μ)."
e. Couper, 1958, p. 102: "Papillate—apex more or less rounded to truncate, trunk not markedly tapering." [*see* ORNAMENTATION].

PARAPORAL CREST, n., pl. -s. (It. imper. of parare, to shield).
Wodehouse, 1928, p. 934: "The ridges bounding the germinal furrows in lophate and sublophate grains. Examples: *Vernonia wrightii, V. texana.*"

PARAPORAL LACUNA, n., pl. -ae. cf. lacuna.
a. Wodehouse, 1928, p. 934: "The lacunae adjacent to the poral lacuna and flanking the germinal furrow or abporal lacuna, generally four—two in each hemisphere—about each poral lacuna. Example: *Vernonia gracilis.*"
b. Wodehouse, 1935, p. 543: "a lacuna adjoining on one side a poral lacuna and wholly within one hemisphere, e.g., in the grains of *Taraxacum officinale* . . . Fig. 765."

PARAPORAL RIDGE, n., pl. -s.
Wodehouse, 1935, p. 543: "the ridges bounding the germinal furrows and extended in a meridional direction, e.g., in the grains of *Tragopogon pratensis* . . . Fig. 762."

PARASPORE, n., pl. -s. (Gr. para, beside). cf. seirospore.
Melchior and Werdermann, 1957, p. 16: *Tr.* Spores appearing in dichotomous branching rows or in somewhat undefined cell masses (parasporangia), e.g. *Rhodophyta.*

PARASYNCOLPATE, adj. cf. colpate, furrow.
Erdtman, 1952, p. 466: "colpate pollen grains are parasyncolpate if the colpi (or their extensions) are bifurcate and the branches meet ± close to the poles, leaving intact apocolpia of regular shape. Examples: . . . Fig. 449–452."
sdb. Pike, 1956, p. 51.

PARS DISTALIS, n. cf. distal face, distal hemisphere, distal surface.

PATTERN, n., pl. -s. cf. ornamentation, sculpture, sculpturing, structure, texture.
Potonié, 1934, p. 11: *Tr.* A term for structure and sculpture is pattern or ornamentation.
sdb. Faegri and Iversen, 1950, p. 162.
Cranwell, 1953, p. 17.

PAULOSPORE, n., pl. -s. (Gr., a pause).
Jackson, 1928: "Klebs' term for CHLAMYDOSPORE." [*see* Melchior and Werdermann, 1954, p. 14, under SPORE].

PAXILLA, n., pl. -ae. (L. paxillus, a peg, small stake). cf. bacularium.
Erdtman, 1948, p. 267: "set of rods." [*see* SPORODERMIS and Fig. 541].

PAXILLATE, adj. cf. baculate, intrabaculate.
Faegri and Iversen, 1950, p. 162: "(Erdtman, 1948) = intrabaculate (+ baculate?)." [*see* Erdtman (1948) under SPORODERMIS].

PECTEN, n. (L. a comb). cf. comb, crista.
I. Jackson, 1928: "± = sterigma."
II. Thomson and Pflug, 1953, p. 21: *Tr.* Parting in the ektexine by elongation of the rodlet elements is . . . not related to a germinal apparatus e.g. . . . [*Pityosporites credoides* (Thomson and Pflug, 1953]. [*see* DISSENCE, PECTEN].

PENTATREME, adj. cf. aperture.
Erdtman and Straka, 1961, p. 65: "(5-treme) spores: with five apertures." [*see* GROUPS OF SPOROMORPHAE].

P:E RATIO, n. cf. dimension, figura.
Harris, 1955, p. 26: "the ratio polar to greatest equatorial axis, taking the longer of the two as 8."

PERFORATE, adj. (L. perforatus, pierced). cf. cribellate, foramen, foraminate, forate, oligoforate, ornamentation, polyforate.
a. Jackson, 1928: "pierced through, or having translucent dots which look like little holes, as in *Hypericum perforatum,* Linn."
b. Traverse, 1955, p. 95: "Having structure with holes, either in individual units, for example, the clubs of clavate (intectate) sculpture, or with holes in and beneath the tectum of a tectate form. The latter is the most common use of the term."

PERICOLPATE, adj. (Gr. peri, all around, about). cf. colpate, furrow, oligopericolpate, polypericolpate.
a. Faegri and Iversen, 1950, p. 24: "the apertures are evenly distributed over the whole surface of the grain (*isometric* distribution, Wodehouse, 1935), *pericolpate periporate.*"
b. Faegri and Iversen, 1950, p. 128: "some or all furrows not meridional." [*see* GROUPS OF SPORO-

MORPHAE and Fig. 62–63].
sdb. Norem, 1958, p. 671.

PERICOLPORATE, adj. cf. oligopericolporate, polypericolporate.
Norem, 1958, p. 671: ". . . with colpori uniformly
distributed over surface . . ."

PERIHETEROCOLPATE, adj.
Norem, 1958, p. 672: ". . . with pori and colpi uniformly distributed over surface."

PERINE, n. cf. perinium, perisporium.
I. Jackson, 1928: "the outermost layer of sculpturing on pollen."
II. a. Erdtman, 1943, p. 52: "perine; the outermost layer, outside of the exine, in certain
spores."
b. Faegri and Iversen, 1950, p. 162: "(Erdtman,
1948): that part of the ektexine which is pre-
sumed to be deposited by a periplasma."
c. Erdtman, 1952, p. 466: "L. perinium (Strasburger, 1882, p. 135): perisporium: the outermost, extra-exinous sporoderm layer in some
spores (in certain mosses, ferns, etc.). It seems
to be due to the activity of a periplasmodium
. . ." [see table under SPORODERMIS].
sdb. Harris, 1955, p. 15–17; [see PERISPORIUM].
Beug, 1961, p. 12.
d. Potonié and Kremp, 1955, p. 18–19: [see
SPORODERMIS].

PERINIUM, n. (L.). cf. perisporium.
Jackson, 1928: "the outermost of the three coats of
a Fern spore, the epispore."

PERINO-CAVATE, adj.
Harris, 1955, p. 25: [see CAVATE].

PERIPORATE, adj. (Gr. peri, all around, about).
cf. aperture, cribellate, Groups of Sporomorphae,
porate, pore.
a. Ingwersen, 1954, p. 40: "Pollen type with several free pores not equatorially distributed."
[see PERICOLPATE, Faegri and Iversen (1950) and
Fig. 71].
b. Norem, 1958, p. 670: "(Forate)—with pores
uniformly distributed over the surface . . . [Fig.
501–1] . . . Oligoperiporate (4–8 pores) . . .
Polyperiporate (>8 pores)".
sdb. Traverse, 1959, p. 95.

PERISACCATE, adj. cf. monosaccate.
Erdtman, 1952, p. 466: "surrounded by a saccus."

PERISPORE, n. cf. perisporium.
Jackson, 1928: "the membrane or case surrounding
a spore;" the mother-cell of spores in Algae;" "=
perigynium;" "an incrustation containing much silica, outside the exospore of Isoetes (Fitting)."

PERISPORINIUM, n. cf. perisporium.
Jackson, 1928: "the outermost membrane of pollen
in Angiosperms (Fitting)."

PERISPORIUM, n. cf. epispore, episporium, perine,
perinium, perispore, perisporinium.
a. Jackson, 1928: "the membrane or case surrounding a spore;" "the mother-cell of spores in
Algae;" "perigynium;" "an incrustation containing much silica, outside the exospore of Isoetes
(Fitting)."
b. Bower, 1935, p. 429 and Fig. 548–550: "The
plasmodium intrudes between the separating
spore-mother-cells, forming a rich nutritive medium, which is absorbed in the more primitive
ferns in the developing spores; but in certain advanced types it may in part remain as a deposit
on the outside of the wall, and is called the
'perispore' . . . The wall of the spore itself is
often marked by characteristic sculpturing which
at times gives a basis for systematic comparison; but in this the perispore is more important.
Ferns may in fact be divided into two groups according to the presence or absence of a perispore.
None is seen in the Eusporangiate Ferns, nor in
the Osmundaceae, Schizaeaceae, Hymenophylla-
ceae, Cyatheaceae, Davallieae, or in Ceratopteris.
In fact, it is absent from all the more primitive
ferns, and of the remaining Leptosporangiates it
is wanting in the Vittarieae, Gymnogrammeae,
Polypodieae, and Pterideae; but is present in the
Asplenieae and Aspidieae. The perispore thus
possesses a certain value for comparison; but con-
fidence in it as a safe criterion is shaken by the
fact that while it is present in Blechnum and
Woodwardia it is absent in the closely related
Brainea and Doodia. It is clearly a feature
adopted late in descent, and restricted to certain
circles of affinity."
c. Dijkstra-van V. Trip, 1946, p. 23: Tr. The perispore
in the case of Carboniferous megaspores appears
only in one case as a closed membrane. In the
Triangulati, a perispore surrounds the entire exospore which is often completely scultureless . . .
[Triletes triangulatus Zerndt, Triletes tricollinus
Zerndt, Triletes triangulatus Zerndt]. In some
other megaspore types the perispore adds a lot
to the outside of the spore.
We think (in contrast to Zerndt's opinion, 1931,
p. 173 and 1934, p. 10 and 19), that Ibrahim
(1932, p. 27), probably by accident, is right
when he designates Triletes brasserti (his spore
type Zonales-sporites saturnoides) as being provided with a perispore and we completely agree
with Sahabi (1936, p. 23) when he believes the
zonal outgrowth and hairs of the Triletes brasserti, T. superbus, and T. dentatus to originate
from the perispore. Sahabi describes the ornamentation which originates from the perispore as
being fixed to the exospore, but not belonging to
it; removing it from the perispore does not result

in a fracture at the surface of the perispore (compare Figs. . . . [567–569, *Triletes brasserti* Stach and Zerndt]); sometimes they fit exactly with somewhat larger or smaller outgrowths of the exospore, as for instance with the triradiate ridges, the arcuate ridges, possibly with the equatorial rims, swellings, thorns, or other places of attachment.

Zerndt (remarks in the description of *Triletes circumtextus*, 1934, p. 19 and 20) appears to contradict his statement that only *Triletes angulatus* shows a perispore: "the different color of the zonal outgrowth and its strongly shiny aspect permits the supposition that the substance from which it is formed is not the same as that of the exospore . . . the zonal outgrowth can be detached without difficulty from the body of the spore . . . one may conclude perhaps that its origin may be different and more recent than that of the exospore." Participation of the perispore in the outer sculpture of the spore is, according to my opinion, also noticeable in *Triletes rotatus,* and *T. mucronatus* and probably in *T. radiatus* and *T. tenuicollatus.*

d. Erdtman, 1943, p. 52: [*see* PERINE].

e. Pokrovskaya, et al., 1950: *Tr.* Perisporium—perispore or perinium—an external membrane of the spore which is formed by the covering tapetum cells of the sporangium. It is usually decomposed in the fossil condition. When preserved it is a good diagnostic indication permitting determination of spores to species (e.g., *Onoclea sensibilis*).

f. Erdtman, 1952, p. 466: "(Russow, 1872, p. 70): Perine." [*see* EPISPORIUM].

g. Harris, 1955, p. 15–17: "The perispore or *perine* was defined by Erdtman (1943) as 'the outermost layer, outside of the exine, in certain spores.' In doubtful cases it may be referred to as the *sculptine.* For reasons which will be explained, the term *winged* is sometimes used in this manual for spores showing an obvious perispore. It is applicable when the outline or profile of the spore, irrespective of elevations or projections which form the sculpture pattern or ornamentation, is modified by the presence of a perine. The presence of a cavity or of cavities between the spore proper and this outermost layer is implied.

"The perine or perispore—"It will be appropriate to introduce a brief discussion of this feature by quoting a concise and authoritative summary from Bower . . . [*see* PERISPORIUM, Bower (1935)].

"It is a common observation that whereas the mature spores of a given species may exhibit a characteristic ornamentation, spores which are not fully mature lack this ornamentation. Between the smooth immature spores and the fully coated mature spores others may be observed which show a partial accretion of the ornamental layer . . . [Fig. 558–559]. If this arises by the gelation of plasmodial substance on the exterior of the spores, it is a perispore as defined by Moll (1934) and other writers. It is possible, then, that the occurrence of a perispore is more general than was supposed by Bower.

"Erdtman, (1943, Pl. 27) figures the spores of *Hemitelia grandifolia* (Cyatheaceae), of which he says 'the contour lines of the endexine form a triangle with rounded corners. The outer part of the spore coat is closely attached to the endexine at these corners, but in other places it is somewhat raised above the surface of the endexine . . .' This loose-fitting envelope does not differ in appearance from a perispore, though it may differ in its mode of origin. The detailed investigation of the formation of spore walls lies outside the scope of the present inquiry, but the fact may be readily verified, as shown by the writer's preparations, that spores of the related *Hemitelia (Cyathea) smithii* lack this outer envelope when slightly immature. One cannot be sure, apparently, of the correct application of such terms as ektexine and perine without reference to the ontogeny of the spore, which, in the case of fossil spores, would not be known.

"Erdtman (1947) stated that 'Many laesurate spores possess a perine, or perisporium, outside the exine. It is sometimes, even when dealing with recent spores, difficult to decide whether a perinous coating is present or not (Hannig, 1911).' There follow suggestions for the appropriate designation of 'fossil spores with a distinct perine' (loc. cit.), the first being applicable to a trilete (i.e., tetrahedral) spore. As an example of the type from living species he gave *Pilularia globulifera* (Marsileaceae, cf. Bower above).

"The paper by Hannig cited by Erdtman has not been accessible to the present writer, but the results of Skottsberg (1942), referred to in a later section, show a graduated series in *Asplenium* from spore types which are 'broadly winged' to types in which the wing is 'very narrow, sometimes quite inconspicuous.' Evidence of a similar nature has been obtained from spore preparations from the New Zealand species of *Blechnum.* . . . [Fig. 551] shows a spore of *Blechnum procerum* with an obvious perispore . . .; a spore of *B. membranaceum,* still with an outer coat, though in this case more comfortable to the size and shape of the spore . . . [Fig. 552]; and a spore of *B. banksii* . . . [Fig. 553], in which the presence of an outer coat is not apparent. A spore from the same slide (and negative) . . . [Fig. 554], however, shows a thin outer layer partly fragmented. This series gives rise to the question whether the spores of *Doodia* (see Bower above) might not show a similar condition to that observed in *Blechnum banksii*. . . . [Figs. 555–556] show spores of *Doodia media* from the same slide which appear to present a parallel case with the example from *Blechnum*. Treatment of the spores was the same in all cases. Even if the results are not acceptable as conclusive in regard to the question of a perispore, they suffice to show that, pending a thorough investigation, there are cases where the ordinary techniques do not enable one to decide with certainty 'whether a perinous coating is present or not.'

"This leads to a difficulty in securing uniformity of description. The term 'sculptine' has been proposed (Erdtman, 1948), which resolves the difficulty when a layer which is doubtfully a perine has to be referred to. Skottsberg (1942) and Selling (1946) used the rather non-committal term 'winged,' and the writer has adopted this term, as explained above, for spores whose general contour is modified by the presence of this layer.

"The term *'cavate'* is used by Faegri and Iversen (1950) for pollen grains in which the ektexine is loosened from the endexine, and might be similarly used of spores, though the terms saccate or vesiculate, where appropriate, may be taken to imply this condition. The prefix perino- might be used if it is necessary to indicate that the layer which is cavate is regarded as a perispore."

sdb. Thomson and Pflug, 1953, p. 17; [*see* LA-MELLA].

Couper, 1958, p. 103.

PEROBLATE, adj. (L. per-, very). cf. shape classes.
a. Erdtman, 1943, p. 44: "<0.50 (<4:8)." [*see* SHAPE CLASSES].
b. Selling, 1947, p. 77: "ratio P:E <0.5; P = polar diameter, E = equatorial diameter."
c. Erdtman, 1952, p. 466: "very flattened. The term is used exclusively in describing radiosymmetric, isopolar spores where the ratio between polar axis and equatorial diameter is <0.50 (<4:8)."

sdb. Faegri and Iversen, 1950, p. 162.
Pokrovskaya, et al., 1950.
Pike, 1956, p. 51.
Norem, 1958, [*see* SHAPE CLASSES and Fig. 167c].
Beug, 1961, p. 12.

PEROBLATOID, adj. cf. shape classes.
Erdtman, 1952, p. 466: [*see* OBLATOID and PERO-BLATE].

PERPROLATE, adj. cf. shape classes.
a. Selling, 1947, p. 77: "ratio P:E (. . . .):>2.00. P = polar diameter, E = equatorial diameter."

b. Faegri and Iversen, 1950, p. 162: "Shape class (q.v.), index more than 2.0." [*see* SHAPE CLASSES, Erdtman, 1943].
c. Erdtman, 1952, p. 466: "(Erdtman, 1943): this term denotes exclusively the shape of radiosymmetric, isopolar spores where the ratio between polar axis and equatorial diameter is >2 (>8:4)."
d. Norem, 1958, [*see* SHAPE CLASSES and Fig. 167g].

PILA, n. cf. pilum.

PILARIUM, n., pl. -ia.
Erdtman, 1952, p. 466: "provisional term used to denote what seems to be groups of ± amalgamated pila. Example: . . . [Fig. 813]."

PILATE, PILATUS, adj. (L. pilum, a pestle). cf. clavate, piliferous, piliformis, subpilate, sympilate.
a. Erdtman, 1947, p. 106: "pilate; with small rods ('pila') with rounded, swollen tops."
b. Faegri and Iversen, 1950, p. 162: "(Erdtman, 1947) = clavate."
sdb. Couper and Grebe, 1961, p. 6; [*see* ORNAMEN-TATION].

PILIFEROUS, adj. (L. pilus, a hair; fero, I bear).
I. Jackson, 1928: "bearing hairs, or tipped with them;" "hair-pointed (Lindley)."
II. a. Erdtman, 1943, p. 51: "with pila." [*see* Fig. 639-C].
b. Pokrovskaya, et al., 1950: *Tr.* pilifera—club-carrying. [*see* ORNAMENTATION].

PILIFORMIS, adj. cf. clavate, pilate.
Kupriyanova, 1948, p. 70: *Tr.* . . . has spines with clublike thickenings (clavata, piliformis) which externally remain granulose. [*see* Fig. 661].

PILUM, n., pl. -a. (L. a pestle). cf. clava, club.
a. Potonié, 1934, p. 11: *Tr.* Pila are very remarkable. In *Pollenites margaritatus*, for example, they stand so close together that the heads of the clubs almost touch . . . [Fig. 654] giving an impression of a network in vertical view. By means of this confluent canopy, there arises below the heads an insulating air layer in which the stems of the heads stand. H. Fischer uses some examples to discuss the fact that the heads or the upper ends of the sculpture elements could be connected (e.g., by cross beams), in some cases they even coalesce to an almost uniform membrane; this is our *exolamella*. By this, it is recognizable in what way the rodlet-impregnated insulation layers of the exoexine are morphologically related to sculptures. Fischer also shows, that sometimes even two insulation layers can lie over each other. [*see* ORNAMENTATION, PILA].
b. Erdtman, 1948, p. 387 and 267: "A pilum con-

sists of a head (caput) and a rodlike pars collaris, or baculum." [*see* Fig. 541].

    c. Faegri and Iversen, 1950, p. 162: "(Potonié, 1934): small rods with rounded, swollen ends = clavae p.p."

sdb. Erdtman, 1952, p. 466.

        Potonié and Kremp, 1955, p. 14; [*see* ORNAMENTATION, PILA].

        Couper and Grebe, 1961, p. 6; [*see* ORNAMENTATION and Fig. 679].

PIT, n., pl. -s.

Jackson, 1928: "a small hollow or depression, as in a cell-wall;" "the endocarp of a drupe containing the kernel or seedstone (Crozier)."

PITTED, adj. cf. aperture, cribellate, foveolate, ornamentation, pore, scrobiculate.

    I. Jackson, 1928: "marked with small depressions, punctate;" "used in a restricted sense for pits in cell walls."

    II. a. Kupriyanova, 1948, p. 70: *Tr.* Pitted (foveolate) sculpture—in which the pits are represented by narrow indentations in the upper layer of the exine. Such sculpture is found in the representatives of the family *Restionaceae.* [*see* ORNAMENTATION and Fig. 665].

       b. Cranwell, 1953, p. 17: "previously used both for a perforate tectum and for a finely reticulate pattern. See foveolate."

    III. Traverse, 1955, p. 94: "The author has used this term in a looser sense than that of Iversen and Troels-Smith to designate sculpturing of isodiametric depressions. This grades over into reticulate—if there is more total area of depression than unmodified surface: Reticulate. If there is less area of depression than unmodified surface: Pitted."

PLAN-APERTURATE, adj. cf. aperture, colpate, furrow, pore.

Erdtman, 1952, p. 459: "the apertures . . . are situated at the midpoints of the sides (sides of amb ± straight." [*see* AMB and Figs. 83 and 499].

PLANITEGILLATE, adj. (L. planus, flat, plane). cf. blotched, flecked, infrapunctate, intrapunctate, laevigate, maculate, maculose, psilotegillate, punctate, punctitegillate.

Erdtman, 1952, p. 467: "with plane tegillum."

PLANOSPORE, n., pl. -s. (Gr. planos, wandering).

    a. Jackson, 1928: "Sauvageau's term for a mobile zoospore."

    b. Melchior and Werdermann, 1954, p. 15, under SPORE.

PLATEA, n., pl. -ae. (L. a street). cf. cuneus, incidence, solution channel, solution meridium.

Thomson and Pflug, 1953, p. 33 and Fig. 410–412: *Tr.* Remarkable is the stripe-like solution process of the endexine, which takes place in a

meridional direction and which proceeds to a different extent in the different groups. We differentiate the following stages: "solution notch" (= Incidence), "solution wedge" (= Cuneus), and the "solution channel" (= Platea). The overall term is "solution meridium." [*see* APERTURE].

PLATEA LUMINOSA, n., pl. -ae.

Iversen and Troels-Smith, 1950, p. 36: *Tr.* Exine surface between the sculpture elements. [*see* ORNAMENTATION].

PLEOTREME, adj. (Gr. pleion, more; trema, hole). cf. aperture, polytreme.

Erdtman and Straka, 1961, p. 65: "Pleotreme spores have more than one aperture." [*see* GROUPS OF SPOROMORPHAE].

PLEOZONOTREME, adj. cf. aperture.

Erdtman and Straka, 1961, p. 66: [*see* ZONOTREME and GROUPS OF SPOROMORPHAE].

PLICA, n., pl. -ae. (L. plico, I fold or plait). cf. furrow.

    I. Jackson, 1928: "a plait or folding;" "the lamella in Fungi;" "a disease of entangled twigs, the buds producing abnormally short shoots."

    II. Thomson and Pflug, 1953, p. 36 and Fig. 413–414: *Tr.* Doublings of the entire exine that run like two parallel garlandlike folds from the pole to the pore region. They are the result of the unequal swelling capacity of the exine. The pore regions expand more strongly than the central part of the side. This type of swelling is systematically important and characteristic for the [*Triatriopollenites*] *Plicatus*-type. [*see* APERTURE].

PLICATE, adj. cf. culpate, furrow.

    a. Jackson, 1928: "folded into plaits, usually lengthwise."

    b. Erdtman, 1952, p. 467: "see Polyplicate."

    c. Norem, 1958, p. 668: "with folds serving as apertures . . . [Fig. 305–1]."

POINTED, adj. cf. shape classes.

Couper and Grebe, 1961, p. 2 and Fig. 179, 183: [*see* SHAPE].

POLAR, adj. cf. figura, polaris.

    I. Jackson, 1928: "relating to the poles of an organ;" "derived from the smaller ends of a flattened rootlet (Lopriore)."

    II. Erdtman, 1947, p. 112: "with distinct polarity." [*see* GROUPS OF SPOROMORPHAE].

POLAR AREA, n., pl., -s. cf. apocolpium, apoporium, ap-orium, area polaris, figura.

    I. a. Faegri and Iversen, 1950, p. 28–29: "As polar area we define that part of the pollen grain which is situated in higher latitudes than all apertures, annuli, or margines. The relative size of this area is of considerable diagnostic value and can be expressed by the angles formed by the latitudes. For practical reasons we have, however, measured the greatest distance between the ends of two furrows, and expressed the *'polar area index'* as the ratio between this measure and the greatest breadth of the pollen grain."

b. Iversen and Troels-Smith, 1950, p. 34: *Tr.* These terms are only used with bipolar pollen grains:

polar = area polaris = polar area: the area which surrounds a pole and which is limited by the edges of intercolpi or interpori.

c. Ingwersen, 1954, p. 40: "area surrounding a pole and delimited by lines connecting neighboring furrow ends or touching the nearest margins of neighboring pores or, if annuli are present, their nearest annulus margins."

II. Erdtman, 1952, p. 467: "this expression is used in the present book to denote areas in radio-symmetric spores ± comparable with the polar areas of the globe (the areas between the poles and the arctic and antarctic circles respectively). 'Polar areas' in Iversen's and Troels-Smith's (1950) sense are referred to as apocolpia or apoporia."

sdb. Pike, 1956, p. 51.

POLAR AREA INDEX, n., pl. -indices. cf. dimension, figura.

a. Faegri and Iversen, 1950, p. 29: ". . . expressed the 'polar area index' as the ratio between this measure and the greatest breadth of the pollen grain." [*see* POLAR AREA].

b. Iversen and Troels-Smith, 1950, p. 42: *Tr.* The relative size of the polar area can be expressed by the ratio of the polar area dimension (polar-M) to largest transverse dimension of the pollen grain (Lt, +; or Lt, max, +). One may differentiate the following classes:

polar-I (0): polar area absent
polar-I (0.25): polar area small
polar-I (0.25–0.5): polar area medium
polar-I (0.5–0.75): polar area large
polar-I (0.75): polar area very large.

c. Ingwersen, 1954, p. 40: "the ratio of the polar area measurement to the greatest equatorial diameter."

POLAR AREA MEASUREMENT, n., pl. -s. cf. dimension, figura.

Ingwersen, 1954, p. 40: "polar area measurement: the longest diagonal or the longest side of a polar area."

POLAR AXIS, n., pl. -es. cf. axis, main axis, pole axis.

a. Potonié, 1934, p. 8: [*see* MAIN AXIS].

b. Wodehouse, 1935, p. 159: "For purposes of discussing the symmetry relations of these spheres it is convenient to speak of their polar axes as lines extending through the centers of the spheres and directed toward the center of the tetrad, where they would all four meet, if so extended, as stated by Fischer (1890). Thus each sphere comes to have an inner and an outer pole, a proximal and distal polar hemisphere, and the equator is the boundary between the two polar hemispheres."

c. Faegri and Iversen, 1950, p. 15: ". . . the line between the proximal and the distal pole of the grain is called the *polar axis*."

sdb. Iversen and Troels-Smith, 1950, p. 31.
Pokrovskaya, et al., 1950, p. 1.
Erdtman, 1952, p. 467; [*see* POLARITY, SYMMETRY].
Thomson and Pflug, (1953) and SACCATE, Erdtman (1957).
Potonié and Kremp, 1955, p. 10.

POLAR FIELD, n., pl. -s. cf. apocolpium, apical field.

Beug, 1961, p. 12: *Tr.* (J. Iversen and J. Troels-Smith 1950) Area surrounding a pole and delimited by the lines connecting the tops of neighboring colpi or margins.

POLAR HEMISPHERE, n., pl. -s.

Wodehouse, 1935, p. 544: "See Equator and Pole." [*see* POLE and APERTURA, Potonié (1934)].

POLARIS, adj. cf. figura, polar.

POLARITY, n., pl. -ies. cf. figura, Groups of Sporomorphae, morphography, shape classes.

a. Jackson, 1928: "the condition of having distinct poles;" "the assumption of a direction pointing to the poles, as the compass-plant, *Silphium laciniatum*, Linn."

b. Erdtman, 1952, p. 11: "Some spores are—as far as their walls are concerned—apolar, i.e., poles or polar regions cannot be distinguished in individual spores after dismemberment of the tetrads. Cryptopolar spores have much the same appearance as apolar spores, but on closer examination they reveal a more or less distinct polarity (*Larix, Psuedotsuga, Equisetum*, and others).

"Polar spores are isopolar, subisopolar, or heteropolar. In isopolar spores an even plane (situated half way between the poles and cutting the polar axis at right angles) divides the spores into equal halves. The surface of the proximal half is known as the proximal face, the surface of the distal as the distal face. In heteropolar spores the two faces are pronouncedly dissimilar to each other with regard to apertures, etc. Subisopolar spores are ± intermediate between isopolar and heteropolar. Their equatorial plane is usually more or less curved."

117

POLAR LACUNA, n., pl. -ae. cf. lacuna.
  a. Wodehouse, 1928, p. 934: "The lacuna at the pole or center of symmetry, in lophate grains, in which the pattern formed by the ridges is radio-symmetrical or nearly so. Examples: *Vernonia jucunda, V. gracilis.*"
  b. Wodehouse, 1935, p. 544: "the one or more lacunae at the pole or center of symmetry in lophate grains in which the pattern is radiosymmetrical or nearly so. When there are more than one at each pole they are polar lacunae, unless, by definition, they are interporal or abporal lacunae. Example: the grain of *Barnadesia trianae* . . . [Fig. 749] and *Barnadesia venosa* . . . [Fig. 748]."
  sdb. Potonié, 1934, p. 11; [*see* ORNAMENTATION, LACUNEN].

POLAR POSITION, n., pl. -s. cf. equatorial position, figura, proximal, shape classes.

POLAR SPORES, n.
  Erdtman, 1952: "Polar spores are isopolar, subisopolar, or heteropolar." [*see* POLARITY].

POLE, n., pl. -s. (L. polus; Gr. polos, a pivot). cf. apex, figura, polus, shape classes.
  I. Potonié, 1934, p. 5: *Tr.* In the usual tetrahedral arrangement of pollen grains in the pollen mother cell, each pollen grain has an inner and an outer pole, i.e., a proximal and a distal polar-hemisphere (Wodehouse, 1929).
  II. Wodehouse, 1935, p. 544: "one of the extremities of the axis of symmetry of radiosymmetrical pollen grains. If there is more than one such axis of symmetry, the word applies only to the extremities of the axis which is directed toward the center of the tetrad or was so directed during the grain's formation. From these tetrad relations the two poles and two hemispheres may be designated as inner and outer or proximal and distal, though in mature pollen grains that are not shed in tetrads the two hemispheres are rarely distinguishable."
  sdb. Iversen and Troels-Smith, 1950, p. 31.
      Pokrovskaya, et al., 1950.
      Erdtman, 1952, p. 467; [*see* POLARITY; SYMMETRY, Potonié and Kremp, (1955), p. 10; *see* APEX].
      Traverse, 1955, p. 91.
      Beug, 1961, p. 12.

POLE AREA, n., pl. -s. cf. contact marking.
  Krutzsch, 1959, p. 37: *Tr.* The area between torus and apex is called the pole, or apex field; also it is designated by R. Potonié (1934) as pole area.

POLE AXIS, n. cf. axis, figura, main axis, polar axis.

POLE CAP, n., pl. -s. cf. shape classes.
  Thomson and Pflug, 1953, p. 38: *Tr.* The long symmetry axis which runs parallel to the colpi pierces the exine at the poles. The equatorial plane cuts the polar axis in half, is perpendicular to it, and at the same time is a plane of symmetry and the plane of the pores. Parallel to the equatorial plane, at a distance of three-fourths of h [height], the "polar cap planes" are situated cutting off the polar caps; h is the height of the poles above the centrum. The outline is designated as the "meridional-contour;" that part of the meridional contour lying between the two polar caps is called the "side-line." The "width/length index" is derived from the ratio of "the largest equatorial diameter to the length of the polar axis;" the "polar cap index" from "the largest diameter of the polar cap plane to the largest equatorial diameter."
  The "figura" is established by approximate comparisons with geometric bodies: spheres, blunt and pointed double cones, ellipsoids, spindles, cylinders, etc.
  The contour is derived from the contours of the polar cap and the sideline. The shapes of the polar cap can be: flattened, hemispherical, sub-hemispherical, or pointed; of the sidelines: straight, weakly convex, strongly convex, or weakly concave . . . [Fig. 509–515].

POLE FIELD, n., pl. -s. cf. contact marking.
  Krutzsch, 1959, p. 37: *Tr.* The area between torus and apex is called the pole, or apex field; also it is designated by R. Potonié (1934) as area.

POLLEN, n. (L. fine flour). cf. Groups of Sporomorphae, miospore, palynomorph, pollen grain, pollenospore, spore, sporomorph.
  a. Jackson, 1928: "the fertilising dust-like powder produced by the anthers of Phanerogams, more or less globular in shape, sometimes spoken of as 'Microspores';" "the antherozoids of Mosses (Hooker and Taylor)."
  b. Erdtman, 1943, p. 43: "as to the terms 'pollen' and 'spore,' it is necessary to keep in mind that they are not homologous—the microspore is the immediate product of division of the mother cell, the pollen grain contains within its wall the microgametophyte developed from the microspore. In view of the fact that pollen analysis commonly deals also with spores, there would seem to be no need for a separate term, 'spore analysis'."
  c. Hyde, 1944, p. 145–149: "The word *pollen* is so well established in English that one discovers with a shock of surprise that, as an English word with its present meaning, it dates from the late eighteenth century. Its original meaning, in classical Latin, was 'fine meal.' The first to use it, still in Latin, in the sense of the dust which is shed from the male organs of flowers and which brings about fertilization, was the great Linnaeus

118

in his *Philosophia Botanica* (1751). His example was followed by other botanical writers from whose pages it has become gradually diffused and internationalized."

d. Faegri and Iversen, 1950, p. 15: ". . . meaning originally 'fine flour' (cp. Skeat, 1910) signifies the substance and ought not to be used in the plural, 'pollen types' and 'pollen grains' being the correct terms." [*see* TETRAD].

POLLEN ANALYSIS, n., cf. palynology.

Hyde, 1944, p. 145: "The technique known as *pollen analysis* has come into vogue during the past twenty years. Essentially it is a simple process, consisting in the microscopic examination of a sample of mixed pollen, whether contained in or mixed with other substances or not, and a determination of the different kinds present and of their relative proportions. Detailed pollen analysis must evidently rest on a wide knowledge of the microscopic appearance of the pollen of various plants.

"Systematic study such as this knowledge presupposes has in the past been carried out by very few workers; in fact it has only caught on since botanists have realized that it led to anything.

"Pollen analysis was originally applied to peat deposits. It was found that, owing to the highly resistant nature of their outer walls, pollen grains of certain kinds were preserved naturally in this medium and that in a given bed of peat the numbers of grains of various kinds—notably of tree pollen—varied widely from one layer to another. . . ."

POLLEN AND SPORE FORMULA, n., pl. -ae. cf. aperture, figura, formula of spores and pollen symbols, Groups of Sporomorphae, morphography, ornamentation, shape classes.

POLLEN CLASSES, n. (L. classis, class). cf. figura, Groups of Sporomorphae, morphography, shape classes.

POLLEN FIBER, n., pl. -s. (L. fibra). cf. aggregation.

Potonié, 1934, p. 14: *Tr.* Likewise still unknown as fossils are fibers in which pollen are arranged in a row (*Halophila*). [*see* APERTURA].

POLLEN FORM INDEX, n., pl. -indices. cf. figura, Groups of Sporomorphae, morphography, shape classes.

Iversen and Troels-Smith, 1950, p. 41: *Tr.* In bipolar pollen grains, the shape of the pollen grains can be expressed by the ratio length (Lg) to largest transverse dimension (Lt, +; or Lt, max, +). The following shape classes may be distinguished (Erdtman, 1943):
perprol: pollina perprolata: Lg/Lt, + >2.0;
prol = pollina prolata: Lg/Lt, +2.0 −1.33;

subsph = pollina subsphaeroidea[1]: Lg/Lt, +1.33 −0.75;
[1] The group subsphaeroidea corresponds to the groups subprolate + spheroidal + suboblate of Erdtman.
obl = pollina oblata: Lg/Lt, +0.75 −0.5;
perobl = pollina peroblata: Lg/Lt, + <0.5.
sdb. Beug, 1961, p. 12.

POLLEN GRAIN, GRANULE, n., pl. -s. cf. pollen, spore, sporomorph.

Jackson, 1928: "the small bodies which compose the entire mass; the latter term is also used for the contents of the grain."

POLLEN MASS, n. (L. massa, a lump, mass). cf. massula.

Jackson, 1928: "pollen-grains cohering by a waxy texture or fine threads into a single body."

POLLENOSPORE, n., pl. -s. cf. palynomorph, spore, sporomorph.

De Jekhowsky, 1958, p. 1392: *Tr.* Reproductive organs of plants, generally measuring from 10 to 100 μ (extreme values: a few microns for certain fungi, a few millimeters for certain Carboniferous macrospores) . . . Essentially of terrestrial origin (at least for those which are fossil forms), they are also found in several types of sediments, not only continental, but also brackish and marine: on the order of one to three (in the case of sorting carried out entirely at random, which will be the only one considered), and often as many as some hundred or thousand per gram of rock (we found as many as 100,000 per gram). Their presence has been reported since the upper Precambrian by the Russians, and we personally have identified them as early as the upper Ordovician.

POLLEN-TETRAD, -TETRAHEDRON, n. cf. aggregation, spore, tetrad.

Jackson, 1928: "the shape of certain groups consisting of four grains cohering in a pyramid, as in *Oenothera*."

POLLEN TUBE, n., pl. -s. (L. tubus).

Jackson, 1928: "the tube emitted by a pollen-grain passing down from the stigma to the ovary and ovules."

POLLINA BISULCATA, n.

Kupriyanova, 1948: *Tr.* . . . having two symmetrically arranged furrows intersecting the polar axis at a right angle. [*see* GROUPS OF SPOROMORPHAE and Fig. 242:3].

POLLINA OCCLUSA, n. cf. aggregation, pseudomonad, spore, tetrad.

Kupriyanova, 1948, p. 75: *Tr.* . . . in which the pollen is closed within a membrane of the maternal pollen cell and is arranged in the form of a tetrahedral tetrad. [*see* GROUPS OF SPOROMORPHAE and Fig. 242:14].

POLLINARIUM, n., pl. -ia. cf. aggregation, Groups of Sporomorphae, spore, tetrad.
I. Jackson, 1928: "=Androecium;" "=Cystidium."
II. Potonié, 1934, p. 8: *Tr.* In many orchids and the Asclepiadaceae all grains of an anther sac are connected to a *pollin(ar)ium* (a single pollen mass).

POLLINATION, n.
I. Jackson, 1928: "the placing of the pollen on the stigma or stigmatic surface."
II. Webster, 1960: "the transfer of pollen from the stamen to the pistil."

POLLINA TRICHOTOMA FISSURATA. cf. trichotomocolpatus, trichotomofissurate, trichotomous.
Kupriyanova, 1948, p. 71: *Tr.* Pollen grains with a three-slit opening. [*see* Groups of Sporomorphae and Fig. 242:1].

POLLINIUM, n., pl. -ia. cf. aggregation, Groups of Sporomorphae, spore, tetrad.
a. Jackson, 1928: "a body composed of all the pollen-grains of an anther-loculus, a pollen-mass."
b. Selling, 1947, p. 77: "a unit of all grains produced by one theca. Cf. Massula."
sdb. Potonié, 1934, p. 8: [*see* Pollinarium].
Beug, 1961, p. 12.

POLOSPORES, n. (Pollen and/or spores). cf. palynomorph, pollen, pollenospore, spore, sporomorph.
Grayson, 1956, p. 71: "is a new term here proposed as a convenient general name used to designate pollen and/or spores."

POLUS, n., pl. -i. cf. pole, figura.

POLYAD, n. (Gr. polys, many). cf. aggregation, Groups of Sporomorphae, pollen mass, spore, tetrad.
Faegri and Iversen, 1950, p. 128: "pollen grains united in groups . . ." [*see* Groups of Sporomorphae and Fig. 6].
sdb. Beug, 1961, p. 12.

POLYBROCHATE, adj. cf. homobrochate, reticulate.
Erdtman, 1952, p. 467: "with a very large number of brochi (>45 in an amb in ±· homobrochate grains). Cf. . . . [Fig. 815–822]."

POLYFORATE, adj. cf. aperture, cribellate, foraminate, forate, Groups of Sporomorphae, oligoforate, ornamentation, perforate, porate, pore.
Erdtman, 1952, p. 467: "with >12 foramina. Example: . . . [Fig. 493]."

POLYGONAL, adj. (Gr. gonia, angle). cf. multilateral, shape classes.

POLYPERICOLPORATE, adj. cf. pericolpate, oligopericolpate.
Norem, 1958, p. 671: ". . . >8 colpi . . . uniformly distributed over surface . . . [Fig. 501–2]."

POLYPERICOLPATE, adj. cf. pericolporate, oligopericolpate.

Norem, 1958, p. 671: ". . . >8 colpori . . . uniformly distributed over surface . . ."

POLYPERIPORATE, adj. cf. periporate, oligoperiporate.
Norem, 1958, p. 670: ">8 pores . . . uniformly distributed over the surface . . . [Fig. 501–1]."

POLYPLICATE, adj. cf. colpate, colpus, furrow, longitudinal furrow, longitudinal rib.
Erdtman, 1952, p. 467: "with ± colpoid grooves; a provisional term used in descriptions of the pollen grains of *Spathiphyllum* . . . [Fig. 298–300], *Ephedra,* and *Welwitschia.*" [*see* Colpus] . . . Steeves and Barghoorn (1959), and Fig. 301–304.

POLYPORATE, adj. cf. aperture, cribellate, Groups of Sporomorphae, ornamentation, porate, pore.
Kupriyanova, 1948: *Tr.* Multiporous pollen grains (pollina polyporata) have very many pores distributed on the entire surface of the pollen. [*see* Groups of Sporomorphae].

POLYRUGATE, adj. cf. colpate, furrow.
Erdtman, 1945, p. 191: "with many rugae." [*see* Rugate and Sulcate].

POLYSPORE, n. cf. spore.
a. Jackson, 1928: "a multicellular spore composed of Merispores (Bennett and Murray)."
b. Melchior and Werdermann, 1957, p. 16, under Spore.

POLYSPOROUS, adj.
Jackson, 1928: "containing many spores, used of Cryptogams, as in asci when more than four or eight spores occur."

POLYSULCATE, adj. cf. colpate, furrow.
Erdtman, 1947, p. 113: "with many sulci." [*see* Groups of Sporomorphae].

POLYTREME, adj. cf. aperture, pleotreme.
Erdtman and Straka, 1961, p. 65: "(-treme) spores: with more than six apertures." [*see* Groups of Sporomorphae].

PONTOPERCULATE, adj. cf. colpate, furrow, operculum, porate, pore.
Erdtman, 1952, p. 467: "said of ± elongate apertures where the apical parts of the opercula merge with the exine surrounding the apertures. Example: . . . [Fig. 416–418]."
sdb. Beug, 1961, p. 12.

PORAL, adj.
Jackson, 1928: "relating to a pore."

PORAL LACUNA, n., pl. -ae. cf. germinal lacuna, lacuna.
a. Wodehouse, 1928, p. 934: "the lacuna enclosing the germ pore. Example: *Stokesia laevis.*"
b. Potonié, 1934, p. 11–12: *Tr.* If the meshes or lumina of a network become larger and if, more-

over, they are very regularly arranged, then Wodehouse (1928) speaks of lacunae; according to their position, he gives the lacunae special names. The *polar lacuna* is located at the pole; the *circum polar lacunae* surround the polar lacuna. In the *poral lacuna* is the germinal; therefore, we prefer to use the general term "germinal lacuna." [*see* ORNAMENTATION].

   c. Wodehouse, 1935, p. 544: "a lacuna enclosing a germ pore. It may be open through a cleft, as in the grain of *Taraxacum* . . . [Fig. 765], or closed, as in those of *Scolymus* . . . [Fig. 764]."

PORATE, adj. (L. poratus). cf. anaporate, angulaperturate, annular, annulate, aperturate, aperturidate, arcuate, arcuatus, aspidate, aspidoporate, aspidorate, aspidote, brevissimicolpate, brevissimirupate, colpato-colporate, colpoid, colpodiporate, colpoidorate, colporate, colporatus, colporoidate, cribellate, cribellatus, demicolporate, dicolporate, diorate, diplodemicolpate, diporate, distaloaperturate, distaloproximaloaperturate, disulcate, diverse porate, extraporate, foraminate, foraminoid, foraminose, foraminosus, forate, fossaperturate, halonate, heterocolpate, hilate, infundibuliformis, mesoporate, monocolporate, monoporate, monopored, nonaperturate, N-stephanoporate, obinfundibuliformis, oligoforate, orate, oroid, oroidiferous, panporate, pentacolporate, perforate, periporate, plan-aperturate, polyforate, polyporate, pontoperculate, poratus, poroid, poroletoid, pororate, protrudent, proximaloaperturate, sinuaperturate, stephanaperturate, stephanocolporate, stephanoporate, tenuate, tetracolporate, tri-colpodiporate, tricolporate, triporate, tumid, ulcerate, vestibulate, zonaperturate, zoniporate, zonorate.

  I. a. Erdtman, 1945, p. 190: "provided with rounded apertures in the general surface of the exine in absence of germinal furrows. Used with the same numerical prefixes as 'colporatus;' 'polyporatus' may replace 'cribellatus'."

   b. Erdtman, 1947, p. 112: "with nonaspidate pores . . ." [*see* GROUPS OF SPOROMORPHAE].

  II. a. Selling, 1947, p. 77: "pollen with one or several pores, not in combination with colpae or rugae. Prefixes: see *Colpa*. Cf. *Colporate* and *Rugoporate*."

   b. Pike, 1956, p. 51: "Pollen grains are porate when they have equatorial ± isodiametric apertures. The limit between a porus and a colpus is defined by the length/breadth ratio of 2:1."

  III. Thomson and Pflug, 1953, p. 20: *Tr.* . . . The ratio of the length to width usually does not exceed 5:1 and the maximum diameter does not reach one-fifth. [*see* PORE].

PORE, n., pl. -s. (L. Porus, channel). cf. annulus (anulus), aperture, arc, arcus, aspid, atrium, canalus pori, centrum pori, colpus, conclave, costa pori,

cover lid, crassimarginate, edge of the annulus, edge of the pore, endannulus (endanulus), endoplica, endopore, endoporus, equatorial pore, equatorial ruga, exit, exitus, exopore, exoporus, fissure, foramen, foramina, formation of lips, fovea, geniculus, germinal aperture, germinal apparatus, germinal lid, germinal point, germinal pore, germ pore, halo, hilum, knee, labrum, lacuna, lid, ligula, ligule, limes annuli, limes marginis, limes pori, lip, margo, margo arcuata, membrana pori, oculus, oncus, opening, open pore, operculum, operculum pori, os, pore canal, pore canal index, pore membrane, poreplug, pore ring, porus, porus annularis, porus collaris, porus simplex, porus vestibuli, postatrium, postcaverna, postvestibulum, praecaverna, praevestibulum, prevestibulum, primary fovea, protrudence, pseudopore, pseudoporus, rimula, tenuimarginate, tenuitas, tenuity, vestibulum, vestibulum apparatus, vestibulum pori.

  I. Jackson, 1928: "any small aperture, as in anthers, for the emission of pollen in the pollen grains themselves, in the epidermis as stomata or water-pores;" "in *Polyporus,* any of the tube-like openings, forming the hymenium;" "large pitted vessels or tracheids in wood;" "an opening in the prickles of *Victoria regia;*" "cavities in soils not occupied by solid substances (Warming);" "minute canals in certain diatom-valves, which pass through the cell-wall (West)."

  II. Wodehouse, 1935, p. 544: "see germ pore."

  III. a. Potonié, 1934, p. 16: *Tr.* Indeed, here the foveae have no doubt already changed into the porus simplex . . . [Fig. 390], viz. not only the exoexine but the entire exine is pierced. According to H. Fischer (1890) only those germinal apparatus called porus simplex may be precisely designated as *germ pore*. Fischer suggests namely ". . . to designate only the true holes of the 'Aussenhaut' (i.e., probably the entire exine) as germ pores . . ." Later authors have repeatedly deviated from this suggestion as well as in regard to porus vestibuli, which will be discussed later. The latter does not pierce the exine; nonetheless, Fischer labeled it as a germinal pore too because its structure was not clear at that time. [Potonié, 1934, p. 18: *Tr.*] *Pollenites simplex* was mentioned as a fossil example. In this fossil type, the exoexine as well as intexine is pierced so that an effectively true porus, a *porus simplex,* is present. But it need be said that it is not always easy to ascertain whether we are dealing with a true porus or with a sharply depressed fovea in which the floor consists only of a thin membrane of intexine. [Potonié, 1934, p. 19: *Tr.*] The question is posed as to whether such a thin layer of intexine is generally or always extant, or is removed during fossilization or the maceration of the fossil material. Then too, in the living grain, the pollen contents beneath the exitus would never be isolated from the outside exclusively by intine. Nevertheless, certain conditions noticed in prepared fossil material satisfy the definition of *porus simplex,* a definition which has been related to the terms *germ pores* and *exitus opening* by previous authors. However, the same authors, in part unconsciously, have from the beginning used these terms in a broader sense. Particularly they have included porus vestibuli . . . [*see* APERTURA].

b. Erdtman, 1943, p. 52: "the rounded apertures which frequently occur in the general surface of the exine in the absence of germinal furrows (*colpae*). There is a gradual transition from a *porus simplex,* which does not lead into a *vestibulum,* to a *porus vestibuli,* which forms the entrance to such a vestibule . . . [Fig. 407]. Sometimes the aperture of a pore leads into a more or less elongated exine collar rising above the general surface of the grain (*porus collaris*)." [*see* ARCUS].

c. Selling, 1947, p. 77: "a ± rounded or only slightly transverse aperture in the general exine surface for the emergence of the pollen tube (cf. *pits,* of the sculpture), with either the absence of furrows or enclosed in a colpa or ruga (see above). Cf. *Furrow* and *Transverse furrow.*"

d. Kupriyanova, 1948, p. 70: *Tr.* The germination pore (porus) is a special outlet place for the pollen tube.

Fischer (1890) whose work has been up to the present time a valuable reference book, recommends to apply the treatment of pollen with a concentrated sulphuric acid in order to determine the pore structure. In such instances, when we deal with a thinning of the exine located on its surface, the protuberances of the exine will remain after the application of sulphuric acid; but if they fall off, it will indicate the absence of the exine, and in this case the membrane of the pore is made up of the intine only.

The thin filmy membrane which covers the furrow or pore is usually composed of endexine elements, it is so thin and transparent that frequently it is very difficult to distinguish it from the intine because both layers are not tinted with fuchsin or other anilin stains.

In contrast with the Dicotyledoneae, the pore of the Monocotyledoneae is never located in the furrow. Therefore, both furrows and pores are not present on the surface of the grains simultaneously.

Pores can be simple (porus simplex) and annular (porus annularis). The former among the monocotyledons are found in representatives of *Pandanaceae, Sparganiaceae, Centrolepidaceae, Palmaceae,* and the latter in representatives of *Gramineae, Restionaceae, Flagellariaceae,* and others. [*see* GROUPS OF SPOROMORPHAE].

e. Pokrovskaya, et al., 1950: *Tr.* A pore of germination, germination pore or embryonic pore. It is an opening in the exine which is the place of the exit of the pollen type. They develop in the membrane of furrows or on the grain surface. Pores may be chambered (p. vestibulum) or non-chambered (p. simplex).

f. Cranwell, 1953, p. 17: "in monocotyledons, an aperture occurring independently of a furrow; serves for emission of pollen tube, as a rule."

g. Oldfield, 1959, p. 21: "The occurrence, shape, size, regularity, visibility and emplacement of apertures in the endexine are noted. In the tetrads the pores are usually situated near to the double walls of contiguous grains, bisecting the double furrow. The pores may however be consistently or inconsistently more or less remote from these double walls."

IV. Erdtman, 1946, p. 71: "the pores (*pori*) are rounded and more or less distinct apertures, as a rule equally distributed on the equator or over the general surface of the grain. '1-po' denotes a grain with one pore. To avoid misunderstandings and the introduction of a new term, '2-po,' '3-po,' '4-po,' etc. should exclusively denote grains with equatorially arranged pores. Any other porate grains should be referred to as 'polyporate' (po-po). This term would thus, according to the definition, include any grains, except monoporate, with pores not confined to the equator. In practice 'polyporate' would mean the same as 'cribellate,' proposed by Wodehouse (1935)."

V. Erdtman, 1947, p. 105: "the term pore (porus) is applied only to non-equatorial pores . . . Equatorially arranged pores are termed 'ora'."

VI. a. Faegri and Iversen, 1950, p. 20: "Most pollen grains possess openings in, or thin parts of the exine, through which the pollen tube emerges. Two different types of apertures can be recognized, which are generally called *pores* and *furrows* (colpi, Wodehouse, 1935).

"In contrast to the furrow the pore is generally isodiametric and, if it is somewhat elongated (not boat-shaped!), the ends are rounded. For practical purposes the limit between pore and furrow may be defined by a length/breadth ratio of 2:1. The endexine of the pore is thinner than that of the rest of the grain; we are not competent to state whether it in any case really is absent in the living grain. In many fossil grains pores appear as genuine holes in the wall, but this may be a secondary phenomenon. It is obvious that as soon as a pollen grain is wetted and its contents swells out, the very thin membrane of the pore is in danger of being ruptured. The pore is the natural place of emergence of the pollen tube. In pollen grains that possess furrows, but not pores, and where the pollen tube must force its way through the thin membrane of the furrows, it frequently leaves an irregular hole with tattered edges.

"Pores and furrows may appear as simple holes or slits in the exine, but generally they are surrounded by distinctive parts of the exine. A pore is frequently surrounded by an annular area (*annulus*), the exine of which is characterized by differences in the outer layer, the endexine." [*see* FURROW and Fig. 350–358].

b. Iversen and Troels-Smith, 1950, p. 33 and Fig. 350, et seq.: Tr.

P = porus = pore: area which serves as the normal exitus place of the pollen tube and whose given length-width ratio is less than two. The pore is designated in three different ways in relation to the surrounding exine.

P(ex = O) = diaporus = open pore: by absence of the exine (e.g., by casting off an operculum).

P(mb): by thinning of the exine.

P(mb), ekt = O): ektexine elements are absent.

P(mb), ekt: ektexine element present.

P(op): by the demarcation of a part of the normal exine by a furrow or suture.

P, 1 = limes pori = edge of the pore: demarcation line of the pore; therefore either the above mentioned furrow or suture or—in the case of a thinned or missing exine—the outer limit of the lighter spot which results from it.

P, mb = mebrana pori = pore membrane: the thinned exine of a pore.

P, op = operculum pori = pore-operculum: the thicker part of a pore which is situated within a furrow or suture. The structure of the operculum is similar to that of the remaining exine of the pollen grain.

P, vest = vestibulum pori = small "vestibulum" within the pore which originates by the exine having two-prong aspect at the edge of the pore in optical cross section.

anl = annulus: area which surrounds the pore as a ring and which is differentiated from the remaining exine of the pollen grain by deviations within the ektexine, e.g., by a greater or lesser thickness of the ektexine. In the case where an annulus contains concentric zones with different structure, they may be named, from the interior to the exterior, anl (1) and anl (2) . . .

anl, 1 = limes annuli = edge of the annulus: outer demarcation line of the annulus.

cost P = coste pori: a thickening of the endexine surrounding the pore.

P, cent = centrum pori: center of the pore.

Pseudo P = pseudoporus = pseudopore: differs from a true pore in that it is not the normal exitus place of the pollen tube.

lac = lacuna = lacuna: probably includes pseudopores as well as pseudocolpi.

c. Ingwersen, 1954, p. 41: "area on pollen grains forming, or surrounding, the normal place of emergence of the pollen tube and whose length-breadth index is lower than 2."

d. Pike, 1956, p. 51: "*Porate*—Pollen grains are porate when they have equatorial, ± isodiametric apertures. The limit between a porus and a colpus is defined by the length/breadth ratio of 2:1."

sdb. Beug, 1961, p. 12.

VII. Thomson and Pflug, 1953, p. 20: *Tr.* Pores are faceted apertures through one or more lamellae and preformed passage for the pollen tube. The ratio of length to width usually does not exceed 5:1 and the maximum diameter does not reach one-fifth the maximum pollen diameter. [*see* SOLUTION and Fig. 368–371].

PORE CANAL, n., pl. -s. cf. canalus pori.

I. Jackson, 1928: "the passage through a pit between neighboring cells."

II. Pokrovskaya, et al., 1950, p. 1: *Tr.* The canal which is extended from the outside of the exine (ektexine) to the intine. The canal may be cylindrical (tubulosus), cone shaped (infundibuliformis), reverse cone shaped (obinfundibuliformis), and disclike (discoideus).

sdb. Takhtadzhyan and Yatsenko-Khmelevskiy, 1945, p. 38.

PORE CANAL INDEX, n., pl. -indices.

Thomson and Pflug, 1953, p. 32: *Tr.* The annulus encloses the pore canal like a beak. The pore canal may be cylindrical or it may widen conically toward the exterior or the interior. Its length is measured from the outer edge of the ektexine to the inner edge of the endexine. The ratio of "length of the pore canal to pollen diameter" (measured in the axis of the pore canal) is called the "pore canal index." [*see* APERTURE].

PORE MEMBRANE, n., pl. -s.

a. Wodehouse, 1935, p. 544: "a delicate membrane covering a germ pore. It may be flecked or bear an operculum."

b. Kupriyanova, 1948, p. 70: *Tr.* The thin filmy membrane which covers the furrow or pore is usually composed of endexine elements. It is so thin and transparent that frequently it is very difficult to distinguish it from the intine because both layers are not tinted with fuchsin or other anilin stains. [*see* PORE].

sdb. Iversen and Troels-Smith, 1950, p. 33; [*see* PORE].

PORE PLUG, n., pl. -s. cf. lid, operculum, pore lid.

PORE RING, n., pl. -s. cf. annulus, aspis, costa pori, dissence, endannulus, formation of lips, halo, labrum, lip, margo, oculus, operculum, protrudence.

POROID, adj. cf. porelike, poroletoid.
   I. Jackson, 1928: "Poroids n., minute circular dots in diatoms, more than 0.6 μ in diameter, tiny cavities resembling pores, but not actual perforations (O. Muller)."
   II. Erdtman, 1947, p. 113: "Compare: poroletoid."

POROLETOID, adj. cf. aperture, monolete, porate, proximalo aperturate, trilete.
   Erdtman, 1947, p. 113: "proximalo aperturate . . . with one poroid aperture (scar)." [*see* GROUPS OF SPOROMORPHAE].

PORORATE, adj. cf. colpate, colporate, colpus, furrow, Groups of Sporomorphae, os, porate.
   a. Erdtman, 1952, p. 467: "with orate pori."
   b. Erdtman and Straka, 1961, p. 66: "with apertures consisting of a distal poroid part and a proximal part (os). Pororate spores are practically always pleotreme."

PORPHYRITIC STRUCTURE, n., pl. -s. cf. granule, granulum, granum, punctum, sporodermis.
   Potonié and Kremp, 1955, p. 18: *Tr.* We call (along with Haberlandt) porphyritic structure [Moertelstruktur], a situation where *micelle nuclei,* or *puncta,* are imbedded in the groundmass or matrix of the exine. [*see* PRISMATIC STRUCTURE].

PORUS, n., pl. -i. cf. pore.
   Jackson, 1928: "= pore."

PORUS ANNULARIS, n. cf. annulus, operculum, pore.
   Kupriyanova, 1948, p. 70: *Tr.* Pores can be simple (porus simplex) and annular (porus annularis) . . . the latter in representatives of *Gramineae, Restionaceae, Flagellariaceae,* and others. [*see* SPORE].

PORUS COLLARIS, n. cf. annulus, aspis, costa pori, dissence, endannulus, formation of lips, halo, labrum, lip, margo, oculus, operculum, pore ring, protrudence.
   I. Erdtman, 1943, p. 52: "sometimes the aperture of a pore leads into a more or less elongated exine collar rising above the general surface of the grain (*porus collaris*)." [*see* PORE].
   II. Pokrovskaya, et al., 1950, p. 3: *Tr.* One of the chamberless types of a simple pore. The exine is noticeably raised around the pore of this type,

forming a fold—"small collar." (e.g., in pollen grains of *Carpinus*).

PORUS SIMPLEX, n.
   a. Potonié, 1934, p. 16: *Tr.* . . . here the fovea have probably already changed into the *porus simplex* . . . [Fig. 390], i.e., not only the exoexine but the entire exine is pierced. [*see* PORE].
   b. Erdtman, 1943, p. 52: "there is a gradual transition from a *porus simplex* which does not lead into a vestibulum, to a *porus vestibuli,* which forms the entrance to such a vestibule." [*see* PORE and Fig. 407].
   c. Kupriyanova, 1948, p. 70: *Tr.* Pores can be simple. [*see* PORE].

PORUS VESTIBULI, n. cf. vestibulate pore.
   a. Erdtman, 1943, p. 52: "forms the entrance to a vestibule." [*see* PORUS SIMPLEX].
   b. Faegri and Iversen, 1950, p. 23: see Fig. 357.
   c. Pokrovskaya, et al., 1950, p. 3: *Tr.* Pores having a pore chamber formed at the expense of the branching exine. The form of the chambered pores may be different (e.g., the chambered pores of the pollen grains of *Tilia, Alnus,* and *Fagus*).

POSITIO AEQUATORIALIS, n. cf. equatorial position, figura, polar position.
   Pokrovskaya, et al., 1950, p. 3: *Tr.* equatorial position. The view of the pollen grain from the equator.

POSITIO POLARIS, n. cf. equatorial position, figura, polar position.
   Pokrovskaya, et al., 1950, p. 3: *Tr.* Polar position. The view of pollen grain from the pole.

POSTATRIUM, n., pl. -ia. (L. after, behind). cf. atrium.
   Thomson and Pflug, 1953, p. 33: *Tr.* Also the endexine can be interlamellarly split and then enclose a "post-vestibulum" or also form an interlamellar "postatrium" . . . [*see* APERTURE and Fig. 385].

POSTCAVERNA, n., pl. -ae. cf. caverna.

POSTVESTIBULUM, n., pl. -a. cf. vestibulum.

POUCHED, adj. cf. rugulo-saccate, subsaccate.
   Jackson, 1928: "expresses priority in time or place."

PRAECAVERNA, n., pl. -ae. cf. caverna.

PRAEVESTIBULUM, PREVESTIBULUM, n., pl. -a. cf. dissence, vestibulum.

PREPOLLEN, n. cf. spore, sporomorph.
   I. Schopf, 1938, p. 12: "In some primitive gymnosperms the male spore shows considerable advance in organization beyond the cryptogamic microspore and yet lack some significant features of modern pollen. For these it is proposed to apply Renault's term 'prepollen'."
   Schopf, 1938, p. 14: "The term *prepollen* was used by Renault [Bassin Houiller et Permien d'Autum et d-Épinac: Études des Gités miner-

aux de la France, fasc. IV, Paris, 1896] in describing the large monolete pteridosperm spores found in *Dolerophyllum* (now *Dolerotheca . . .*) fructifications and also spores of the same type found in the pollen chambers of some pteridospermic seeds. The cellular organization in these is very different from that of cryptogamic spores. For reasons given below it is believed that a satisfactory distinction may be made between them and other spores or spore equivalents found in maceration residues of coal. Because the points of distinction seem to be botanically significant, prepollen is used as a category coordinate with pollen. Prepollen is considerably more primitive in type of organization but it may represent the Paleozoic equivalent of certain types of modern pollen, in the same way that pteridosperms are considered ancestral to the Cycadales. Most authors have designated these spores as 'pollen.' It is believed, however, that revival of Renault's term will facilitate more precise discussion of such fossils.

"A distinction of fundamental importance serving the distinguish prepollen from modern pollen is the point of germinal exit. The term germinal exit is used to indicate rupture of the spore coat for normal exposure of the gametophytic product whether this be spermatozoids, a pollen tube, or vegetative gametophytic tissue. The existence of pollen tubes in plants of the Paleozoic [Schopf, 1938, p. 15:] is still open to question since they have yet to be demonstrated. Germinal exit in prepollen is probably proximal but in modern pollen typically distal. Germinal exit in cryptogamic spores is so far as known, always proximal. The process of exit (commonly spoken of as germination) is a common feature in all spore bodies and it is physiologically significant since the spore coat, which was previously the dominant feature of the gametophytic environment, is essentially dispensed with at this stage. In the species of prepollen described . . . features indicative of proximal germinal exit are found.

"Cryptogamic spores are similar to prepollen in having proximal exit and it is chiefly due to this similarity that prepollen is considered as primitive among the male spores of seed plants. It seems likely that originally the postulated cryptogamic ancestors of modern seed plants also were proximal in germinal exit and that at some stage of evolution rather abrupt change in the exit position occurred. In the Paleozoic prepollen figured by Zerndt 1930 as 'Pollen of *Dolerophyllum* sp.' there seems to be distal infolding of the coat similar to that seen in cycad and some angiosperm pollen when dry. Such spores as this

may represent a transitional stage in the process whereby distal exit finally became established. However this may have occurred, the prevalence of proximal exit among the primitive land plants shows that this condition is probably a fundamental feature of spore organization and hence it is a point of significance when this condition is indicated in the male pores of ancient seed plants".

II. Dijkstra-van V. Trip, 1946, p. 22: *Tr.* One finds in coal pollen or spores of various types, which Halle (1933) would assign to *Whittleseyinae*. Zerndt has then used them for stratigraphic purposes because their size is such that they could not pass through the sieve after maceration and describes them as type 31. Their nature, however, is not clear. Halle calls them microspores, Florien (1937) calls them pollen. Both authors have studied them thoroughly in fructifications. Ibrahim calls them megaspores and in part microspores, Zerndt pollen, Schopf prepollen. The latter name as first proposed by Renault when he described the spore of *Dolerophyllum* (now *Dolerotheca*). Probably, prepollen grew no pollen tube, rather a straight furrow burst open in the middle of the spore. Very probably a great number of sporangia-genera have had such pollen or spores which were described by several scientists.

PRIMARY FOVEA, n. (L. primus, first). cf. fovea.

PRIMARY GERMINAL, n., pl. -s. cf. meridional area.

PRIMARY MERIDIONAL AREA, n., pl. -s. cf. meridional area.

PRISMATIC STRUCTURE, n., pl. -s.

Potonié and Kremp, 1955, p. 17: *Tr.* The so-called "*prismatic structure*" is shown in the cross section of the exine as radially directed lighter "rodlets" which stand close to one another. [*see* SPORODERMIS and PORPHYRITIC STRUCTURE].

PROCESS, n., pl. -es. (L. processus, a prolongation). cf. baculum, elevation, excrescence, projection, protrusion, protuberance.

Jackson, 1928: "any projecting appendage *Processus Hymenii,* 'the aciculae of certain Fungals' (Lindley);" "see also bands, in fruit of *Zostera minor*."

PROCESSES-PROJECTIONS, n. cf. baculum.

Kosanke, 1950, p. 11: "type of ornamentation," [*see* ORNAMENTATION and Fig. 640:O].

PROJECTION, n., pl. -s. (L. projectus). cf. baculum, elevation, excrescence, mammilla, process, protrusion, protuberance, tuber, tubercle.

Harris, 1955, p. 19–20: "The pattern is regarded as comprising elevations of, or projections from, the general surface, these being not less than $1\,\mu$ in height. The distinction here made between projections and elevations is a matter of convenience and is not necessarily of structural significance . . . A. Projections. irrespective of height, which are less than twice as long as wide. These may be grouped according to the characters of the apex, trunk, and base . . . B. Elevations. These are long in surface view (length more than twice breadth), except where otherwise specified." [see ORNAMENTATION].

PROLATE, adj. (L. prolatus, a bringing forward). cf. shape classes.
I. Jackson, 1928: "drawn out towards the poles."
II. a. Erdtman, 1943, p. 44: "P:E ratio 2–1.33 (8:4–8:6)." [see SHAPE].
b. Erdtman, 1952, p. 467: "this term denotes exclusively the shape of radiosymmetric, isopolar spores where the ratio between polar axis and equatorial diameter is 2–1.33 (8:6–8:7)."
c. Norem, 1958. [see SHAPE CLASSES and Fig. 167f].
sdb. Faegri and Iversen, 1950, p. 162.
   Selling, 1947, p. 77.
   Beug, 1961, p. 12.

PROLATE SPHEROIDAL, adj. cf. shape classes.
a. Erdtman, 1943, p. 44: "ratio P:E 1.01–1.14."
b. Erdtman, 1952, p. 467: "this term denotes exclusively the shape of radiosymmetric, isopolar spores where the ratio between polar axis and equatorial diameter is 1.14–1.00 (8:7–8:8)."
sdb. Selling, 1947, p. 77.

PROMINENT, adj. (L. projecting). cf. protrudent, shape classes.
Jackson, 1928: "standing out beyond some other part."

PROTRUDENCE, n., pl. -s. (L. protrudere, to thrust). cf. annulus, aspis, costa pori, dissence, endannulus, formation of lips, halo, labrum, lip, margo, oculus, operculum, pore ring, swelling.

PROTRUDENT, adj. cf. aperture, aspidate, aspidoporate, aspidorate, aspidote, porate, prominent, protrudence, shape classes.

PROTRUSION, n., pl. -s. cf. protuberance.

PROTUBERANCE, n., pl. -s. cf. baculum, elevation, excrescence, mammilla, process, projection, protrusion, tuber, tubercle.

PROXIMAL, adj. (L. proximus, next, nearest). cf. contact figure, dorsal, internal, proximalis.
a. Jackson, 1928: "the part nearest the axis, as opposed to distal."
b. Potonié, 1934, p. 5: Tr. . . . an inner pole . . . [see POLE].
c. Wodehouse, 1935, p. 159: ". . . line extending through the centers of the spheres and directed toward the center of the tetrad . . . Thus each sphere comes to have inner and outer poles, a proximal and distal polar hemisphere . . ."

sdb. Erdtman, 1943, p. 52.
   Selling, 1947, p. 77.
   Kupriyanova, 1948; [see GROUPS OF SPOROMORPHAE].
   Faegri and Iversen, 1950, p. 15.
   Erdtman, 1952, p. 11; [see POLARITY].
   Thomson and Pflug, 1953, p. 18; [see SYMMETRIE].
   Potonié and Kremp, 1955, p. 10; [see APEX].
   Traverse, 1955, p. 91.
   Couper, 1958, p. 103.
d. Pokrovskaya, et al., 1950: Tr. Proximal (dorsal—or spinal, according to Wodehouse, and internal according to Kozo-Poljansky). The part of the pollen grain or of the spore facing inward in the tetrad. The trilete or monolete crack is formed in spores on this side.

PROXIMAL CREST, n., pl. -s. cf. cap ridge, crista marginalis, crista proximalis, marginal ridge.

PROXIMAL FACE, n., pl. -s. cf. proximal.
Erdtman, 1952, p. 467: "that part of a spore surface which is directed inward in its tetrad." [see POLARITY].

PROXIMAL HEMISPHERE, n., pl. -s.
Potonié and Kremp, 1955, p. 10: Tr. With respect to the arrangement in the mother-cell, each grain has an inner or proximal and an outer or distal pole as well as a proximal and a distal hemisphere. [see APEX].

PROXIMAL SURFACE, n., pl. -s. cf. figura.
Ingwersen, 1954, p. 41: "the surface of a pollen grain or a spore facing the center of the tetrad."

PROXIMALIS, adj. cf. proximal.

PROXIMALOAPERTURATE, adj. cf. aperture, colpate, furrow.
Erdtman, 1947, p. 113: "with an aperture, simple or composite, on the proximal part." [see GROUPS OF SPOROMORPHAE].

PSEUDOCOLPUS, pl. -i. (Gr. pseudes, false). cf. colpus, furrow.
a. Faegri and Iversen, 1950, p. 162: "(pseudopore): differs from a normal furrow (pore) in that it is not an exit for the pollen tube."
b. Erdtman, 1952, p. 467: "colpoid streaks not functioning as apertures. Example: . . . [Fig. 294]."
sdb. Iversen and Troels-Smith, 1950, p. 32; [see COLPUS].

PSEUDOMONAD, n., pl. -s. cf. aggregation, pollina occlusa, spore, tetrad.
Selling, 1947, p. 77: "an apparent monad with walls formed by the PMC, thus homologous to a tetrad."
Selling, 1947, p. 350–351: "The pollen morphology of this family [Cyperaceae] is described in numerous papers. The structures observed are generally

called pollen grains, a somewhat misleading term, however, since they are not homologous to pollen grains in general. I suggest the designation *pseudomonads*. In this the morphological features are chiefly considered. 'Cryptotetrads' would be a term stressing the developmental peculiarities, but in this connection I prefer a morphological term. As emphasized by Wille (1882, 1886), later also by Strasburger (1884), Herail (1889), Biourge (1892), Juel (1900), Ziegenspeck (1938), Wulff (1939), and others, the wall of the pseudomonad is formed by the PMC and contains a pollen tetrad, in which only one component is developed. There are thus close similarities in the development of these structures and that of the juncaceous tetrads (cf. Wille, Ziegenspeck, Wulff). They are generally 30–40 μ in largest diam., subglobular to conical, and provided with a most characteristically granular exine showing a few (1–4) often ± obscure exits. In cross sections of the anthers they can be seen to be arranged in a circle, their acuminate ends towards the center (Juel, 1900, p. 652)."

sdb. Cranwell, 1953, p. 17; [*see* MONAD].

PSEUDOPORE, PSEUDOPORUS, n., pl. -s, -i. cf. aperture, diaporus, endoporus, exitus, exoporus, germinal apparatus, germinal point, germ pore, lacuna, open pore, os, pore, pseudopore, simple pore, ulcus.

I. Jackson, 1928: "in *Sphagnum* leaves, thickened rings without perforations (Russow)."

II. Iversen and Troels-Smith, 1950, p. 34: *Tr.* . . . differs from a true pore in that it is not the normal exitus place of the pollen tube . . . Lacuna: includes pseudopores as well as pseudocolpi. [*see* PORE].

sdb. Faegri and Iversen, 1950, p. 162; [*see* PSEUDOCOLPUS].

PSILATE, PSILATUS, adj. (Gr. psilos, bare). cf. laevigate, smooth, unadorned.

a. Wodehouse, 1928, p. 934: "Unadorned—without spines, ridges, or projections other than germinal apertures. Examples: *Phleum pratense, Artemisia tridentata.*

sdb. Wodehouse, 1935, p. 544; see Figs. 268, 710.
Erdtman, 1943, p. 52.
Erdtman, 1947, p. 106; [*see* GROUPS OF SPOROMORPHAE].
Iversen and Troels-Smith, 1950, p. 46; [*see* ORNAMENTATION].
Erdtman, 1952, p. 467.

b. Faegri and Iversen, 1950, p. 27: "Sculpturing elements s.s, absent; surface even or diameter of pits less than 1 μ (*Betula*)." [*see* ORNAMENTATION].

sdb. Ingwersen, 1954, p. 41.
Traverse, 1955, p. 94.

Beug, 1961, p. 13.

c. Cranwell, 1953, p. 17: "W., . . . perfectly smooth (*Thismia*) or with minute pits or grooves (Faegri and Iversen)."

sdb. Pokrovskaya, et al., 1950, p. 3.

PSILOLOPHATE, PSILOLOPHATUS, adj. cf. lophate.

a. Wodehouse, 1928, p. 934: "With the outer surface thrown into ridges which lack spines or conspicuous adornments. Examples: *Pacourina edulis, Barnadesia spinosa." see* Fig. 750.

b. Wodehouse, 1935, p. 544: "lophate, with the ridges smooth on their crests."

sdb. Erdtman, 1947, p. 106.

PSILOLUMINATE, adj. cf. reticulate.

Erdtman, 1952, p. 467: "with smooth lumina (without bacula, granula, etc.)."

PSILOTEGILLATE, adj. cf. blotched, flecked, infrapunctate, laevigate, maculate, maculose, planitegillate, punctate, punctitegillate.

Erdtman, 1952, p. 467: "with tegillum without adornments on its upper surface. Examples: . . . [Fig. 723]."

PUNCTATE, adj. (L. punctatus). cf. blotched, flecked, infrapunctate, intrapunctate, maculate, maculose, planitegillate, psilotegillate, punctitegillate, scabrate, subpsilate.

I. Jackson, 1928: "marked with dots, depressions or translucent glands; punctata Vasa = dotted vessels."

II. a. Potonié, 1934, p. 10: *Tr.* In all such cases when in the cross section no rodlets are recognizable and where in the vertical view of the exine, punctate to maculate patterns become noticeable without the existence of a sculpture, one speaks of *punctation* or *flecking* (maculation).

If possible the exine is designated as punctate only (in the formula p = punctatus) if it is not possible to measure the size of the individual structure elements with an enlargement of some 400 X; as soon as the elements can be measured with the aforesaid enlargement we speak of a maculose exine (in the formula m = maculatus). [*see* MACULATUS].

b. Potonié and Kremp, 1955, p. 13: *Tr.* An exine with structure can have a smooth surface, that is a smooth outline . . . [*Leiotriletes sphaotriangulus* (Loose)]; it does not have to show a sculpture. However, it can show a distinct design in plan view, for instance a punctation, maculation, or netlike design . . . [*Calamospora pedata* (Kosanke)] or a streaky, cloudy, flowlike structure or chagrenate appearance, etc. These apply only for the inner structure of the exine and also sometimes for its infrasculpture.

Therefore, terms like punctation, etc., have never been used by us for surface sculptures; however, as already mentioned, they have previously been confused with granulation, etc. Therefore, the terms *infrapunctate, infragranulate, infrareticulate* (Iversen and Troels-Smith) appear to be practical to us.

One would speak of infrapunctation, -maculation, etc. only in those cases where one can see punctated, maculated, etc. designs in the plan view of the exine without the existence of an outer sculpture. Therefore, these constructions are called structures of the exine. Those elements which are on top of the exolamella of the exine form the outermost "layer."

Only those form elements which stand out in relief on the surface of the exine can be called sculpture.

c. Krutzsch, 1959, p. 39: *Tr.* The word "sculpture" is not needed when using the word punctate. But if one wants to indicate a punctate structure then one must append the word structure, since otherwise it would refer to sculpture . . . [*see* ORNAMENTATION].

III. a. Faegri and Iversen, 1950, p. 162: "punctatus (Potonié, 1934): structure, the elements of which are too small to be measured by 400 X magnification."

b. Kosanke, 1950, p. 11: [*see* ORNAMENTATION and Fig. 640:D].

sdb. Zaklinskaya, 1957; [*see* SACCATE and Fig. 223].

PUNCTATE-RETICULATE, adj. cf. reticulate.
Kosanke, 1950, p. 11; [*see* ORNAMENTATION and Fig. 640:E].

PUNCTITEGILLATE, adj. (L. punctum, point). cf. punctate.
Erdtman, 1952, p. 467: "with a tegillum with minute perforations (puncta). Example: . . . [Fig. 716 d and e]."

PUNCTUM, n., pl. -a. cf. granule, granulum, granum.
I. Jackson, 1928: "the marking on the valves of Diatoms."
II. Erdtman, 1952, p. 467: "With a tegillum with minute perforations (puncta)." [*see* PUNCTITEGILLATE].
III. Potonié and Kremp, 1955, p. 18: *Tr.* Usually in thin sections [of spores], one does not notice any rodlets, but only a feature called *porphyritic structure* (*Mörtelstrucktur*); often this is best visible in the intexine we call. As porphyritic structure (along with Haberlandt), a situation where *micelle nuclei* or *puncta* are imbedded in the groundmass or matrix . . . [*see* SPORODERMIS].

PUSTULATE, adj.
I. Jackson, 1928: "as though blistered."
II. Krutzsch, 1959, p. 40: *Tr.* The wall is interspersed with roundish to ovule lacunae of diverse sizes, which give the wall a vesicular (pustulate-like) structure (character). [*see* ORNAMENTATION and Fig. 563].

PYCNIDIOSPORE, n., pl. -s.
Jackson, 1928: "a spore produced in a pycnidium."

PYCNIDIUM, n., pl. -a. (Gr. pyknos, thick; idion, dim. suffix).
Jackson, 1928: "a cavity resembling a Pyrenocarp in Lichens, etc., containing gonidia (pycnoconidia or stylospores)."

PYCNIOSPORE, n., pl. -s. cf. pycnidiospore.

Q

QUADRANGULIS, adj. cf. shape classes.
  Pokrovskaya, et al., 1950: *Tr.* Used to classify the outline of certain microspores. [*see* CIRCUMSCRIPTIO].
QUADRANGULUS, adj.
  Jackson, 1928: "having four angles, which are usually right angles."

QUINQUANGULIS, adj. (L. quinque, five). cf. shape classes.
  I. Jackson, 1928: "five-cornered."
  II. Pokrovskaya, et al., 1950: *Tr.* Used to classify the outline of certain microspores. [*see* CIRCUMSCRIPTIO].

R

RADIAL, adj. (L. radialis, radius, the spoke of a wheel). cf. contact figure.

Jackson, 1928: "radiating, as from a center; belonging to the ray, as in the flowers of Composites; = actinomorphic."

RADIAL CREST, n., pl. -s. cf. contact figure, crista radialis, radial comb.

Dijkstra-van V. Trip, 1946, p. 20: *Tr.* . . . e.g. . . . [*Triletes superbus* Bartlett] the radial crests are high and folded over.

RADIAL EXTREMITY, n., pl. -ies. cf. contact figure.

Dijkstra-van V. Trip, 1946, p. 20: *Tr.* End of the radius, abstract conception, the spot where rays meet the two arcuate rims . . . (also defined by others, as Schopf, 1935, as the "radial extremity;" by this is meant the extreme end of the radial crest . . . continuing into the equatorial zone . . .). This spot is especially marked because in cases of a certain spore type ear-lobelike processes occur there . . . [*Triletes auritus* Zerndt], whereas in other types the radial part of the equatorial zone . . . grows more than the interradial part . . . [*Triletes triangulus* Zerndt, *Microsporites karczewskii* Zerndt, *Triletes brasserti* Stach and Zerndt] among others, whereby the whole spore obtains a triangular shape, or as in case of another type, the hairs of the arcuate rims . . . are better developed . . . [*Triletes praetextus* Zerndt] and similar ones.

RADIAL POSITION, n., pl. -s. cf. spore.

General term used in orientation and description of spores. Couper and Grebe, 1961 Fig. 1: *see* Fig. 160.

RADIAL RIDGE, n., pl. -s. cf. contact figure.

Dijkstra-van V. Trip, 1946, p. 20: *Tr.* . . . for example in . . . [*Triletes auritus* Zerndt] one sees two of the radial ridges from the side and one from the front. In *Aphanozonati* they are tubelike in the beginning . . . [*Triletes mamillarius* Bartlett]; later on they split with round lips, compare . . . [*Triletes auritus* Zerndt] triradiate germination cleft. Also in *Auriculati,* examples occur with sharpened radial ridges which are completely split, but which are not opened. The lips which are just in the process of opening are also discernible in . . . [*Triletes horridus* Zerndt]. There already· is a suture to be seen at the part of the radial ridge which does not yet seem to be opened . . . [*Triletes mamillarius* Bartlett]. Such a suture can also be seen in . . . [*Triletes brasserti* Stach and Zerndt].

RADIALLY SYMMETRICAL SPORE, n., pl. -s.

Kosanke, 1950, p. 14: "Diagrammatic drawing of spore." *see* Fig. 109.

RADIOSYMMETRIC, adj. cf. shape classes.

Erdtman, 1947, p. 112: "with more than two vertical planes of symmetry, or, if provided with two such planes, always with equilong equatorial axes."
sdb. Erdtman, 1952, p. 468.
    Pike, 1956, p. 51.

RADIUS, n., pl. radii. (L. a ray). cf. contact figure, ray.

Jackson, 1928: "the ray of Compositae, the outermost florets when distinct in form from those composing the disk; a partial umbel in Umbelliferae; the structures known as medullary rays."

RAISED COMMISSURE, n., pl. -s. cf. contact figure, vertex.

Couper, 1958, p. 102: "the actual line of dehiscence carried on a narrow ridge above the general surface of the spore."

RAMIBACULATE, adj. (L. ramus, branch). cf. baculate.

Erdtman, 1952, p. 468: "with ± branched bacula. Cf. . . . [Figs. 648, f and g, 728] Iversen-Troels-Smith (1950): 'provided with columellae digitatae.' An example of pollen grains with a special kind of branched bacula ('columellae conjunctae') mentioned by the same authors, has not yet been published." [*see* LO-ANALYSIS].

RAY, n., pl. -s. (L. radius). cf. contact figure, radius, tecta.

a. Jackson, 1928: "the marginal portion of a Composite flower, when distinct from the disk; a branch of an umbel, a partial umbel."

130

b. Potonié, 1934, p. 15: [*see* APERTURA].

c. Dijkstra-van V. Trip, 1946, p. 20: *Tr.* [it] used *in abstracto* . . . occasionally used in a description, without it, referring to a certain shape.

d. Kosanke, 1950, p. 14: see Fig. 109–110.

e. Potonié and Kremp, 1955, p. 10: *Tr.* The three rays of tecta originate at the apex . . . [Fig. 132–133]. In plan view, they form angles of approximately the same size. [*see* Y-MARK].

RECTANGULATE, RECTANGULATUS, adj. (L. rectus, right + angulus, angle). cf. shape classes.

RECTIMURATE, adj. (L. rectus, straight). cf. reticulate.

Erdtman, 1952, p. 468: "with ± straight muri. Example: . . . [Fig. 751–753]."

RENIFORM, adj.

Norem, 1958; [*see* SHAPE CLASSES and Fig. 167k].

RESTING SPORE, n., pl. -s.

Jackson, 1928: "a spore with a thick integument, needing time before germinating, usually passing the winter or dry season in a dormant state."

RETICULA COMPOSITA SEPIBUS EVALLIS, n. cf. reticulum cristatum.

Kupriyanova, 1948, p. 71: *Tr.* Complex netlike sculpture having rodlike walls. [*see* ORNAMENTATION and Fig. 668].

RETICULA SEPIBUS DENSIS, n. cf. reticulum simplex.

Kupriyanova, 1948, p. 71: *Tr.* Simple net sculpture having unbroken walls. [*see* ORNAMENTATION and Fig. 666].

RETICULA SEPIBUS EVALLIS, n. cf. reticulum cristatum.

Kupriyanova, 1948, p. 71: *Tr.* Simple netlike sculpture having rodlike walls. [*see* ORNAMENTATION and Fig. 667].

RETICULATE, adj. (L. reticulatus, rete, a net). cf. augustimurate, carinimurate, curvimurate, extrareticulate, heterobrochate, homobrochate, infrareticulate, intrareticulate, intectate-reticulate, latimurate, lophoreticulate, mesobrochate, microreticulate, oligobrochate, polybrochate, psiloluminate, punctate-reticulate, rectimurate, subreticulate, suprareticulate, tectate-reticulate.

a. Jackson, 1928: "netted like network, as in certain cell-thickening.

b. Wodehouse, 1935, p. 544: "with the surface thrown into anastomosing ridges enclosing lacunae, generally smaller than in lophate grains, e.g., in the grains of *Ligustrum* . . . [Fig. 782]."

c. Erdtman, 1947, p. 106: "reticulate; consisting of a network, reticulum, formed by anastomosing ridges ('muri'), enclosing small, frequently more or less irregular, spaces ('lumina'). The muri of

a reticulum are usually lower and have a less intricate construction than the ridges of lophate pollen grains."

d. Traverse, 1955, p. 94: "Sculpture of elements joined to a network, so that the total area of depression exceeds that of unmodified surface."

e. Kupriyanova, 1948, p. 71: *Tr.* The most widespread sculpture element (among pollen of monocotyledonous plants) is represented by the netlike sculpture (reticulata). The netlike state can be of two types: (1) a little net of rodlike forms (sepes e vallis)—when the walls of the net consist of joined rods providing the upper surface of the walls with a beadlike character. This sculpture is characteristic of pollen of the Lily, the Amaryllis, Iris family, Orchid family, and to a smaller degree of the Palm family. Walls of the net can be simple (reticulum simplex) and complex (reticulum compositum). In the first case, the walls of the net consist of one row of rods, while in the second, it consists of several rows, usually two, joined together; sometimes the rod rows diverge, in turn forming cell-like walls. (2) When the walls of a net do not consist of rods and are unbroken (sepes densi), the upper surface does not show a beadlike pattern. An example of such sculpture is the family *Aponogetonaceae* and *Aracea*.

Discontinuous walls of the net or pollen of representatives of the family Potomogetonaceae must be considered as the result of the destruction of the walls related to the general process of the reduction of exine. [*see* ORNAMENTATION and Fig. 658–669].

f. Thomson and Pflug, 1953, p. 23: *Tr.* Muri . . . arranged into nets . . . The ridges (muri) are as wide as, or not as wide as, the space between them. The ridges have the same overall crest height. [*see* ORNAMENTATION and Fig. 673].

sdb. Potonié, 1934, p. 11; [*see* RETICULUM SIMPLEX and ORNAMENTATION].

Faegri and Iversen, 1950, p. 27; [*see* ORNAMENTATION and Fig. 688–689].

Iversen and Troels-Smith, 1950, p. 46; [*see* ORNAMENTATION].

Kosanke, 1950, p. 11; [*see* ORNAMENTATION and Fig. 640].

Pokrovskaya, et al., 1950; [*see* ORNAMENTATION].

Cranwell, 1953, p. 17.

Harris, 1955, p. 27.

Beug, 1961, p. 10.

RETICULOID, adj. cf. fossula, negative reticulum, ornate, reticulum fossulare.

Potonié and Kremp, 1955, p. 14: *Tr.* If the verrucae are close together, they become polygonal in plan view. In this case, their mosaic consists of elements which are separated by more or less straight clefts . . . [Fig. 739]. The cleft system forms a negative *reticulum* or *reticuloid*. (We distinguish the reticulum = extra-reticulum . . . from the infrareticulum and from the reticuloid or negative reticulum.) [*see* ORNAMENTATION].

RETICULUM, n., pl. -a. (L. a little net). cf. endoreticulum, extrareticulum, infrareticulum, intrareticulum, microreticulum, reticulate.
I. Jackson, 1928: "a membrane of cross-fibres found in Palms at the base of the petiole; applied to the network of lining in the nucleus."
II. Potonié, 1934, p. 11: *Tr.* These are fences, or combs, that are built up from pila, rodlets, etc., which sometimes may even be laterally connected and which join to form networks. [*see* ORNAMENTATION].
sdb. Erdtman, 1943, p. 52.
Faegri and Iversen, 1950, p. 27; *see* Fig. 698–700, 707–709.
Iversen and Troels-Smith, 1950, p. 36; [*see* ORNAMENTATION].
Pokrovskaya, et al., 1950; [*see* ORNAMENTATION].
Erdtman, 1952, p. 468; *see* Fig. 751–760.
Potonié and Kremp, 1955, p. 14–15; [*see* ORNAMENTATION].
Zalinskaya, 1957; for ornamentation of disaccate pollen grains [*see* SACCATE; *see* Fig. 224–225].
Beug, 1961, p. 13.

RETICULUM CRISTATUM, n., pl. reticula cristata. (L. crista, a crest). cf. reticula composita sepibus evallis, reticula sepibus evallis.
a. Potonié, 1934, p. 11: *Tr.* A frequent type of arrangement of the above-named sculpture elements is that of *cristae*. These are fences or combs that are built up from pila, rodlets, etc., which sometimes may even be laterally connected and which can join to form networks (reticula cristata). [*see* Fig. 657].
b. Erdtman, 1943, p. 52: "crests arranged in networks."
sdb. Pokrovskaya, et al., 1950.

RETICULUM FOSSULARE, n. (L. fossula, a little ditch). cf. fossula, negative reticulum, ornate, reticuloid.
Kupriyanova, 1948, p. 70: *Tr.* If the tubercles have a form of areas (areolate) surrounded by narrow ditches (fossula) a negative reticulate structure is created (areolate, reticulum fossulare). *Dracaena aurea* . . . , *Lysichiton kamtschatkense* . . . , and *Chamaedora concolor* . . . , as well as some others can serve as examples of the negative reticulate structure. [*see* ORNAMENTATION and Fig. 662].

RETICULUM SIMPLEX, n. (L. simples, simple, of one piece or series, opposed to compound). cf. reticula sepibus densis.
a. Potonié, 1934, p. 11: *Tr.* Often the "reticula" consist only of smooth stripes or *lists* (muri). We then have the *reticulum simplex.* [*see* ORNAMENTATION].
b. Erdtman, 1943, p. 52: "a network of the ordinary type, consisting of low and smooth ridges (muri)."
c. Kupriyanova, 1948, p. 71: *Tr.* . . . the walls of the net consist of one row of rods. [*see* ORNAMENTATION and Fig. 666–667].
sdb. Pokrovskaya, et al., 1950.

RETIPILARIATE, adj. cf. comblike, cristate, cristoreticulate, sepes a vallis.
Erdtman, 1952, p. 468: "with pilaria arranged in a ± reticuloid pattern."

RETIPILATE, adj. cf. retipilariate, sepes a vallis.
Erdtman, 1952, p. 468: "with a reticuloid pattern with pila instead of muri."
sdb. Beug, 1961, p. 13.

RHOMBOHEDRAL, adj. (Gr. rhombos, rhom ± hedra, base). cf. shape classes.

RHOMBOHEDRAL TETRAD, n., pl. -s. cf. aggregation, spore, tetrad.
Potonié, 1934, p. 14: *Tr.* An additional special case of the arrangement of pollen tetrads is that of the *rhombohedral tetrad* . . . [Fig. 3]. *Naegeli* denotes this as half tetrahedral. It may be clearer to speak, together with Wodehouse (1929), of a rhomboidal arrangement. The rhomboidal tetrad makes it understandable how the one and same mother-cell might supply pollen with different numbers of germinals.

RHOMBOIDAL, adj. (Gr. rhomboeides, rhomboidal). cf. shape classes.
a. Jackson, 1928: "approaching a rhombic outline, quadrangular, with the lateral angles obtuse."
b. Zaklinskaya, 1957: for outline of disaccate pollen see Fig. 227.

RHOMBOIDAL TETRAD, n., pl. -s. cf. aggregation, spore, tetrad.

RIB, n., pl. -s. cf. fissura dehiscentis, monolete aperture, monolete mark.
I. Jackson, 1928: "a primary vein, especially the central longitudinal or midrib."
II. Dijkstra-van V. Trip, 1946, p. 20: *Tr.* Abstract concept; in the *Monoletes* it occurs instead of the apex; in fig. . . . [pollen of *Whittleseyinae,* type 31 Zerndt] it is perhaps recognizable as a middle line; in the illustrations of Schopf (1938, Pl. 6, figs 2, 5, and 6) it seems to be . . . a slit; in general, folds near the contact areas will diminish the possibility of recognizing the ribs . . . see Schopf (1938, pl. 6, figs 1 and 3).

RIBBED, adj. cf. canaliculate, cicatricose, costate, striate.
Jackson, 1928: "furnished with prominent ribs."

RIDGE, n., pl. -s. cf. lira, list, murus, rugula, stripe, vallum, wall.
a. Jackson, 1928: "an elevated line on the fruit of Umbelliferae; either primary or secondary."
b. Special outline at the dorsal part of disaccate pollen grains. (Zaklinskaya, 1957.) See Fig. 231–235 and Pokrovskaya, et al., 1950, Fig. 204.

RIM, n., pl. -s. cf. arc line, arcuate ridge, arcuate rim, crista arcuata, curvatura, margo arcuata.

RIMULA, n., pl. -ae. (L. rima, a cleft, crack). cf. colpus, furrow, pore.
a. Potonié, 1934, p. 20–21: *Tr.* A peculiarity in the differentiation of the germinal apparatus is there in given—that in *Corylus* the pore is sometimes not completely round but is somewhat elongated in the meridional direction. As in meridional folds, compression of the margins of the pores ensues somewhat more in the equatorial direction. The small, incidental elongation of the pore in *Corylus* points to the distinctly stretched rimula being developed in other species.

In *Alnus* there is already established in clear form a small, longish slit (a rimula) . . . [Fig. 398]. If one looks perpendicularly at the exine, it is even seen that in the meridionally elongated rimula it extends a little beyond the inner circle on both sides.

The rimula is to be understood as a special case of porus vestibuli. For that reason, the perimeter of the rimula should likewise be called a *pore ring* . . .

The vestibulum and rimula are especially well developed in *Tilia*.

In the plan view of the germinal, the elongated rimula of *Tilia* extends far over the double circle enclosing the vestibulum. (That is not to say that the double circle need be considered a boundary of the vestibulum as in the above types. In those instances too, focused in a particular way, the lip formation produces such an optic cross-section.) In *Tilia* the rimula is quite clearly an open slit only within the double circle, functioning as an exit from the vestibulum. Outside the double circle the rimula continues in both directions as only a small, narrow furrow, the *sulcus,* cutting into the exoexine . . . [Fig. 315]. The sulci, pointed at their ends and pointing toward the poles, terminate quite far from the poles in *Tilia*. However, they are essentially nothing more than the long furrows of many pollen types differentiated on the basis in which the furrows approach the poles and of which are transitions to those broad meridional areas which

appear in fossil pollen grains, now evenly expanded, now sharply in-folded. Consequently, there is no boundary between simple sulci and small in-folded meridional areas. [*see* APERTURE and SULCUS].
b. Faegri and Iversen, 1950, p. 163: "= slightly elongated pore."

RING, n., pl. -s. cf. cingulum, zonal.
Jackson, 1928: [*see* ANNULUS for the various senses in which used].

ROD, RODLET, n., pl. -s. cf. bacularium, baculum, columella, rodlet-layer.
a. Potonié, 1934, p. 9: *Tr.* Generally these structures seen in a cross-section of the exine are in the form of many closely packed, radially arranged rodlets, which in contrast to the space between them, have the ability to absorb dye material such as fuchsin, gentian-violet, etc., in strong intensity. H. Fischer and others recognized these rodlets as minute pillars which stand between the lamellae and hold them apart. An air-interspersed *insulation layer* originates in the exine. [*see* ORNAMENTATION].
b. Faegri and Iversen, 1950, p. 163: "= columella."

RODLET-LAYER, n., pl. -s. cf. bacularium, columella.
Thomson and Pflug, 1953, p. 18: *Tr.* Ekt- and endexine are usually formed by alternating structureless and structured lamellae. The structured lamellae usually consist of radially directed individual elements which often have the shape of small rods. Therefore, these lamellae are also called "rodlet-layers." Their thickness varies between fractions of a micron, but can also reach five microns and more in the annulus structure. The structureless lamellae are always thinner-walled than the rodlet layers and form the base and the roof surfaces of the rodlets which stand like small pillars between the two surfaces. [*see* LAMELLA].

ROTUND, adj. (L. rotundus, round).
Jackson, 1928: "rounded in outline, somewhat orbicular, but a little inclined towards oblong."

ROTUNDATUS, adj. (L. round). cf. shape classes.
I. Jackson, 1928: "rounded."
II. Pokrovskaya, et al., 1950: *Tr.* Used to classify the amb of certain microspores. [*see* CIRCUMSCRIPTIO].

ROTUNDUS, adj. (L. round). cf. rotund, shape classes.

ROUNDED, adj.
Couper and Grebe, 1961: [*see* SHAPE].

ROUNDISH, adj. cf. shape classes.
Zaklinskaya, 1957: for outlines of disaccate pollen grains see Fig. 227.

ROUND-LOBATE, adj. cf. rotundo-lobatus, shape classes.

RUGA, n., pl. -ae. cf. colpus, fold, furrow, sulcus.
I. Jackson, 1928: "a wrinkle or fold."
II. Potonié, 1934, p. 21: *Tr*. Here it must be inserted that the sulci and the afore-mentioned, related, plain differentiations of the germinal apparatus will generally be called *folds* or *rugae*. H. Fischer said (1890, p. 16): the exitus places are ". . . simple folds" . . . if . . . "they are elongated and sharply infolded in the dry grain." It will be shown in the following what special terms will be needed to designate more precisely the individual peculiarities of rugae. [*see* APERTURA, RUGA COMPRESSA, AQUATORIAL RUGA].
III. a. Erdtman, 1945, p. 191: "Although originally applied by Potonié (1934) in another sense, the term 'ruga' may be used to denote furrows equally distributed over the surface of the grains. Of the three terms suggested 'ruga' is shortest, 'colpa' intermediate, and 'sulcus' longest. As to the features, which these terms are supposed to denote, the rugae in a polyrugate grain, as a rule, are shorter than the colpae in, e.g., a grain of the common tricolpate type. The colpae, in their turn, are as a rule comparatively shorter than the sulcus in a typical monocotyledonous monosulcate pollen grain."
   b. Erdtman, 1946, p. 71: "A polycolpate pollen grain has many furrows which, according to the definition, are equally spaced in the equatorial, more or less extended zone of the grain. If, on the other hand, the grain is provided with many furrows equally distributed over the general surface of the exine the furrows are referred to as *rugae*. In a hexarugate pollen grain the rugae correspond to the six edges of a tetrahedron. By shortening and contraction the rugae may, in the same way as sulci and colpae, exhibit any transition to pori. Some plants produce both tricolpate and hexarugate grains but it is felt, nevertheless, that the term ruga should be retained as a purely descriptive term. . . . [Fig. 33] shows a polyrugate grain."
   c. Selling, 1947, p. 77: "a furrow, not meridional (cf. *Colpa*) and not occurring singly or just two in each grain (cf. *Sulcus*). Rugae thus denote non-meridional furrows (occasionally transitional to pores) evenly distributed over the surface of the grain."
sdb. Erdtman, 1947, p. 105.
IV. a. Faegri and Iversen, 1950, p. 163: "(Potonié, 1934) = furrow pp. (Erdtman, 1945) = isometrically distributed furrows."
   b. Erdtman, 1952, p. 468: "global, ± regularly arranged apertures with length-breadth ratio >2:1. Examples: . . . [Figs. 75:7 and 436–

438]. Potonié (1934b) defines rugae in a different way: . . ."
V. Thomson and Pflug, 1953, p. 20: *Tr*. . . . colpi of appropriate structure which stretch parallel to the equator. [*see* SOLUTION].

RUGA AEQUATORIALIS, n. cf. aperture, colpate, equatorial ruga, furrow, transverse fold.

RUGA COMPRESSA, n.
Potonié, 1934, p. 21: *Tr*. The vestibulum of *Tilia* is strikingly formed. Differentiation of a vestibulum in *Corylus* and still more in *Alnus* resulted essentially from the formation of lips or else beak; also by deep, inward folding of the inner exine and by especially deep impression of a *compressed fold* (ruga compressa). This arrangement is recognizable in the optical section of the germinal apparatus . . . [Fig. 316]. In cases of more distinct development of compressed folds, one recognizes how at first the intexine extends to the edge of the rimula, then bends over as far as possible, laying entirely against the wall (compressed fold) and then turns over to the other edge of the rimula. If in an optical cross-section, one focuses as accurately as possible on the center of the rimula . . . [Fig. 317], the vestibulum seems to be less deep, but broader. This condition also appears in a vertical view of the germinal apparatus . . . [Fig. 317]. The punctate line of . . . [Fig. 317, lower part] corresponds to the optical cross-section through the middle of the rimula just discussed. . . . [Fig. 317, lower part] shows that the double circle bounding the vestibulum produces two small pointed lobes (tips) opposite each other and which are equatorially directed parallel to it. [*see* APERTURA, RUGA COMPRESSA].

RUGATE, adj. (L. rugatus, wrinkled). cf. pericolpate, rugoporate, triacontarugate.
I. Jackson, 1928: "wrinkled."
II. a. Erdtman, 1945, p. 191: ". . . may be used to denote furrows equally distributed over the surface of the grains." ". . . the rugae in a polyrugate grain, as a rule, are shorter than the colpae in, e.g., a grain of the common tricolpate type," [*see* SULCUS].
   b. Erdtman, 1947, p. 112: "with rugae without pori or poroid apertures." [*see* GROUPS OF SPOROMORPHAE].
   c. Selling, 1947, p. 77: "pollen with rugae, each without a pore. The prefixes hexa-, dodeca-, and poly- denote the number of rugae."
   d. Faegri and Iversen, 1950, p. 163: "rugate (Erdtman, 1945) = pericolpate."
   e. Erdtman, 1952, p. 468: "with rugae." [*see* RUGATE].

RUGOPORATE, adj. cf. rugate.
a. Erdtman, 1947, p. 112: "with poriferous or pori-

diferous rugae." [*see* GROUPS OF SPOROMORPHAE].

b. Selling, 1947, p. 77: "pollen with rugae, each with a pore or transverse furrow. The prefixes septa- and poly- denote the number of rugae."

RUGOSE, RUGOSUS, RUGOUS, adj. (L. rugosus). cf. corrugate, costate, obvermiculate, ribbed, rugulate, rugulose, wrinkled.

I. Jackson, 1928: "covered with or thrown into wrinkles."

II. Potonié, 1934, p. 33: *Tr.* Surface wrinkled (superficies rugosa).

III. Faegri and Iversen, 1950, p. 163: "rugosus (Potonié, 1934) = rugulate p.p."

IV. Kosanke, 1950, p. 11: for illustration see Fig. 640.

RUGULA, n., pl. -ae. (L. a little wrinkle or fold). cf. lira, list, ridge, stripe, vallum.

I. Jackson, 1928: "a longitudinal groove in the upper lip of the flower, which encloses the style of *Justicia* Houst. (Lindau)."

II. Couper, 1958, p. 102: "Rugulae are long and narrow in surface view, whereas verrucae if elongate, are relatively broad." [*see* ORNAMENTATION].

RUGULATE, adj. (L. rugulatus). corrugate, costate, obvermiculate, ribbed, rugose, rugous, rugulate, rugulose, wrinkled.

a. Faegri and Iversen, 1950, p. 27: "Radial projections elongated (length at least twice the breadth). Elements irregularly distributed. (*Nymphoides peltatum.*)" [*see* ORNAMENTATION and Fig. 701–703].

b. Cranwell, 1953, p. 17: "irregular elongate ridges." [*see* Fig. 778].

c. Ingwersen, 1954, p. 41: "sculpturing type with elongated sculpturing elements in irregular or not predominatingly regular arrangement."

d. Harris, 1955, p. 20: "Wrinkled elements irregularly distributed (*Botrychium australe*) N.B. Rugulae are long and narrow in surface view, whereas verrucae, if elongate, are relatively broad. Intermediates can only be indicated by hyphenating the two terms." [*see* ORNAMENTATION].

sdb. Iversen and Troels-Smith, 1950, p. 46; [*see* ORNAMENTATION].

Thomson and Pflug, 1953, p. 21–23; [*see* ORNAMENTATION].

Traverse, 1955, p. 94.

Couper, 1958, p. 102 [*see* ORNAMENTATION].

Beug, 1961, p. 11.

Couper and Grebe, 1961, p. 7; [*see* ORNAMENTATION and Fig. 685].

RUGULO-SACCATE, adj. cf. pouched, sub-saccate.

Harris, 1955, p. 20: "Pouched, cavate, elevations, 'long' (length more than twice breadth). Example: *Asplenium falcatum.*"

RUGULOSE, adj. (L. rugulosus). cf. rugulate.

Jackson, 1928: "somewhat wrinkled."

RUGULO-STRIATE, adj. cf. striate.

Harris, 1955, p. 20: [*see* ORNAMENTATION].

RUPATE, adj. cf. colpate, furrow.

Erdtman, 1952, p. 468: "with rupi."

RUPUS, n., pl. -i. cf. furrow, colpate.

Erdtman, 1952, p. 468: "equatorial colpoid apertures converging in pairs."

S

SAC, n., pl. -s. cf. aerifera, air sac, air sack, bladder, cavea, saccus, sack, vesicula, wing.
  Jackson, 1928: "a pouch, as air sac . . ."

SACCAL, adj. (L. saccus, a bag). cf. bisaccate, disaccate, monosaccate, perisaccate, saccate, saccatus, subsaccate, vesiculate, vesiculosus.
  Jackson, 1928: "relating to a sac, as the embryo sac."

SACCALE AREA, n., pl. -s.
  Erdtman, 1957, p. 3: "area . . . forming the floor of the sacci;" [see SACCATE].

SACCATE, adj. (L. saccatus, bag-shaped). cf. aerostatic umbrella, air sac, air sack, aperture, bisaccate, cavate, cavea, central cell, comb, corpus, crest, cristae, dimensions, disaccate, discus, disk, dorsal root of the bladders, form, furrow, Groups of Sporomorphae, limbus, marginal frill, marginal ridge, mesosaccium, mesosaccale area, monosaccate, ornamentation, pecten, perisaccate, pouched, rugulosaccate, sac, saccal, saccale area, saccate, saccatus, saccus, sack, shape, shape classes, subsaccate, sulcoid groove, sulcus, tenuitas, velum, ventral root of the bladders, vesicula aerifera, vesiculate, vesiculosus, wing, zona.
  I. a. Jackson, 1928: "bag-shaped."
  b. Erdtman, 1947, p. 106: "Saccatus; with air sacs ('sacci')."
  c. Faegri and Iversen, 1950, p. 163: "(Erdtman, 1947) = vesiculate."
  d. Erdtman, 1957, p. 3–4 and Fig. 211: "The saccate pollen grains in recent gymnosperms are heteropolar, bilateral or radiosymmetric (sometimes slightly asymmetric). They consist of a body (corpus) and a varying number of air-sacks or bladders (sacci). The aperture is distal, and should often perhaps more appropriately be referred to as a tenuitas (i.e., a thin aperturoid area functioning as an aperture and gradually merging into the surrounding exine). It has earlier, as a rule, been described as a sulcus or a sulcoid groove.
    "The surface of the corpus of a pollen grain with n bladders can be divided into the following areas: n saccale areas, forming the floor of the sacci, n mesosaccale areas (mesosaccia), i.e., areas between the sacci and in the same latitude as these, and finally two aposaccale areas (aposaccia), one at the distal pole, and the other, usually much larger than the former, in the proximal face of the grain with the proximal pole in its center.
    "With respect to the thickness of the exine of the corpus certain pollen types . . . exhibit two distinct exine areas: a proximal, crassi-exinous (referred to as cap, cappa), and a distal, tenui-exinous (referred to as cappula). The non-saccale exine of the cappa consists of comparatively thick sexine and thin nexine. The outer, ectosexinous part of the sexine is usually thin, tegilloid, and as a rule connected to the nexine—except within the sacci—by baculoid, densely spaced endosexinous elements.
    "The sacci are separated from the interior, non-exinous parts of the corpus, by saccale nexine. Their outer wall consists of thin ectosexine which is often perforated (shown in electron micrographs—not published—by Erdtman and Thorsson in 1950). The small holes (micropuncta) are usually difficult to observe through an ordinary light microscope. In the majority of the *Tsuga* species the ectosexine of the sacci (as well as that of the corpus . . .) is studded with spinules or small spines. Attached to the inner surface of the outer wall of the sacci are endosexinous elements protruding into the lumen of the bladders (in *Pherosphaera fitzgeraldii* stray endosexinous rods are also found on the saccale nexine). These elements are more widely spaced than those of the body. Branched or unbranched, single or combined in different ways, they tend to produce an array of patterns which are difficult to draw and hard to describe. Microtome sections (cf. particularly Afzelius in *Grana palynologica*, 1:2, 1956) make these subtle details of pollen construction easier to observe and safer to interpret.
    "Near the proximal root of the sacci are often found slight, sexinous ridges or frill-like projec-

tions (proximal crests, cristae proximales, also referred to as cristae marginales) varying in appearance in different species. At the distal root of the sacci, where these merge into the distal aposaccium, the characteristic pattern of the bladders comes abruptly to an end.

"The height of the corpus coincides with the polar axis (i.e., the perpendicular line connecting the poles); the breadth is identical with its maximum horizontal extension in grains in equatorial longitudinal view (marginal crests extending beyond the general surface of the corpus not included); and the depth (in bilateral grains) is equal to the transverse ('non-sacciferous') diameter of the corpus. It is often preferable and, at the same time easier to calculate the inner dimensions of the corpus.

"The height of a saccus is the shortest distance from the highest point of the saccus (or from a line drawn through this point parallel to the saccale nexine) to the underlying nexine of the corpus. Its breadth is equivalent to the 'tangential' diameter of the saccus in pollen grains in polar view. In radiosymmetric grains the breadth can also be measured in pollen grains seen in equatorial view; bilateral grains must be in a transverse equatorial position if the breadth shall be measured. In microscope slides this, however, is seldom the case. Its depth—in bilateral grains—is equal to the maximum diameter of the saccus in grains in equatorial longitudinal view. In radiosymmetric grains the depth is calculated in a similar way. Height, breadth, and depth of corpus and sacci are illustrated in . . . [Fig. 211].

"As shown by Afzelius the inner part of the nexine (the endonexine) in *Cedrus* is laminated . . . In acetolyzed pollen grains of this genus, and of *Abies,* etc., it can often be seen, even by means of an ordinary microscope, that the nexine consists of two distinct layers, which often split apart as a result of the chemical treatment. So striking is this feature that it seems extraordinary that it has not been mentioned until now.

"The morphology of bisaccate pollen grains has been dealt with by numerous botanists, among whom was Strasburger, who believed that the floor of the sacci was formed by intine. This opinion has often been echoed right up to recent years, although Strasburger himself soon corrected his mistake.

"In conclusion, it ought to be mentioned that, according to Ciguriaeva, the thin ends (with sexine and nexine slightly separate from each other) of the pollen grains in *Ephedra* and *Welwitschia* may be interpreted as the remainder of the true sacci."

e. Wodehouse, 1935: see Fig. 205–206.

sdb. Pokrovskaya, et al., 1950: *see* Fig. 204.
Zaklinskaya, 1957: *see* Fig. 208–210, 219–221, 224–225.

II. Harris, 1955, p. 19–20: "projections which appear to be flaccid and hollow . . . Pouched, cavate, elevations approximately isodiametric. N.B.—The walls of saccate elevations tend to be flaccid and hollow looking, whereas verrucate projections are firmer and appear to be solid."

SACCUS, n., pl. -i. cf. air sac, air sack, bladder, cavea, sack, vesicula aerifera, wing. For the dimension of saccate grains cf. also dimension and saccate.
a. Erdtman, 1952, p. 468: "air sac (sexine loosened from nexine: bacula or ± baculoid elements usually sticking to the under surface of the tegillum). Saccate grains are termed 'vesiculate' by Iversen and Troels-Smith (1950)."
b. Potonié and Kremp, 1955, p. 19: *Tr.* In the development of *sacci* the basis of the columellae detach from the intexine, and the air space between the intexine and exoexine expands. [*see* LIMBUS, VELUM, SPORODERMIS, Potonié, 1934, p. 9].
sdb. Couper and Grebe, 1961, p. 6: *see* Fig. 196.

SACK, n., pl. -s. cf. saccus.
Jackson, 1928: "= sac."

SCABRA, n., pl. -ae. cf. scabrate, micro-.
Beug, 1961, p. 12. [*see* MICRO-].

SCABRATE, adj. (L. scabratus). cf. baculate, flecked, maculate, micro-, microbaculate, microclavate, microgemmate, papillate, punctate, subpsilate.
a. Jackson, 1928: "made rough or roughened."
b. Faegri and Iversen, 1950, p. 27: "Radial projection of sculpturing elements more or less isodiametric. No dimension of 1 µ or more (*Artemisia campestris*)." [*see* ORNAMENTATION].
c. Harris, 1955, p. 18: "Flecked: with minute pits or elevations, the psilate in part and scabrate of Faegri and Iversen (1950) (*Paesia scaberula*)."
d. Traverse, 1955, p. 92: "With sculpturing of projections, no dimension of which equals or exceeds 1 µ. The writer has used this term wherever the small projections were not amenable to closer study—usually because of small size. Where closer description was possible it has been done—for example: 'scabrate, of tiny beads'."
sdb. Iversen and Troels-Smith, 1950, p. 46; [*see* ORNAMENTATION].
Cranwell, 1953, p. 17.
Ingwersen, 1954, p. 41.
Couper, 1958, p. 102; [*see* ORNAMENTATION].
Beug, 1961, p. 13.

SCAR, n., pl. -s. (Gr. eschara, hearth, scab). cf. contact figure, laesura, monolete mark, tetrad mark, Y-mark.
a. Jackson, 1928: "A mark left on a stem by the

separation of a leaf, or on a seed by its detachment, a cicatrix."

b. Traverse, 1955, p. 93: "A remnant, in a pollen grain or spore, of contact with the other members of the original tetrad. In this monograph all of these are the single scars of monolete spores."

SCLERINE, SCLERINIUM, n. (Gr. skleros, hard). cf. sclerosporium.

Erdtman, 1952, p. 468: "sporoderm except for intine. Cf. . . . Tab. 2 . . ." [see SPORODERMIS for Erdtman's Table 2].

SCLEROSPORIUM, n. = sclerine.

SCROBICULATE, adj. (L. scrobiculatus, a little trench). cf. aperture, cribellate, foveolate, ornamentation, pitted, pore, scrobiculatus.

    I. Jackson, 1928: "marked by minute or shallow depressions, pitted."

    II. Erdtman, 1943, p. 52: "Scrobiculatus: see Ornamentatio." [see ORNAMENTATION and Fig. 639].

    III. Faegri and Iversen, 1950, p. 163: "scrobiculate (Potonié, 1934) = foveolate."

    sdb. Pokrovskaya, et al., 1950; [see ORNAMENTATION].

SCROBICULUS, n., pl. -i. cf. foveola, pit.

    I. a. Potonié, 1934, p. 12: *Tr.* If the lumina become very small and the stripes between them broader than the lumina, then the lumina become *scrobiculi.* Frequently it takes some effort in order to discover whether it is a true net sculpture or a punctation, maculation, granulation, or verrucation. The space between the puncta, macules, granules, and warts often gives the impression of stripes or lists by which more or less big net lumina seem to be separated (especially if these elements are arranged very regularly and if one does not look too closely. [see ORNAMENTATION].

    b. Potonié and Kremp, 1955, p. 15: *Tr.* If the lumina of the reticula become very small and the spaces between the lumina larger than the lumina, then the lumina are called *scrobiculae;* larger scrobiculae are called *foveolae.* [see ORNAMENTATION].

    II. Erdtman, 1952, p. 468: "very small, ± circular lumina separated by sexinous streaks several times as wide as the average diameter of a single scrobiculus. Examples: . . . [Figs. 634 and 788–789]."

SCULPTINE, SCULPTINIUM, n.

a. Erdtman, 1948, p. 387: "may be used as a provisional, neutral term, embracing any strata, or fragments of strata, belonging to the exine (n.b., the sexine) the perine, or to both." [see SPORODERMIS].

b. Faegri and Iversen, 1950, p. 163: "sculptine

(Erdtman, 1948) seems to correspond more or less to ektexine."

c. Erdtman, 1952, p. 468: "(Erdtman, 1948 . . . , L. sculptinium): sclerine except for nexine. Cf. Tab. 2, . . ." [see SPORODERMIS for Erdtman's Table 2].

d. Harris, 1955, p. 27: "a neutral term for the sculptured layer, useful when there may be doubt whether the pattern belongs to the exine or perine." [see PERISPORIUM].

SCULPTURE, n. (L. sculptura). cf. ornamentation, pattern, sculpturing.

a. Kupriyanova, 1948: *Tr.* The sculpture (ornamentation) of the exine must be differentiated from the texture (texura); the latter is created by partial separation of layers of the exine, in this way forming a slight roughness. on the surface of the exine. [see ORNAMENTATION].

b. Traverse, 1955, p. 93: "External modifications of the form of the ektexine, giving it recognizable and often distinctive character."

SCULPTURING, n., pl. -s. cf. sculpture.

a. Potonié, 1934, p. 11: *Tr.* Only those form-elements which stand out in relief on the surface of the exine are to be called *sculptures.*

    A term for structure and sculpture is *pattern* or *ornamentation.*

    An exine equipped with structure accordingly may have a smooth surface, i.e., it does not have to show sculpture. [see STRUCTURE, Potonié, 1934, p. 9 (Skulptur)].

b. Faegri and Iversen, 1950, p. 18: "In accordance with Potonié (1934) we distinguish between the structure (texture) and the sculpturing of the exine.

    "The term sculpturing on the other hand comprises the external geometrical features without reference to their internal construction. Thus a spine may consist of a single granule, but it may also be a highly complex structure comprising a number of granules. We shall later in this chapter deal with the sculpturing in more detail, but we want to emphasize here that unless one is very careful, it is easy to confuse structure and sculpturing. Eg., in *Fritillaria* structure and sculpturing are identical (reticulate), whereas *Galeopsis tetrahit* . . . [Fig. 698–700] possess tectate-reticulate pollen grains, in which the sculpturing only is reticulate, whereas the structural elements are evenly distributed. On the other hand it is self-evident that sculpturing is in any case due to the arrangement and form of ektexine elements, e.g., the striation of the pollen grains of most *Rosaceae* are just as much a structural feature as a sculpturing type." [see ORNAMENTATION].

sdb. Kupriyanova, 1948, p. 70; [*see* ORNAMENTA-
TION].

    Cranwell, 1953, p. 17.

    Beug, 1961, p. 13.

SCULPTURING ELEMENTS, n. cf. ornamentation.

SCUTULUM, n., pl. -a. (L. a small shield).

Potonié, 1956, p. 37–38: *Tr.* On the equator, between the two Y-rays is found a shieldlike dome,
a scutulum. [*see* Fig. 161].

SECONDARY FOLD, n., pl. -s. (L. secundus, second). cf. furrow.

  a. Potonié, 1934, p. 16: *Tr.* Probably favored by
the subequatorial position of the pori; *Pollenites
simplex* shows three secondary folds . . . [Fig.
390] which have been produced by the exine
being pressed together from pole to pole while
in the sediment. Such folds, related to the state of
preservation, are to be called *secondary folds.*
In *P. simplex* they consist of arcs which begin
between the pori at the equator and which are
concave toward the pores. [*see* APERTURA].

  b. Potonié and Kremp, 1955, p. 13 and Fig. 597:
Tr. Folds which result from compression in the
sediment are called *secondary folds* . . . [*Calamospora multabilis* (Loose)]. The thinner the exine,
the more frequently the secondary pleats or folds
appear; especially in the case of initially ± spherical grains.

SECONDARY GERMINAL, n. cf. meridional area.

SECONDARY MERIDIONAL AREA, n. cf. meridional area.

SEED MEGASPORE, n., pl. -s. cf. spore, sporomorph.

Schopf, 1938, p. 11: *"Lepidocarpon* shows a minimum of heterangy and in a great many particulars
agrees favorably with the fructifications of heterosporous but free-sporing lycopods. It is well established that we are justified in speaking of the large
spore within this and similar seeds as seed megaspores." [*see* SPORE].

SEIROSPORE, n., pl. -s. (Gr. seira, a rope).

Jackson, 1928: "a spore produced in a branched
row resulting from the division of terminal cells of
particular branches in certain Ceramiaceae."

SEPES A VALLIS, SEPIBUS EVALLIS, n. (L.
sepes, a hedge; densus, thick). cf. cristate, cristo-
reticulate, retipilariate, retipilate.

Kupriyanova, 1948, p. 71 and Fig. 667: *Tr.* Little
net of rodlike forms (sepes a vallis)—when the
walls of the net consist of joined rods providing
the upper surface of the walls with a beadlike
character. This sculpture is characteristic of pollen
of the Lily, Amaryllis, Iris, and Orchid families, and
to a smaller degree of the Palm family. [*see* ORNA-
MENTATION].

SEPES DENSI; SEPIBUS DENSIS, n. (L. sepes, a
hedge; densus, thick). cf. ornamentation, reticulate,
reticulum simplex, retipilate.

Kupriyanova, 1948, p. 71 and Fig. 666: *Tr.* When the
walls of a net do not consist of rods and are unbroken (sepes densi), the upper surface does not
show a beadlike pattern. An example of such sculpture is the family *Aponogetonacea* and *Araceae.*
[*see* ORNAMENTATION].

SETACEOUS, adj. (L. setaceus, seta, bristle). cf.
setose.

  I. Jackson, 1928: "bristle-like; applied to a stem it
means slender, less than subulate."

  II. Kosanke, 1950, p. 11: [*see* ORNAMENTATION and
Fig. 640].

SETOSE, adj. (L. setosus, bristly). cf. echinate,
echinulate, spinose, thornlike.

  a. Jackson, 1928: "bristly, beset with bristles;"
"having setae usually ending in glands (Babing-
ton)."

  b. Erdtman, 1947, p. 106: "setose; with bristles
('setae')."

  c. Faegri and Iversen, 1950, p. 163: "(Erdtman,
1947): with bristles = echinatus p.p."

SETULA, n., pl. -ae. cf. spinula, spinule.

Jackson, 1928: "the stipe of certain Fungi (Lindley);" "a minute bristle."

SETULATE, adj.

Harris, 1955, p. 19: "Trunk (of the projection) not
tapering. Apex not sharp. Example: *Polystichum
sylvaticum.*" [*see* ORNAMENTATION].

SEXINA, SEXINE, n. cf. ectexina, ektexina, ektexine, ektexinium, exoexine.

  a. Erdtman, 1948, p. 387: "the sculptured part of
the exine." [*see* SPORODERMIS].

  b. Faegri and Iversen, 1950, p. 163: "(Erdtman,
1948) seems to correspond more or less to ektexine."

  c. Erdtman, 1952, p. 468–469: "the outer, sculptured part of the exine . . ." [*see* SPORODERMIS
and Fig. 648].

  d. Potonié and Kremp, 1955, p. 18: [*see* SPORO-
DERMIS].

sdb. Pike, 1956, p. 51.

    Kuprianova, 1956; [*see* SPORODERMIS].

SHAGREEN, cf. chagranate.

SHAPE, n., pl. -s. cf. figura, Groups of Sporomorphae, morphography, shape classes, spore.

  a. Zaklinskaya, 1957, for shape of disaccate pollen
grains *see* Fig. 227, 236–241.

  b. Couper and Grebe, 1961, p. 2 and Fig. 168–
183: *Tr.* It is recommended that the shape of a
spore should be described first as seen in polar
view and secondly as seen in equatorial view, if
both views are obtainable.

    (a) *Polar view*

Equatorial contour: the shape of the equatorial outline of a spore as seen in polar view. The equatorial contour can be of three basic shapes: *circular, triangular* or *oval* . . . [Fig. 168–170]. In the case of triangular equatorial contours, the sides can be further described as *convex, straight* or *concave* . . . [Fig. 171–173], and the angles as *sharply rounded, rounded* or *flat* . . . [Fig. 174–176]. Flattened angles may be due to a non-structural feature or due to *structural differentiation* (e.g., a *radial crassitude*).

(b) *Equatorial view*

Proximal profile: the shape of the outline of proximal surface as seen in equatorial view, and should include any shape feature due to the *tetrad mark*. The basic profiles are *flat, convex, pointed* or *concave* . . . [Fig. 177–180]. In monolete spores, the shape of the proximal profile should be qualified by the view in which it is seen (i.e., whether in *longitudinal* or *transverse* view).

Distal profile: the shape of the outline of the distal surface of a trilete or monolete spore as seen in equatorial view. The basic profiles are *convex, pointed* or *flat* . . . [Fig. 181–183].

SHAGREEN. cf. chagrenate.

SHAPE CLASSES, n. cf. angular, angular-elliptical angular-oval, angulatus, apiculate, apolar, bean-shaped, bilateral, circular, conical cryptopolar, dimension, elliptical, fabaeformis, fabiformis, figura, figure, flattened, form, global, Groups of Sporomorphae, hemispherical, hexeropolar, lalongate, lolongate, morphography, multilateral, oblate, oblatoid, oblete, orbicular, orbiculate, orbiculatus, oval, ovatus, perprolate, peroblate, peroblatoid, pointed, polar cap, pollen classes, pollen form index, polygonal, prolate, prolate-spheroidal, protrudent, prominent, quadrangulis, quinquangulis, radiosymmetric, rectangulate, rectangulatus, rhombohedral, rhomboidal, rotundatus, rotundo-lobatus, rotundus, roundish, round-lobate, shape class index, short, side line, size classes, size range, sphaeroideus, spheroidal, subisopolar, suboblate, subprolate, subspheroidal, trapeziform, trapeziform-roundish, triangular, triangulatus, trilobate, trilobatus, triquete.

a. Erdtman, 1943, p. 44–45 and Fig. 521: "The shape of these grains is that of an ellipsoid of revolution with the polar axis as axis of rotation. In polar view, the outline of the grain is circular, in some cases somewhat triangular, with furrows at the angles. In equatorial view, the outline is elliptical. With the polar axis comprising the major axis of the ellipse, the pollen grains with decreasing eccentricity may be termed perprolate, prolate, subprolate, prolate spheroidal, and spherical, while those—in case the polar axis com-

prises the minor axis of the ellipse—with increasing eccentricity may be termed spherical, oblate, spheroidal, suboblate, oblate, and peroblate . . . [Fig. 521] where also the suggested relations between polar axis and equatorial diameter have been given. The details of this terminology have been discussed with Dr. Paul Richards (Trinity College, Cambridge) and some outstanding English mathematicians to whom the writer is indebted for valuable suggestions."

Shape classes and suggested relations between polar axis (P) and equatorial diameter (E). (From Erdtman, 1943, Table 5).

Shape Classes	P:E	
Perprolate	>2	(>8:4)
Prolate	2–1.33	(8:4–8:6)
Subprolate	1.33–1.14	(8:6–8:7)
Spheroidal	1.14–0.88	(8:7–7:8)
prolate spheroidal	1.14–1.00	(8:7–8:8)
oblate spheroidal	1.00–0.88	(8:8–7:8)
Suboblate	0.88–0.75	(7:8–6:8)
Oblate	0.75–0.50	(6:8–4:8)
Peroblate	<0.50	(<4:8)

sdb. Erdtman, 1946, p. 72.

b. Norem, 1958, p. 673–676 and Fig. 167a–167n: "Shape classification (Applied as subclasses in master key . . .)

G. Shape more or less spheroidal or elliptical

H. Orientation of polar axis indeterminate or vertical (polar view)

1. *Ellipsoidal* . . . ratio of major to minor axis >1.25

2. *Globoid* . . . ratio of major to minor axis about equal to 1.(<1.25)

HH. Polar axis oriented in traverse view (parallel to slide)

1. *Peroblate* . . . ratio of polar axis to equatorial diameter <0.50

2. *Oblate* . . . ratio of polar axis to equatorial diameter between 0.50 and 0.80

3. *Spheroidal* . . . ratio of polar axis to equatorial diameter between 0.80 and 1.25

4. *Prolate* . . . ratio of polar axis to equatorial diameter between 1.25 and 2.0

5. *Perprolate* . . . ratio of polar axis to equatorial diameter >2.00

GG. Shape more or less triangular

H. Polar axis vertical

1. *Deltoid* . . . equilateral triangle with more or less straight sides and sharp apices

2. *Subtriangular* . . . equilateral triangle with more or less straight or convex sides and rounded apices

3. *Triquete* . . . equilateral triangle with concave sides and rounded apices

4. *Trilobate* . . . clover-leaf shaped with more or less sharp angles between lobes

GGG. Shapes more or less equilateral with four or or more sides

  H. Polar axis vertical

    1. *Tetragonal* . . . , *Pentagonal* or *Polygonal*

GGGG. Grains with bilateral symmetry

    1. *Reniform* (phaseolate) . . . kidney or bean shaped

    2. *Saccate* . . . with wings or bladders

    3. *Lenticular* . . . lens shaped

    4. *Crescent* . . . crescent shaped

    5. *Spatulate* . . . spoon or spatula shaped

    6. *Drop*-shaped like a falling drop.

SHAPE CLASS INDEX, n., pl. indices. cf. dimension, figura, shape classes.

Faegri and Iversen, 1950, p. 163: "is expressed by the relations between the length of the polar axis and that of the greatest equatorial diameter (Erdtman, 1943)."

sdb. Ingwersen, 1954, p. 41.

SHARPLY ROUNDED, adj.

Couper and Grebe, 1961: [*see* SHAPE].

SHORT, adj. cf. shape classes.

SIDE LINE, n., pl. -s. cf. polar cap.

SIMPLE PORE, n., pl. -s. cf. aperture.

  I. Jackson, 1928: "with only a slight enlargement at the center, where it meets the neighboring cell."

  II. Kupriyanova, 1948, p. 71: *Tr.* Pollen grain with a simple opening or pore. [*see* GROUPS OF SPOROMORPHAE].

SIMPLIBACULATE, adj. (L. simplex, simple). cf. baculate, oligobaculate.

Erdtman, 1952, p. 469: "muri- etc., supported by a single row of bacula are simplibaculate. Examples: . . . [Figs. 630 and 724–727]."

SINUAPERTURATE, adj. cf. aperture, colpate, furrow, pore.

Erdtman, 1952, p. 459: ". . . the apertures are situated equally halfway between the angles (sides of amb ± concave . . .)." [*see* AMB and Fig. 500].

SIZE, n., pl. -s. cf. dimension, figura.

SIZE CLASSES, n. cf. dimension, figura, shape classes.

  a. Erdtman, 1943, p. 48: "When quoting the dimensions of pollen grains and spores, all possibility that may lead to a misinterpretation must be avoided. In radiosymmetrical grains, the size is expressed simply by quoting the length of the polar axis and the equatorial diameter. In monocolpate grains, on the other hand, the length may be expressed as the distance between the extreme points of a central longitudinal section running in the same directions as the *colpa* (furrow). The maximum breadth is usually equal to the distance between the extreme points of a central trans-

versal section through the grain or spore. When speaking of winged conifer pollen grains, a special terminology should be used. The width of a fully expanded grain (a figure which, incidentally, usually does not seem to be of much diagnostic value) may be defined as the distance between the extreme parts of the two opposite wings. The width of the body (i.e., the distance between the two points where the proximal root of the bladders meet the body) is more reliable as a diagnostic character. The breadth of the body and wings can only be measured in grains in polar view. Their height is measured in grains in end view with both bladders fully exposed. The height of the body is identical with the length of the polar axis, while the height of the bladder is identical with the length of a perpendicular line stretching from the convex extremity of the bladder to the endexinous floor constituted by the body. Both figures are of minor importance. Several measurements concerning winged conifer pollens hitherto published are of no value since there are no precise descriptions regarding the way the actual measurements were made."

  b. Erdtman, 1946, p. 72: "The following size classes, based on the length of the longest pollen or spore axis, have been suggested (Erdtman, 1945, p. 191):

PI = very small grains or spores (*pollina perminuta; sporae perminutae*) . . . . $<10.0\,\mu$

MI = small do. (*p. minuta; sp. minutae*) . . . . 10– 25 μ

ME = medium size do. (*p. media; sp. mediae*) . . . . 25– 50 μ

MA = large do. (*p. magna; sp. sp. permagnae*) . . . . 50–100 μ

PA = very large do. (*p. permagna; magnae*) . . . . 100–200 μ

GI = gigantic do. (*p. gigantea; sp. giganteae*) . . . . >200 μ

"In the smallest size class the size should be expressed in tenths of μ, in the second smallest class in one and a half μ, in the other classes in μ. The sculptural elements, such as spines, etc., should not (particularly if they are more or less prominent) be included in the general dimensions of the grains or spores but should be dealt with independently. The same points of view are also adopted in mycology (cf. Malencon, 1929) . . . A nonarbitrary distinction between prominent and not prominent sculptural elements will, however, sometimes be difficult or even impossible. In such cases, the methods used in obtaining the measurements should be clearly stated. Finally, it may sometimes be profitable to quote the dimensions of the exine lumen."

SIZE RANGE, n., pl. -s. cf. dimension, figura, Groups of Sporomorphae, morphography, shape classes, size classes.

Couper, 1958, p. 103: "unless otherwise stated the size range given in descriptions of both associated and dispersed spores and pollen grains is based on at least 25 specimens. The bracketed figure is the mode."

SLIT, n., pl. -s. cf. rimula.

SMALL SPORE, n., pl. -s. cf. microspore, miospore, sporomorph.

Schopf, 1938, p. 11: "Spores which cannot be assigned to the isospore, microspore, prepollen or pollen categories will be merely termed small spores."

SMOOTH, adj. cf. laevigate, psilate.

Jackson, 1928: "not rough, opposed to scabrous, free from hairs;" "glabrous, as opposed to pubescent."

SOLUTION, n., pl. -s. (L. solutio, to loosen). cf. colpus, pore, ruga, solution area.

I. Jackson, 1928: "the detachment of various whorls normally adherent;" "the opposite of adhesion."

II. Thomson and Pflug, 1953, p. 20: *Tr.* The solutions are usually localized and often mark the germinal apparatus. The following types are frequent in pollen:

(1) "Pores" (pori) are broad apertures [flächige Durchbrüche] through one or more lamellae and preformed paths for the pollen tube. The ratio of length to width usually does not exceed 5:1 and the maximum diameter does not reach one-fifth the maximum pollen diameter . . . [Fig. 368–370].

(2) "Colpi" are linear gaps [Auflösungen], therefore slits in the lamellae, without widening, or are more than five times longer than wide, which extend in meridional direction.

(3) "Rugae" are forms which correspond to colpi, however they are equatorial stretched. (Slits which run obliquely to the equator are considered to be colpi) . . . [Fig. 371].

(4) Large areal solution zones of different shape are here called solution areas; they exceed the dimensions which are mentioned in 1 and 2. Here, for instance, the "atrium" of the Triatriates and the "solution meridal" of some Extratriporates must be named. Solution areas can extend far beyond the real germinal apparatus. Finally, entire lamellae can be entirely absent. Iversen separates exines which consist only of one lamella as "intectate" from the "tectates" which have more lamellae (Faegri and Iversen, 1950). However, this criterion can be called upon only to a certain degree for systematic classification. An "intectate"

phenomenon need not only be the result of lamellar solution, rather it can also result from a marked thinning of the exine so that its fine lamellar structure cannot be resolved with the light-microscope (lamellae oppressae). Therefore, one can only call the "intectates" an "apparently one-layered" exine and use this term only in connection with a certain microscope aperture as a systematic criterion.

One finds dehiscence marks on spores. They are developed as Y-marks or as Y-fissure marks in the trilete, zonal, and triplanate spores; in the monolete spores, they are linear or split linear . . . [Fig. 113–114].

SOLUTION AREA, n., pl. -s.

Thomson and Pflug, 1953, p. 20: *Tr.* Large areal solution zones of different shape are here called solution areas. [*see* SOLUTION].

SOLUTION CHANNEL, n., pl. -s. cf. platea.

SOLUTION MERIDIUM, n. cf. cuneus, incidence, platea, solution channel, solution notch, solution wedge.

Thomson and Pflug, 1953, p. 33: *Tr.* The solution process of endexine is remarkable; it extends as a stripe in a meridional direction, to a different degree in different groups: We differentiate the following conditions the solution notch (= incidence), the solution wedge (= cuneus), and the solution channel (= platea). The overall term is "solution meridium." [*see* Fig. 410–412].

SOLUTION NOTCH, n., pl. -es. cf. incidence.

SOLUTION WEDGE, n., pl. -s. cf. cuneus.

SPARSE, adj. (L. sparsus, spread open). cf. pattern, sculpture.

a. Jackson, 1928: "scattered."

b. Harris, 1955, p. 21: "The practice is to describe the sculpture pattern as *dense* when the space separating the projections is less than one-twentieth the equatorial diameter of the spore and as *sparse* when in the main it is greater than one-tenth the equatorial diameter."

S-PATTERN, n. cf. LO-pattern, suprategillar pattern.

Erdtman, 1952, p. 468: "S-pattern: designations usually used in descriptions of tegillate (or tectate) spores only; connote the superficial pattern produced by suprategillar (or supratectal) elements."

SPATULATE, adj.

Norem, 1958, [*see* SHAPE CLASSES and Fig. 167n].

SPERM CELL, n., pl. -s. (Gr. sperma, a seed). cf. antherozoid, sporomorph.

Jackson, 1928: "a male reproductive cell, as (a) an antherozoid, (b) a pollen grain; usually a minute active cell whose function is that of fusion with a large resting cell (oösphere), to form a zygote;"

"sperm cell, sometimes restricted to the spermatozoid mother-cell."

SPHEROIDAL, adj. (L. sphaera, a sphere). cf. shape classes, spheroidal, sphaeroideus, subspheroidal.
  a. Jackson, 1928: "globular, any solid figure approaching that of a sphere."
  b. Erdtman, 1943, p. 45: "In spheroidal spores the ratio between polar axis and equatorial diameter is 0.88–1.14 (7:8–8:7)."
  sdb. Erdtman, 1952, p. 469.
  c. Selling, 1947, p. 77: "P:E ratio (polar and equatorial diameters) 1:00 (cf. Erdtman, 1943, p. 45, where the term is used in a somewhat different sense). Note: 'Globular' used to denote grains isodiametrical throughout."
  d. Zaklinskaya, 1957; for outline of disaccate pollen grains see Fig. 227–241:B.
  e. Norem, 1958. [see SHAPE CLASSES and Fig. 167e].
  f. Beug, 1961, p. 13: Tr. Pollen grain with a ratio of length (polar axis) to greatest diameter (pollen form index) = 1.33–0.75 (subspheroidal according to J. Iversen and J. Troels-Smith, 1950).

SPHAEROIDEUS. cf. sphaeroidal.

SPINA, n., pl. -ae. (L. a thorn). cf. echinus, spine, thorn.
  I. a. Potonié, 1934, p. 11: Tr. Sculpture elements consisting of regularly arranged, more or less broadly projecting, irregularities of the exoexine. [see ORNAMENTATION and Fig. 652, 656].
    b. Erdtman, 1952, p. 469: "long, conspicuous, and generally sharp, pointed excrescences; length exceeding 3 μ. Examples: . . . [Fig. 712–714, 721–722, 729]."
    c. Faegri and Iversen, 1950, p. 163: "spina (Potonié, 1934) = echina."
  II. Potonié and Kremp, 1955, p. 14: Tr. The spines or spinae . . . [Fig. 740] narrow more slowly than the coni. Also they often show a thickened base. In this case, the length of that part of the spinae which is above the basal thickening is more than twice the dimension of the diameter of the spinae as measured above the thickening. Sometimes the spinae are distinctly pointed; occasionally the points are broken off. The spinae are more or less straight. [see ORNAMENTATION].
  sdb. Kupriyanova, 1948, p. 70, Fig. 658; [see ORNAMENTATION].

SPINE, n., pl. -s. (L. spina, a thorn). cf. spina.
  I. Jackson, 1928: "A sharp pointed woody or hardened body, usually a branch, sometimes a petiole, stipule, or other part."
  II. Ueno, 1958, p. 176: "long, conspicuous, and generally sharp, pointed excrescences; length exceeding 3 μ."

SPINOLOPHATE, SPINOLOPHATUS, adj. cf. echinolophate.
  Erdtman, 1947, p. 106: "(Wodehouse, l.c.: echinolophate): spinolophate; with crests bearing spines."

SPINOSE, adj. (L. spinosus). cf. echinate, echinulate, setose, thorn-like.
  a. Jackson, 1928: "spiny, having spines."
  b. Erdtman, 1947, p. 106: "Spinosus: spinose; with spines . . ."
  c. Kosanke, 1950, p. 11; [see ORNAMENTATION and Fig. 640(M].
  sdb. Kupriyanova, 1948, p. 70; [see ORNAMENTATION and Fig. 658].
    Couper and Grebe, 1961, p. 7; [see ORNAMENTATION and Fig. 681].

SPINULA, n., pl. -ae. (L. dim. of spina). cf. setula, spinule.
  Erdtman, 1952, p. 469: "small spines, not exceeding about 3 μ in length. Examples: . . . [Fig. 795–797, 803–806]."

SPINULE, n., pl. -ae. cf. spinula.
  I. Jackson, 1928: "a diminutive spine."
  II. Ueno, 1958, p. 176: "small spines, not exceeding about 3 μ."

SPIRAPERTURATE, adj. (Fr. Gr. speira, a coil). cf. colpate, furrow.
  a. Erdtman, 1952, p. 469: "with one (sometimes several) spiral apertures. Examples: . . . [Fig. 439–442]."
  b. Norem, 1958, p. 672: "Syncolpate—with furrows fused into spirals."

SPIRATE SULCATE, adj. cf. colpate furrow.
  Kupriyanova, 1948, p. 75: Tr. Pollen grains with a spiral furrow. [see GROUPS OF SPOROMORPHAE].

SPIROID. cf. colpate, furrow.
  I. Jackson, 1928: "a delicate thickening in the cells of the tentacles Drosera (Kerner)."
  II. Erdtman, 1952, p. 469: "spiroid apertures are ± similar to the apertures in spiraperturate grains although less distinct."

SPONGY, adj.
  I. Jackson, 1928: "Having the texture of a sponge, cellular and containing air, as in many seed coats."
  II. Krutzsch, 1959, p. 40: Tr. The wall is stuffed with alternating meshlike little lacunae. [see ORNAMENTATION and Fig. 769].

SPORA, n., pl. -ae. (Gr. spora, a seed). cf. pollen, spore, sporomorph.
  Jackson, 1928: "= spore."

SPORA DISPERSA, n., pl. -ae. cf. dispersed spore, sporomorph.
  Potonié and Kremp, 1955, p. 2: Tr. We designate as sporae dispersae such spores and pollen as are

not found in situ, that is, in fructifications; rather they are found dispersed in sediments.

SPORANGE, n., pl. -s. (L. sporangium; Gr. spora, a seed; aggeion, a vessel).

Jackson, 1928: "a sac endogenously producing spores;" "sometimes applied to the volva among Fungals (Lindley)."

SPORANGIOPHORE, n., pl. -s. (L. sporangiophorum; Gr. angeion, vessel; phoreo, I carry).

Jackson, 1928: "a sporophore bearing a sporangium, such as the sporophyll in *Equisetum,* or the columella in Ferns."

SPORANGIUM, n., pl. -a.

Jackson, 1928: "cf. sporange."

SPORE, n., pl. -s. (Gr. spora, a seed). cf. aecidiospore, akinete, arthrospore, aplanospore, autospore, ascospore, auxospore, azygospore, basidiospore, carpospore, chlamydospore, conidiospore, conidium, cyst, cystospore, endospore, exospore, gemma, heterospore, homospore, isosporae, isospore, macrospore, meiospore, microcyst, microspore, miospore, monospore, ooidium, palmellastadium, palynomorph, paraspore, planospore, pollen, pollen grain, pollen granule, pollen mass, pollenospore, polospore, polyspore, prepollen, pychidiospore, pychiospore, resting cell, resting spore, small spore, spora, spora dispersa, sporomorpha, swarm spore, teleutospore, teliospore, tetraspore, urediniospore, uredospore, zyospore, zoospore.

a. Jackson, 1928: "a cell which becomes free and capable of direct development into a new bion; in Cryptogams the analogue of seed in Phanerogams, understood by Saccardo as a BASIDIOSPORE; further particularized by C. MacMillan . . ."

b. Schopf, 1938, p. 10: "The relation of spores to life cycles of their respective plants is a basic consideration in the systematic study of such bodies. For further information pertinent to the subject, the reader is referred to discussions given by Bowers, 1923, and by Eames, 1936, and others which discuss the gametophyte generation. the present section deals principally with the different kinds of spores and equivalent structures which may be isolated from coal.

"The term spore is used in the present paper to designate those propagative bodies which consist essentially of the gametophyte (derived from a single cell) enclosed and protected by more or less waxy non-cellular layer or membrane, the spore coat. In coal ordinarily the spore coat alone is distinguishable although in some cases the gametophyte may also be present as a carbonized lamina.

"The kinds of spores differ according to the type of plant reproduction and may best be discussed from that standpoint. The two major reproductive types which concern us are the free-sporing and the seed-bearing. The former are more primitive; some plants of this organization show diversification in the sexual specialization of their spores. The spores or spore equivalents of seed plants are always highly differentiated with respect to the sexes.

"Free-sporing plants are those in which the spores are liberated from the sporangia which produced them and continue development, under favorable conditions, separately from their parent sporophyte. Seed-bearing plants retain and nourish the female gametophyte (spore equivalent) within the sporangium during growth and the male spore is generally transported to it via the air after being shed in the usual way. Such a reproductive method is much more complex than that found in free-sporing plants. The latter, however, were much more abundant in the ancient flora than they are at present, although the free-sporing habit is still found in our common ferns and other present day Pteridophytes.

"Sexual differentiation among spores of the primitive free-sporing type may be recognized by differences in spore size but not all sex differentiation involves size. Examples of this sort occur among modern heterothallic bryophytes but these plants have not been recognized in the Paleozoic. Plants showing spore size differentiation are said to be heterosporous, the male spores being generally small and easily dispersed by air, the female spores relatively [Schopf, 1938, p. 11]: large. Free-sporing plants whose spores are not thus differentiated are said to be isosporous or homosporous. The spores of isosporous plants are all the same size, generally small, serve both male and female functions, and are termed isospores. Only the small male spores of heterosporous free-sporing plants are true microspores; the larger female spores are called megaspores. The larger size of the megaspore is presumably an adaptation for the storage of food which nourishes the embryo arising later.

"Since the isospores of homosporous plants and the microspores of heterosporous plants are much alike in size and form, discrimination between the two is usually difficult for isolated specimens. On the other hand, since in many specific instances representatives of each type are distinctive, there is reason to believe that many forms may eventually be assigned correctly through inference from similar spores being identified in known fructifications. The practice generally followed by coal microscopists of calling all smaller varieties of spores 'microspores' should therefore be avoided, since it will tend toward confusion when more precise information becomes available. The author suggests that, where there is doubt, spores of the isospore and microspore types should be referred to merely as small spores.

"As was stated, the seed-bearing mode of reproduction is characterized throughout by marked male and female sexual differentiation of the spore structures. Since these structures are often found in maceration residues along with spores of free-sporing plants they must be accorded some consideration particularly in regard to the terminology employed.

"Thomson [Evolution of the seed habit in plants: Trans. Roy. Soc. Canada, ser. 3, v. 21, p. 229–272, 1927] has pointed out that sexual differentiation in seed plants is expressed to a considerable degree through modification of the sporangia, the manifestation of which he calls heterangy. He does not believe that size differentiation of spores, i.e., heterospory, *necessarily* preceded the seed habit in the course of evolution. Consequently some caution must be exercised in calling the male and female spore structures of plants microspores and megaspores respectively . . .

"At least one group of plants is known, however, in which there is no question that seeds contained true megaspores. These are the isolated spores described herein as new genus, *Cystosporites* known to be allied with *Lepidocarpon,* a seed belonging to the lycopod order. *Lepidocarpon* shows a minimum of heterangy and in a great many particulars agrees favorably with the fructifications of heterosporous but free-sporing lycopods. It is well established that we are justified in speaking of the large spore within this and similar seeds as seed megaspores. The male spores of *Lepidocarpon* are cryptogamic in character and no doubt are truly microspores. [Schopf, 1938, p. 12]: "In several cases gymnospermic seeds have likewise been isolated from coal and in addition to the female spore membrane show adhering remnants of integumentary cuticle. Although these are not described in the present paper it is convenient to mention them here. Gymnosperms are markedly heterangious and attained the seed habit wholly independently of *Lepidocarpon.* It seems that these large gymnospermic female spores ought also to be called seed megaspores. However, since no free-sporing heterosporous ancestry is *known* to have preceded the seed habit in the gymnosperm line complete justification for this designation is lacking. In some primitive gymnosperms the male spores show considerable advance in organization beyond the cryptogamic microspore and yet lack some significant features of modern pollen. For

these it is proposed to apply Renault's term 'prepollen.' A more detailed discussion of prepollen is given on pages 14–15. [*see* PREPOLLEN for Schopf, 1938, p. 14–15].

"For completeness the spore equivalents of modern gymnosperms and angiosperms should be mentioned although no angiosperm ancestors are recognized in the Paleozoic flora. The male spore equivalents in these are fairly similar in organization and are called pollen. Modern pollen is especially characterized by the formation of a pollen tube. The female gametophytes of modern gymnosperms appear to resemble the fossil forms of the Paleozoic in all significant characteristics of spore organization. The female gametophytes of angiosperms on the other hand show little to aid in comparison. There is no suggestion of a

"Table 1

Type of Reproduction	Kinds of Spores or Spore Equivalents
A. Free-Sporing Mode of Reproduction: 1. Spores not distinguishable as to sex: Plants *Homosporous* or *Isosporous*	For Both Sexes *Isospore* (=homospore)
2. Spores distinguishable according to sex: Plants *Heterosporous*	Male *Microspore* Female *Megaspore*
B. Seed-Bearing Mode of Reproduction: Always distinguishable as to sex: not necessarily heterosporous in the sense given above. Plants *Heterangious* 1. Primitive Seed Plants	Male *Microspore* *Prepollen* or *Pollen*(?) Female *Gametophyte of Primitive Seed* (May, or may not, be properly called a *seed megaspore,* depending on whether plants are Heterosporous, as well as Heterangious.)
2. Modern Seed Plants	Male *Pollen* (produces pollen tubes) Female *Gametophyte of Modern Seed*"

spore coat and the gametophyte as a whole is so extremely reduced that spore homologies pertinent to the above discussion cannot be confidently drawn.

"The accompanying table 1 shows in a condensed form the general relationship of spores in vascular plants and summarizes the preceding discussion. [Schopf, 1938, p. 13: on p. 145].

c. Dijkstra-v. V. T., 1946, p. 21–22: [*see* MEGASPORE]
d. Erdtman, 1947, p. 104: ". . . The spores have, as a rule, been subjected to acetolysis. In material so treated, the intraexinous parts have dissolved away leaving only the more resistant extra-intinous parts of the spores (the term 'spore' indicates spore or pollen or both according to the context)."
e. Kosanke, 1950, p. 14: *see* Fig. 109–110.
f. Melchior and Werdermann, 1954, p. 14. *Tr.*: Asexual germ cell formation. . . . Special germ cells are formed and isolated (spores, agametes), which develop directly into new organisms. Gonospores include all those spore types in whose development a reduction-division takes place and whose mother cell is the Gonotocont.

1. *Resting cells* (resting spores, paulospores): one vegetative cell (sometimes more) develops directly into the resting cell.
    1.1 *Microcysts* (Cystospores): a myxoflagellate becomes rounded and surrounds itself with a thin membrane, e.g. Myxomycetes.
    1.2 *Chlamydospores*: thick or double-walled resting spores originating by destruction or in persistence of a hypha, e.g. *Fungi.*
    1.3 *Oidia* (Oidium chains, oöspores): the tip of the hypha is destroyed by simultaneous cross division (oidium division) in a row of thin-walled, spore-like cells, which mostly remain together in chains for a period of time, e.g. *Fungi.*
    1.4 *Gemmae* (resting conidia): terminal cells which become especially thick-walled conidia, e.g. *Fungi.*
    1.5 *Akinetes* (resting spores, arthrospores, resting akinetes): The complete cell including its membrane contracts by alteration of the membrane and by simultaneous accumulation of reserve substance and often also of pigments: very wide-spread in the algae groups, resting cells of the *Cyanophyta.*
    1.6 *Teleutospores* (compare p. 15 [*see* under Exospores]).
    1.7 *Brandspores*: resting cells surrounded with resistant membrane; initially multi-nucleate: *Ustilaginales.*
2. *Exospores* (exogene spores, acrogenous spores, conidia, exoconidia, stylospores): the origin and

separation of spores occurs at the end of certain cells (Conidiophores).
Special types of exospore formation:
Single conidia: formation and pinching off of spores singly at the end of the mother cells, cf. *Fungi.*
    Row conidia (conidia chains): formation and pinching off of spores in a row directly at the conidiophore or at special stalks (sterigmata). The formation generally occurs at the base in such a way that the hyphen end at first forms a conidium, the second conidium formed by the next cell, etc.; [Melchior and Werdermann, 1954, p. 15. *Tr.*] the lowermost conidium is therefore the youngest one. Or the formation is acrogenous, in which case from the first formed conidium a second one grows, etc.; the uppermost one is the youngest, cf. *Fungi.*
    Basidiospores (sporidia): At the conidiophore on awl-like sterigmata generally in a special arrangement and often formed in tetrads, thin-walled spores. The conidiophore, promycelium or basidium is divided (protabasidium) or undivided (autobasidium): *Basidiomycetes, Basidiolichenes.*
    Aecidiospores (Aeciospores): multinucleate spores, pinched off in rows in basipetal succession. They are formed in containers (Aecidium, Aecium) which are surrounded by a coat: *Uredinales.*
    Uredospores (summer spores, protospores): nomads, multi-nuclear, often covered with verrucae, spines, or lists, germinating immediately by a simple germ tube after becoming mature: *Uredinales.*
3. *Endospores* (endogene spores, endo-conidia, sporangiospores, conidia, sporangiospores, conidia, gonidia): the spores are formed in a spore container (sporangium, gonidangium) which often has a special shape, and from which they become liberated. The spore formation occurs successively in such way that at each nuclear division the protoplast immediately becomes divided, or simultaneously in such way that at first the nucleus is repeatedly divided and only then the protoplast is subdivided. As sporangioles are designated secondary sporangia with few spores, which originate in the form of vesicula on the outside of reduced primary sporangia which no longer produce spores. Upon further reduction of one conidium the wall of the sporangiole fuses with the conidium wall: the sporangiole then becomes a conidium, and the sporangiaphore becomes a conidiophore (*Mucorinales* . .). As single spored sporangioles are also considered the basidiospores (see above), as a consequence of that opinion, the

basidium is an ascus with exogene spore formation . . . .

3.1 *Isospores* (Homospores): All spores are morphologically similar, they are of mixed sexuality or sexually determined.

3.11 *Zoospores,* planospores (Zoogonidia, Schwärmspores, Flagellates . . .): without membrane, with active motion by one, two or more flagella, sometimes by a ring of flagella. They originate, multiple or singly, in a zoosporangium (flagellate sporangium). If smaller zoospores occur along with the usual ones, one speaks of macrozoospores and microzoospores: e.g. *Ulothrix.*

Synzoospores: Multiflagellate, giant zoospores which correspond to numerous diflagellate zoospores, hence constituting a zoospore unit: *Vaucheria.*

Carpozoospores: Diflagellate zoospores, which originate in groups of 8–16 during germination of the oospore: *Coleochaete.*

3.12 *Aplanospores* (often designated only as "spores"): Without flagella, therefore not self-propelled, but generally surrounded by a resistant membrane. Origin in the sporangium (Aplanosporangium). [Melchior and Werdermann, 1954, p. 16: *Tr.*].

Special types of aplanospore formation:

Autospores: Aplanospores, which while in the mother cell acquire the same form and appearance as the latter: e.g. *Chlorophyta.*

Palmellastadium: Aplanospores which are kept together in nests by filament production of the mother cell membrane: e.g. *algae.*

Ascospores: Formation of 8 (seldom fewer or up to 32) spores in the ascus: *Ascomycetes, Ascolichenes.*

Tetraspores: Formation of each four spores during reduction-division in the tetrasporangium: *Dictyota, Rhodophyta.*

Polyspores: Makes cells homologous to tetraspores, originating by many-fold cell division in the polysporangium: e.g. *Florideae,* especially the *Ceramiaceae.*

Paraspores (seirospores, seirogonidia): spores appearing in dichotomous branching rows or in somewhat undefined cell masses (parasporangia): e.g. *Rhodophyta.*

Monospores: Formation of only one spore without reduction-division in the monosporangium: e.g. *Rhodophyta.*

Carpospores (carpogonidia): Formation of only one naked spore in the carposporangium: e.g. *Rhodophyta.* If the carpospores are diploid, they are also called *diplospores.*

Cysts (cystospores, hypnospores, hypnocysts): the protoplast contracts within the cell and forms at its periphery a new membrane which becomes especially thickened; the cysts can maintain therefore a longer resting period before their germination: e.g. *Pyrrophyta, Euglenophyta, Chrysophyceae, Chlorophyta, Bacteriophyta* (here designated as "endospores"). If the new-formed membranes become silicified one speaks of silica cysts. Resting-cells are also very often simply called cysts.

3.2 *Heterospores*: Formation of two kinds of spores which are of different size and of different sex: *heterospory.*

Microspores formed in the microsporangium.

Macrospores (megaspores) formed in the macrosporangium (megasporangium).

The leaflike structures which bear spores and which often are ± metamorphosized are called sporophylls, or microsporophylls and macrosporophylls. Sporophylls can be combined in special stands: sporophyll stand or microsporophyll stand and macrosporophyll stand.

The corresponding forms among the higher plants, in the divisions of the gymnosperms and angiosperms are designated as: microspores—pollen grains, microsporangium—pollen sac or pollen loculus, microsporophyll, male sporophyll—stamen, macrospores—embryosac, macrosporangium—nucleus enclosed in the ovulum, strobilus, conus . . . .

g. Winslow, 1959, p. 11–14: "The spores described and discussed in this report were largely, perhaps entirely, derived from vascular plants, the Tracheophyta. Many of the plants were large and tree-like and are referred to as arborescent. Others grew close to the ground and are called herbaceous. All vascular plants produce spores of some kind and have an alternation of sporophyte and gametophyte generation during their life cycles.

"The mature plants, part of the sporophyte generation, produce spore mother cells, each of which produces four unicellular spores of the gametophyte generation. Some vascular plants are homosporous, producing spores that germinate into multicellular gametophytes, which in turn produce both male and female gametes. Other plants are heterosporous and bear: 1) megaspores that germinate into multicellular female gametophytes that produce female gametes and 2) microspores that germinate into multicellular male gametophytes that produce only male gametes. The union of female and male gametes initiates the sporophyte generation of the life cycle.

"Most spores are enclosed within a protective coat that is resistant to chemical and physical attack. Because of this, the spore coats or exines, referred to generally as spores throughout this paper, commonly are well preserved in coals and many kinds of sediments. The gametophyte inside the spore coat is preserved only under exceptional conditions.

"Different plants produce spores of different sizes and shapes and, as plants evolved through time, the size, shape, and ornamentation of the spores changed, although possibly not at the same rate as other observable changes in the plants. Because of the great variety of spore types, their gradual change in appearance through geologic time, and the resistant nature of the spore coat, the spore assemblages found in any one sediment are likely to be different in composition from those found in older or younger beds.

"Also, because of possible geographic differences in the distribution of plants, the assemblages from the same bed may have a slightly different aspect from one area to another. Spores, especially microspores and isospores, are produced in great numbers, are small, and are widely dispersed by natural agencies such as wind and water currents. Although megaspores are generally larger, sometimes very large, and are produced by only a segment of the floral population—the heterosporous plants—they, too, are sometimes widely dispersed. However, their distribution is more likely to be restricted to an area close to their site of reproduction.

"Because we have very little chance of determining whether a fossil spore performed a male or female function in the life cycle, unless cone studies have provided the knowledge, an arbitrary lower size limit, of $200\,\mu$ has been given for megaspores (Guennel, 1952). Dijkstra-van V. Trip (1946, p. 21) also review the size problem. The size limit suggested follows from using the Tyler sieve, with a mesh opening of about $210\,\mu$ to separate the fine from the coarse residue. Guennel further proposes the use of the term 'miospore' for all spores or sporelike bodies less than $200\,\mu$ in size. These could include isospores, microspores, small megaspores, pollen, and prepollen.

"The spores of *Spencerisporites* are isolated spores similar to those occurring in some of the cones of *Spencerites,* reported as eligulate and homosporous. Thus, on available evidence, the spores referred to *Spencerisporites* could be considered isospores. *Renisporites* spores, rather small in comparison with most megaspores, may be megaspores, isospores, or microspores. *Monoletes* and *Parasporites* are considered prepollen. However, most of the spores described here are megaspores of the arborescent and herbaceous lycopsids and lepidocarps.

"The larger spores show considerable differ-

ences in size. In *Cystosporites* three members of the original tetrad are abortive, much smaller than the fertile specimens and sometimes much different in aspect.

"Spores are characteristically either radially (trilete suture . . . [Fig. 106]) or bilaterally (monolete suture . . . [Fig. 111]) symmetrical, the type of symmetry being controlled by the division of the spore mother cell. For descriptive purposes, spores are oriented with reference to their original position in the tetrad grouping . . . [Fig. 106, 111]. The side of a spore toward, and including, the original areas of mutual contact in the tetrad is designated as proximal (side, surface, or hemisphere). The side of the spore external to, or away from, the center of the tetrad is distal. The axis of a radially symmetrical spore passes through the center of the original tetrad.

"The apices of the four spores at the proximal poles originally touched in the tetrad.

"Bilaterally symmetrical spores have one axis of symmetry through the long dimension of the spore and another through the proximal surface of the spore and the center of the distal surface. The contact areas are the two or three surfaces of mutual contact in the original tetrad.

"The trilete suture . . . or line of dehiscence forms along three radiating lines; the trilete ray refers to one extension of the trilete suture or its expression as lips or as a fold. The monolete suture forms along a single line . . . Some spores may possess very insignificant lips bordering the suture, others may have straplike lips . . . and still others may have lips surmounted by an apical prominence . . .

"Those features influenced by the contact relationship of spores during growth in a tetrad are designated as haptotypic by Wodehouse (1935). The contact areas may be bounded by arcuate ridges, flanges, or ornamentation which may or may not extend over the entire distal surface. Distal ornamentation, in reference to megaspores, is not generally strictly limited to the distal hemisphere, but extends distally from the arcuate ridges or contact areas. Specifically inherited characters, such as distal ornamentation, of which the type, shape, and size is relatively constant for any one species, are designated as emphytic by Wodehouse (1935).

"Some spores . . . possess bladders or membranous air sacs which are attached to the spore coat. Still others possess a wrinkled inner membrane . . . variously ornamented, referred to as endosporal membrane by Schopf (1938), the mesosporium of Dijkstra-van V. Trip, (1946) and Høeg, Bose, and Manum (1955). Original spore shape may vary from sac-shaped, to spherical, to distinctly oblate. The shape is usually more or less distorted by compression, the manner of compression being determined by both the original shape and the presence and kind of ornamentation. Thus, the compressional form is usually characteristic of a species, although other species may also show the same form."

h. Erdtman and Straka, 1961, p. 65: ". . . 'Spore' here and in the following means pollen grain or spore or both according to the context." [see GROUPS OF SPOROMORPHAE].

SPORE ANALYSIS, n. cf. palynology.

SPORE COAT, n., pl. -s. cf. exine, spore wall, sporoderm, sporodermis.

SPOREDISK, n., pl. -s. cf. corpus sporae. = body of the spore.

SPORE WALL, n., pl. -s. cf. exine, spore coat, sporoderm, sporodermis.

SPORIDESM, n. (Gr. desmos, a bond).
Jackson, 1928: "a pluricellular body becoming free like a spore, in which each cell is an independent spore with power of separate germination."

SPORIDIUM, n., pl. -a.
Jackson, 1928: "a synonym or diminutive of spore, or a granule which resembles a spore . . . ."

SPORODERM, n., pl. -s. (L. sporodermis; Gr. derma, a skin). cf. exine, spore coat, spore wall, sporodermis.
I. Jackson, 1928: "the integument of a spore."
II. Erdtman, 1952, p. 469: "(Bischoff, 1833): the wall of a spore."

SPORODERMIS, n. cf. afzelius layer, bacularium, columella, columella conjuncta, columella digitata, columella simplex, columellate, crassexinous, crassinexinous, crassisexinous, crassitegillate, cavea, cavate, caverna, cavium, cell contents, dissence, dissencelumen, ectexina, ectonexine, ectosexine, ektexina, ektexine, ektexinium, ektexospore, ektexospore lamella, ektonexine, endexina, endexine, endexospore, endexospore lamella, endo-cracks, endolamella, endonexine, endosexine, endospore, endosporium, epispore, episporic, episporium, euintina, euintine, exina, exine, exine index, exinium, exintina, exintine, exoexine, exolamella, exospore, exosporinium, exosporium, extine, general exine, ground substance, infratectal, infrategillar, insulating layer, intectate, intectate-reticulate, intectatus, interloculum, intexine, intine, intinium, intratectal, intrategillar, lamella, lamella conspicus, lamella oppressa, lamellar, laemelliform, matrix, matrix pollinis, medine, membrane, mesexine, mesexinium, mesine, mesonexine, mesospore, mesosporium, mesosporoid, mexina, mexine, murus sporae, nexina, nexine, nexinium, oncus, ornamentation, pattern, perine, perinium, perispore, perisporinium, perisporium, porphyritic

structure, prismatic structure, rod, rodlet-layer, sclerine, sclerinium, sclerosporium, sculpture, sculptine, sculptinium, sexina, sexine, spore-coat, spore wall, sporoderm, structura, structure, structure element, sublayer, synexina mihi, synexine, tectate, tectatus, tectoid, tectum, tectum perforatum, tegillum, tenuiexinous, tenui-nexinous, tenuisclerinous, tenuisexinous, tenuitegillate, winged.

a. Potonié, 1934, p 8: *Tr.*

Intine, Exine—The wall of fresh pollen grains consists of two parts which lie closely on top of each other and which are strongly chemically differentiated; the exine and the intine (Fritzsche). The *intine* is fragile and pliable and consists predominantly of pectin-rich cellulose. Hereby the pectin is especially enriched under the germinal apparatus (which see); it is strongly capable of swelling. The *exine* is stronger, less flexible, and strongly cutinized.

In fossil pollen which have been prepared from sediments in the described manner, the intine is no longer identifiable; also, attempts to indicate its presence with cloro-zinc iodide prove unsuccessful.

Lamellae—The exine consists of several layers. Jentys-Szafer (1928) has distinguished 5 in *Corylus* and *Betula* and 3 in *Myrica*. Two distinct layers are often recognized in the fossil material. They partly contrast each other in their color, take on dyes in different intensities, and are often separated from each other by a quite distinct fine line. In 1889, Mangin already recognized in *Spartium junceum* an exceptionally distinct, two-layered exine. Fossil material now demonstrates that in many cases subdivision into two principal layers can be done. If one of the two principal layers (in fossil material; without the usual further preparation) can occasionally be further distinguished as two distinct composite layers, then the line separating these *sublayers* or lamellae from each other is always much less distinct.

Exoexine, Intexine—In fossil pollen, two layers of the exine are frequently well differentiated and separated from each other by sharper lines: the exoexine and the *intexine*.

The term intexine was coined by Fritzsche (1832). H. Fischer considers it as unnecessary (op. cit. p. 15), since only in the *Cucurbita* are there two distinctly distinguishable membranes. The famous lid or stopper of the germinal-apparatus of the pollen of *Cucurbita* is found according to him in the outer membrane. [Potonié, 1934, p. 9: *Tr.*] We will see that the construction of the germinal-apparatus often owes its peculiarity to the existence of the two principal layers of the exine and that we are usually able to decide very clearly what is to be labeled as exoexine or intexine at the germinal-apparatus. A clear description of many relationships of the pollen-exine is hardly possible without conceptual recognition of the two principal layers of the exine.

Consequently H. Fischer was not right when he wrote: "All those cases where former observers, as Fritzsche, Mohl, Meyen, and Sehacht have noted more than two pollen membranes are mistakes."

The exine of *Spartium junceum* according to Mangin (1889) has two layers despite its slight thickness. If the pollen grain swells and the enrichment of pectin under the exitus becomes effective, the exoexine dissociates and makes way for a small hill of intexine.

Often the two layers are also differentiated from each other by the fact that the exoexine has a microscopically discernible structure, but the intexine does not.

Structure, Rodlets, Insulation Layer, Exolamella, Air Sacs—As far as possible, only the characteristics of cell wall expressed by certain designs in optical sections of the exine are called *structures*. Generally these structures seen in a cross section of the exine are in the form of many closely packed, radially arranged rodlets, which in contrast to the space between them have the ability to absorb dye-material, such as fuchsin, gentianviolet, etc., in strong intensity. H. Fischer and others recognized these rodlets as minute pillars which stand between the lamellae and hold them apart. An air-interspersed *insulation layer* originates in the exine. In fossil material, these conditions are often clearly recognizable. Thus, e.g., it is often shown very well in the case of the fossil exine of Pinaceae. This shows a construction in which at a cursory glance one would say that it is here (as an exception) the intexine which shows the just-mentioned rodlet construction. On a rather thin innermost lamella (which sits on top of the intine, which is not preserved in fossils) there is namely a rather strong layer of rodlets which are separated from the outside by a strong *exolamella*. Because of the strong frame of the exolamella, one is tempted to take this alone as the exoexine and to interpret the layer of rodlets, together with the innermost lamella, as endoexine. That this interpretation is incorrect can be shown in the construction of the *air-sac* of Pinaceae pollen which were hitherto not yet clearly understood. These air-sacs originate in the following way: the exolamella [Potonié, 1934, p. 10: *Tr.*] together with the shrinking rodlet-layer separate from the innermost lamella and inflate . . . [Fig. 547]. Note that we want to differentiate among other things an exoexine from an intexine, be-

cause in a cross section of the exine, a tangential separation-line is often clearly displayed (besides possibly an extant, very delicate tangential line which demarcates the sublayers or lamellae). It is not known at the present where one has to imagine this separation line [to be] in Pinaceae pollen. It is probably, however, that the border of the principal layers of exine is where the disconnections usually occur, i.e., below the rodlet-layer in Pinaceae pollen. This theory is supported by the fact that the structure of all fossil exines known so far, is always found in the exoexine. Pinaceae exine is no exception. It is striking only that the unstructured outer lamella of the exoexine is especially thick. By the expansion of the air-sacs, the whole rodlet layer actually becomes detached from the rather thin intexine, which follows from the fact that the wall which separates the air-sacs from the cell of the pollen grain no longer shows any structure and therefore, it consists only of that part of the wall which at the beginning has been called the innermost lamella. Therefore, we must label this part of the wall as intexine. The wall which separates the air-sac from the outside in cross section shows unaltered exolamella; there are small irregularities fastened to its innerside, the shrunken remains of the rodlet layer. The air-sac is nothing more than a considerable expansion of the gas-filled insulation layer. It is homologous to this layer. If one looks at the exine of Pinaceae pollen in plan view, then the cross sections of the rodlets can be seen in an arrangement which gives the impression that they are the lumina of a rather regular-shaped network, which, however, does not stand out in relief. That is why we are dealing with a structure and not with a sculpture. [Potonié, 1934, p. 11: *Tr.*]

Pila—*Pila* are very remarkable. In *Pollenites margaritatus,* for example, they are so close together that the heads of the clubs almost touch . . . [Fig. 654], giving an impression of a network in vertical view. By means of this confluent canopy, there arises below the heads an insulating air layer in which the stems of the heads stand. H. Fischer uses some examples to discuss the fact that the heads, or the upper ends of the sculpture elements, could be connected (e.g., by cross beams); in some cases they even coalesce to form an almost uniform membrane; this is our exolamella. By this, it is recognizable in what way the rodlet-impregnated insulation layers of the exoexine are morphologically related to sculpture. Fischer further shows that sometimes even two insulation layers can lie over each other.

b. Erdtman, 1943, p. 43–44: "The majority of pollen grains have two coats—an outer, the exine, and an inner, the intine (Fritzsche, 1837, p. 28). These terms may also be used in the description of the spores of mosses and ferns instead of the longer terms 'exosporium' and 'endosporium.' The intine and its eventual subdivisions will not be considered here for they are not found in the fossil state, at least not in peat, clay, brown-coal, etc. The exine, as pointed out by Fritzsche (1.c.), frequently consists of two layers, but the subdivision of the exine (*sensu lat.*) into exine (*sensu str.*) and intexine, as proposed by him, may lead to confusion. Therefore, some authors have used the term exoexine for exine *sensu stricto*. To avoid this unwieldy term as well as to have a convenient term in case the exine should exhibit three layers instead of two, we may speak of ektexine (for exine *sensu stricto*, or exoexine), mesexine (or a layer between the outermost and the innermost exine layers), and endexine (for intexine). The outer surface of the exine may sometimes be provided with some kind of sculpturing or ornamentation. The ornamentations of sculptured pollen grains are exceedingly varied: the exines may be echinate, reticulate, etc. . . . Unsculptured grains, lacking spines or projections of any kind, are known as psilate (Wodehouse, 1935)."

c. Dijkstra-van V. Trip, 1946, p. 22: *Tr.*

Structure of the Spore Wall—In the case of the megaspores one distinguishes the following layers (from the inside to the outside): endospore, mesospore (sometimes), exospore, and sometimes perispore. The endospore is never preserved in the fossils: it is a thin cellulose membrane in the case of recent spores. A mesospore, according to Fitting (1900), occurs in the case of *Isoetes* and *Selaginella*. Ergolskaya (1930) declares the small dark band which he notices in the section within the exospore (exine) incorrectly as the endospore (intine). Wicher (1934) described, in the case of megaspores which have been macerated from bituminous coal "for mere descriptive reasons" an "extremely thin, folded, hyaline membrane" as mesospore. This term was extended by Zerndt (1934, p. 11) to an important palaeobotanic concept: the mesospore is supposed to be a completely closed and homogenous spore membrane without a definite opening place; it is fixed to the exospore at the inside of the contact areas and during germination it is supposed to tear along the triradiate ridges of the exospore. Although we do not know the real mesospore, it is not very likely that it is an artificial phenomenon caused by the maceration as Sahabi (1936) thinks. Zerndt found it in the case of *Triletes brasserti* (1934, p. 24). In *Triletes triangulatus* and *T. brasserti* we have very distinctly

recognized the place where a possible mesospore is attached, a somewhat rough surface at the inside of the exospore at the place of the contact areas. Also Wicher mentioned these spore types among the species where a mesospore occurs. Nikitin (1934) believes that the limitation of the "Androtheca" of his Devonian megaspore [Dijkstra, 1946, p. 23: *Tr.*] *Kryshtofovichia africani* toward the true spore body is established by a thick granular mesospore.

The exospore is usually the carrier of the outward sculpture of the spore. It follows from the countless coal petrographic pictures where megaspores are cut in polished section that the exospore is definitely not homogeneous. Slater and Eddy (1932) even described some of their types as "double layered." We illustrate a specimen of *Triletes mamillarius*, in which by the process of peeling off the exospore an interior membrane becomes visible, which probably is nothing else but an interior part of the exospore. A photograph with infrared light by Leclerq (1933) indicates something similar. Also in the case of the specimen of *Triletes hirsutus* . . . it seems that an interior part of the exospore has become visible by too much maceration, it is shiny and partly transparent; we take *Sporites fumosus* Schopf (1938) for something similar, probably it is also the same spore type.

The exospore of a megaspore which has burst open shows mostly a dull, smooth surface at the inside. In the case of *Triletes brasserti* and *T. triangulatus* a wrinkled triangle appears within the contact figure (but much smaller than the contact figure), formed by three slightly bent lines which connect the ends of three seams which are visible at the inside (compare Stach and Zerndt, fig. 28).

The perispore in the case of Carboniferous megaspores appears only in one case as a closed membrane. In the *Triangulati*, a perispore surrounds the entire exospore which is often completely sculptureless . . . . [*Triletes triangulatus* Zerndt, *Triletes tricollinus* Zerndt, *Triletes triangulatus* Zerndt]. In some other megaspore types the perispore adds a lot to the outside of the spore. We think (in contrast to Zerndt's opinion, 1931, p. 173 and 1934, p. 10 and 19) that Ibrahim (1932, p. 27) (probably by accident) is right when he marks *Triletes brasserti* (his spore type *Zonales-sporites saturnoides*) as provided with a perispore and we completely agree with Sahabi (1936, p. 23) when he believes the zonal outgrowth and hairs of *Triletes brasserti*, *T. superbus*, and *T. dentatus* to originate from the perispore. Sahabi describes the ornamentation which originates from the perispore as to be fixed to the exosphere, but not belonging to it; removing it from the perispore does not result in a fracture at the surface of the perispore . . . [compare Fig. 567–569]; sometimes they fit exactly with somewhat larger or smaller outgrowths of the exospore, as for instance with the triradiate ridges, the arcuate ridges, possibly with the equatorial rims, swellings, thorns, or other places of attachment.

Zerndt's remarks at the description of *Triletes circumtextus* (1934, p. 19–20) appears to contradict his statement that only *Triletes angulatus* shows a perispore: "The different color of the zonal outgrowth and its strongly shiny aspect permit to suppose that the substance from which it is formed is not the same as that of the exospore . . . the zonal outgrowth can be detached without difficulty from the body of the spore . . . one may conclude perhaps that its origin may be different and more recent than that of the exospore." Participation of the perispore in the outer sculpture of the spore is, according to my opinion, also noticeable in *Triletes rotatus* and *T. mucronatus* and probably in *T. radiatus* and *T. tenuicollatus*.

d. Erdtman, 1948, p. 387: "The term 'sporoderm' (sporodermis), as suggested by Leitgeb (1883) and others, means the wall, all layers included, of pollen grains and spores . . ."

e. Erdtman, 1948, p. 267: ". . . [Fig. 541]. Outline section through an imaginary pollen (spore) wall to illustrate the tentative terminology used in this paper.

"Only the sclerine (vide postea) is shown. Different kinds of sculpturing are exhibited on the outer side of the wall. The normal complete sequence of layers is as follows (from the inside outwards):

A. Soft (malacodermic) layers (not preserved in fossil spores): intine (endosporium).
B. Entirely, or chiefly, hard (sclerodermic) layers (sclerine), as a rule preserved in fossil spores.
  I. Exine (exosporium).
    a. Nonsculptured exine: nexine (exinium nonsculpturatum).
      1. Endonexine (the lowest stratum in . . . [Fig. 541]).
      2. Mesonexine (marked with cross hatching in . . . [Fig. 541]).
      3. Ectonexine (marked with solid black in . . . [Fig. 541]).
    b. Sculptured part of the exine: sexine (exinium sculpturatum). Among the phanerogams the basic structure of the sexine seems to be small drumstick-shaped rods (pila), projecting at right angles from the nexine surface. Each pilum has a head (caput) and a rodlike pars

collaris, or 'baculum.' A typical pilate exine is shown in . . . [Fig. 541], left of the aperture. On the right side of the aperture are two sets of pila with their heads coalesced laterally ('baculate exine') followed by one set of rods (paxilla) without swollen tops. This exine is 'paxillate' (or 'spino-paxillate'), if the spines (vide postea) are also considered.

II. Perine (perisporium). A perine is formed when a periplasm with perinogeneous tendencies is present at the formation of the spore wall. It is sometimes difficult to decide whether a certain stratum or sculptural element is of a perinous or of an exinous nature. In such cases sculptine may be used as a provisional neutral term, embracing any strata, or fragments of strata, belonging to the sculptured part of the scleroderm i.e., either to the exine (n.b., the sexine), the perine, or to both. The two small triangular spinules in . . . [Fig. 541] on the right side of the aperture, may be referred to as 'sculptinous' if their nature cannot be otherwise ascertained. On the other hand, the two large spines on the same side of the figure differ so widely from the supporting rod layer that there can be no doubt as to their perinous nature."

f. Faegri and Iversen, 1950, p. 16: "The third layer, the *exine,* is formed of one of the most extraordinarily resistant materials known in the organic world. Apparently unchanged spore walls (consisting of the same or a closely related substance) are found in Paleozoic rocks where all other organic remains have been carbonized and distorted. Recent pollen grains can be heated to almost 300°C (Zetzsche, 1929, p. XXIX) or be treated with concentrated acids or bases with very little effect on the exine. According to Zetzsche (1.c.) and Vicari (1936) the substance is less resistant to oxidation. As a consequence of this great resistance the chemical composition of the exine is very inadequately known. According to Vicari (1.c.) the so-called sporopollenins are N-free substances, which are specifically different, as is the quantity of sporopollenin in different pollen species. It is interesting to note that the very resistant *Pinus* pollen also possess the greatest quantity of sporopollenin.

"The exine is highly variable. In some few pollen types (hydrogamous species) it seems to be absent altogether. Apart from these the simplest cases are those where the grain is enveloped in a thin, continuous cover, apparently consisting of one homogeneous sheet (*Larix*).

"Where the exine is more complex, it is possible to distinguish between two layers (Fritzsche, 1837), an inner and an outer, which are called, respectively, *endexine* and *ektexine.* The inner layer forms a continuous, homogeneous membrane, corresponding to the simple exine of the above-mentioned pollen type.* In contrast the outer layer always seems to consist of small elements (*granula,* Fritzsche, 1.c.), the development and distribution of which cause the extreme variability of the structure of the exine.

* In some grains the pores (cp. later) seem to form true holes in the endexine. However, it is very difficult to ascertain this; nevertheless, the continuous endexine may have thinner areas equivalent to fully perforated pores.

"In the very simple grains the ektexine is confined to isolated small knobs, [Faegri and Iversen, 1950, p. 17]: "which are scattered on the inner membrane (*Juniperus*). In other types these projections are more densely crowded (*Populus*) or have the form of rather conspicuous clubs (*Ilex*). More elaborate structures are formed when the granules are grouped into various patterns and fused laterally more or less completely, forming an open network (*reticulum, Armeria*). A contrast to the open structure of the above-mentioned exines is formed by those pollen grains in which the outer ends of the granules fuse, forming another, outer sheet which envelopes the whole grain covering all the other exine elements. If the granules assume the form of radially placed prismatic elements that are fused along their whole length, the ektexine is compact, but in most cases a very careful examination (preferably of sections) discloses that the granules are fused at their tips, or at their tips and bases, thus forming an outer membrane which is separated from the inner one by a cavity and is borne by small columns (*columellae*) which may in some cases fuse to form an inner reticulum (*Alisma*). In some pollen types the outer part of columellae is branched, forming intricate patterns (*Stellaria,* cf. . . . [Fig. 693–697])." [*see* ORNAMENTATION].

g. Erdtman, 1952, p. 18: "The spore wall (sporoderm) usually consists of two main strata, an inner, soft (malacodermatous) layer, intine, and an outer, hard (sclerodermatous) layer, sclerine.

"Sclerine is usually synonymous with exine. In the spores of certain plants (certain mosses, ferns, etc.), however, it also comprises an outer layer, perine. The presence of this layer seems to be due to the activity of a periplasmodium with perinogeneous properties. The physicochemical qualities of the perine differ more or less from those of the exine. It is, however, still impossible, particularly when dealing with pollen grains, to decide—without undertaking cytological investigations—whether a certain sporoderm layer or sporoderm element is exinous or ± perinous. As long as this cannot be definitely proved

'perine' can be classified under the same noncommittal heading—'sculptine'—as the outer, sculptured part of the exine. This has, however, not been observed in the present book in order to reduce the terminology: 'sclerine' is described as 'exine' even if it should happen to be ± perinous. Likewise 'sculptine' is always replaced by 'sexine.' [Erdtman, 1952, p. 19]: "Sexine (from 's' in sculptured, and 'exine') is the outer sculptured part of the exine. It often exhibits two layers, an inner (endosexine) and an outer (ectosexine). These terms are further explained below under 'Sexine Patterns.'

"The inner, nonsculptured part of the exine is known as nexine (from 'n' in nonsculptured, and 'exine'). It usually exhibits an outer, ± thick, not very refractive zone (ectonexine) and an inner, ± thin, more refractive zone (endonexine).

Table 2. *Sporoderm stratification—*

S P O R O D E R M	S C L E R I N E	E X I N E	PERINE		
			SEXINE	ECTOSEXINE	SCULPTINE
				ENDOSEXINE	
			NEXINE	ECTONEXINE	NEXINE
				ENDONEXINE	
			INTINE		

h. Erdtman, 1952, p. 463: "Exine (Fritzsche, 1837, p. 28); exinium (Strasburger, 1882, p. 135); exosporium; tegmen exterius; extine: the main, outer, usually resistant layer of a sporoderm. . . .

i. Thomson and Pflug, 1953, p. 17–18: [*see* LAMELLA].

j. Van Campo, 1954, p. 252: *see* Fig. 546.

k. Potonié and Kremp, 1955, p. 17: *Tr.* The individual *layers of the spore wall (sporodermis)* are not always distinguishable in the case of fossil material. It is established as a fact, however, that sometimes different layers are observable and that authors do not always have the same thing in mind when they use terms which have been created for the individual layers of the spore wall.

Intine, Exine, Exoexine, Intexine—One only occasionally recognizes more than one layer in coal thin sections. In the interior, there sometimes is a very thin layer which has been changed to

153

vitrinite, and it is interpreted as the *intine* which originally was composed of cellulose. Then the *exine* follows which appears as ± amber-yellow exinite. The exine again can show different layers, the *exoexine* and *intexine*. In thin section, one layer can be more distinctly punctate, maculose, streaked, cloudy, or even prismatic than the other layer . . . Toward the exterior the exoexine usually forms the sculptures of the spore surface.

The so-called *"prismatic structure"* is shown in the cross section of the exine as radially directed lighter "rodlets" which stand close to one another (see also below under columella). One particularly recognizes light-yellow, delicate, approximately radially directed elements in the area of the folding of the exine (about at the equator of the spore) and especially in the exoexine. This structure has never been found clearly visible on a spore, not even on all parts of the same polished cross section. So, for example, only the phorphyritic structure [Mörtelstruktur] is recognizable in the case of some photos of cross sections of megaspore exines in transmitted light. (This structure however is not shown in the sculpture elements which are attached to the walls . . . [*Zonalisporites ovalis* Stach and Zerndt]).

A distinct limit between intexine and exoexine is often not seen; however, the structural elements in the intexinous part of the exine are more tangentially arranged and in the exoexinous part more radially or irregularly arranged. [Potonié and Kremp, 1955, p. 18: *Tr.*]

The chart shows the ideal stratification of the sporoderm. Where possible, the oldest of the specific terms have been chosen in order that the palaeontologic publications of earlier authors remain readable.

On the other hand, structures are sometimes recognizable in the macerated material which are not visible in polished sections of a coal. For instance, when one observes the exine in plan view, the rodlet structure on the inside of the exoexine here becomes recognizable as punctation.

Porphyritic Structure—Usually, in thin sections one does not notice any rodlets, but only a phenomenon called porphyritic structure [Mörtelstruktur]. Often this is best visible in the intexine. As porphyritic structure we call (along with Haberlandt) a phenomenon where *micelle nuclei* or *puncta* are imbedded in the groundmass or matrix of the exine. As previously mentioned, the limit between int- and exoexine is only sometimes discernible in coal thin sections in transmitted light and then may be marked by a fine dark line. It is as if a delicate fissure had been formed here which is filled with a flaky dark-brown (that is vitrinitic) substance. This substance is similar to the substance occurring in breaks where the exine has folded (see above) [Potonié and Kremp, 1955, p. 19: *Tr.*] and is also similar to the substance occurring in the inner space of the spore, which appears as a line in cross section, and also in the intratectum.

In some cases the thin section shows different colors for the exo- and intexine. The intexine is somewhat more brownish or lighter.

The terms *endospore, mesospore,* and *exospore* are generally not used in this paper; occasionally, the term *perispore* is used (= epispore Russow 1872 = perine Strasburger 1882). For simplicity, we use only the terms appearing on the chart on page 18 since these can be used in the description of both pollen and spores.

Endospore—The *intine* corresponds to the *endospore* (it must be mentioned here that use of the term endospore is avoided, among other reasons, because it is sometimes quite differently defined).

Exospore—The exine includes the entire cutinized part of the spore or pollen membrane, that is the exospore with its mesospore (and also the perispore if present).

Perispore—The *perispore* (perine) often cannot be differentiated from the exoexine in fossil material. It is a secondary, additional, subsequent formation. In paleontological considerations, the exterior parts of the exine which are more or less easily lost, leaving no breaking-points but

		Perispore cf. BISCHOFF 1842, p. 594 (= Perine STRASBURGER 1882, ERDTMAN 1950)	
Sporodermis cf. BISCHOFF 1842, p. 595	Exine FRITZSCHE 1837 (= Exospore)	Exoexine POTONIE 1934 (= Ektexine ERDTMAN 1943 = Sexine ERDTMAN 1948)	Exolamella POTONIE 1934 (= Ectosexine ERDTMAN 1950)
			Isolierschicht POTONIE 1934 (= Endosexine ERDTMAN 1950)
		Intexine FRITZSCHE 1832 (= Endexine ERDTMAN 1943 = Nexine ERDTMAN 1948)	
	Intine FRITZSCHE 1837 (= Endospore)		

rather natural separation surfaces, are considered as the perispore. We use the term perispore only in this sense.

In using the term exospore, one has to remember that this term does not include the conception of the perispore; the perispore lies on the exospore.

Mesospore, Mesexine—The *mesospore* is an interior part of the exospore and might correspond with the intexine. In Zerndt, 1934, p. 10, the mesospore does not belong to the exospore. Other authors use the term mesexine, which is supposed to lie between intexine and exoexine. Van Campo's mesexine is the inner part of our exoexine. The *mesexine* separates from the intexine during formation of the saccus. Therefore, the mesexine is the insulation layer which consists of columellae; the endosexine of Erdtman.

Insulation Layer, Exolamella—As previously mentioned, the exine consists of intexine and exoexine. The exoexine is composed of an inner *insulation layer* and an outer *exolamella*. The sum of the more or less loosely standing, radially directed rodlets or *columellae* which support the exolamellae, and their interspaces, is to be called the insulation layer. The columellae form a forest of pillars with air space between them.

l. Harris, 1955, p. 25: "*Sporoderm*: all layers of the spore wall taken together."

m. Traverse, 1955, p. 92: "*Exine*: This is the 'wax'-impregnated outer coat of pollen grains. It has usually been referred to as *the* outer layer of pollen grains, by comparison with the internal layer, or intine. However, it is typically of complex structure, with at least two layers:

(1) *Endexine*: The distinct inner layer of the exine. The endexine is generally of simple, homogenous structure. Sometimes, the exine consists only of endexine.

(2) *Ektexine*: The outer layer of the exine; it is composed of small grains or elements. According to the disposition of these elements on the inner layer, or endexine, the ektexine may be:

(a) *Intectate*: If the ektexinous elements are not united in such a way as to form a continuous cover over the endexine. For example, the ektexinous papillae that cover many pollen grains.

(b) *Tectate*: If the ektexine forms a continuous coat outside of the endexine. This coat is formed by the fusion of ektexinous elements and may be rather homogenous or very complex in structure, as for instance when the coat is made of columns fused at the top to give a columellate structure. The

outer, continuous layer of ektexine in tectate types is the *tectum*. In the instance of columellate structure, the tectum is the layer on top of the columns."

n. *Kupriyanova, 1956*, p. 1212–1216 : *Tr.* [*see* Fig. 540]. It is impossible not to mention here about another layer, covering not only individual pollen grains but the whole tetrad, and which can be observed in pollen tetrads of Juncaceae, Cyperaceae, Thurniaceae, Orchidaceae, Ericaceae, and in certain other families. The layer mentioned is nothing but the membrane of the maternal cells. To distinguish it from the exine of the singular pollen grain it can be called the general exine or synexine (synexina mihi).

The synexine is well stained by basic fuchsin and perhaps it can be referred to the group of ektexine layers.

In summarizing the above discussion we have to conclude that it is impossible to apply the names ekt- and endexine to concrete layers, because under the name of ektexine is combined a whole group of layers, (1) the uppermost granulated, easily soluble layer, first described by Afzelius in 1955, (2) one or several columnar layers (sexina), (3) the lowest coarse-grained loose layer (mexina), and (4) the general exine of tetrads (synexina).

The names, proposed by Erdtman, sexine and nexine are proper and must be retained, but the content given by them in our opinion must be somewhat changed.

The scheme given by us in . . . [Fig. 540] shows differences in views of various palynologists on the structure of the membrane of pollen grains.

With a model such as this . . . [Fig. 540] in the margins are proposed the following terms.

For *acetolysis-resisting ektexine* group of layers:

1) synexine (synexina)—it is a membrane of the maternal cell membrane covering the tetrads of some pollen grains.

2) mexine (mexina)—it is a layer of exine, situated under the sexine, which is stained by fuchsin most intensively and found in pollen of many Dicotyledoneous plants. The term sexine must be used in designating the upper layer of the exine consisting of structural elements (it includes tectum as well as bases of columns in some instances). The term nexine must be retained for the innermost homogenic layer of exine that is not stained with fuchsin.

For the *intine group of layers which is unstable during the acetolysis*:

1) exintine (exintina)—the outside layer of the intine, frequently having subapertural thick-

enings (the term proposed by Fritzsche).

2) euintine (euintina)—internal layer of the intine that participates in the formation of the pollen tube.

o. Krutzsch, 1959, p. 37: *Tr.* The designation of the inner layer as, e.g., endexospore and of the outer one as ectexospore* is really inopportune because these layers are not entirely *homologous* in the different groups of spores, rather be *analogous*. One must logically assume that similar designations of different structure elements among diverse groups of *Sporites* (and not only here!) are homologous.

* or also as tectum or exolamellae, etc.; cf. R. Potonié and Klaus, 1954, p. 525.

Therefore, it is much more useful to speak, when needed, at this time, only of an outer, middle, inner layer. If in the description of the spore coat, still other layers or groups of lamellae have to be determined, their position can be determined from the outside to the inside according to their position, or perhaps be designated by a letter, such as a, b, or c (cf. Thomson and Pflug, 1953, p. 58). This is valid in a still higher degree for the essentially more complicated walls of the pollen. Here "morphological unity designations," which *presume* homology are much less useful than in the strongly less differentiated spore walls.

Layer stratification: Cavities may be present between the different layers of the spore wall. They may be more or less complete, that means they are developed everywhere or they may occur only locally. In respect to the degree of the dissociation of the layers or lamellae one can dissociate analogous to the situation in the pollen:

a) a solid unit

b) a combined layer

c) an interloculum

In spores the latter, if restricted at the equatorial region, is in part called a cingulum.

p. Rowley, 1959, p. 19–20 and Fig. 588a: "Afzelius' 1955 study on the fine structure of the pollen wall in *Clivia miniata* has provided the only high resolution electron micrographs which give information on both the intine and exine. Evidence is provided by Afzelius for a morphologically distinct layer between intine and exine. By careful light microscope analysis, Erdtman, (1948 . . .) had indicated that such a layer existed as a thin, highly refractile zone of the nexine and termed it endonexine. In *Clivia* this endonexinous layer is laminated and shows a greater electron density than either the exine or intine. Ehrlich [Exptl. Cell Res. 15 (3), 463 (1958)] has obtained electron micrographs of the *Saintpaulia ionantha* pollen wall which clearly

show that the thin endonexine of the extracolpate region widens under the pore in this species to form a thick lens-shaped layer.

I have used the term *mesine* (endonexine, Erdtman; mexine, Kuprianova, 1956) for this layer. *Mesine* reflects an integral association with both the intine and exine without suggesting that it is one or the other. In keeping with Fritzsche's use of -ine as a common ending for *exine* and *intine* the intermediate, lamellated, electron-dense layer may be termed the mesine, the prefix being derived from Kreek *mesos,* in the middle.

In my Commelinaceae material no unequivocal example of the mesine was found. Sections of *Tradescantia paludosa* and *Commelina coelestis* had a thin (50 mμ or less) electron-dense layer which could be interpreted as mesine overlying the intine, under and occasionally between apertural insulae and spines (. . .). Since the mesine is described as a laminated layer and observation here reveals only one electron-dense lamina, it seems best to interpret this layer as a thin endexine.

I have found a prominent mesine in the pollen wall of several dicotyledonous genera fixed and embedded by the same methods used for pollen of the Commelinaceae; thus it is likely that the absence of an easily identifiable mesine in pollen of the Commelina Family is not due to technique. It can be tentatively concluded that the mesine in pollen of this family is either minimal or absent.

In *Clivia* Afzelius has described and illustrated mesine-like laminae above the endexine. The aperture region of *Commelina coelestis,* in my material, also shows a mesine-like laminated material between the spines of the exine."

q. Larson, Skvarla, and Lewis, 1962, and Fig. 545: "Sculptured exines, as observed in the electron microscope, have been found, with exceptions, to be composed of two layers which are homologous to the ektexine and endexine as defined by Faegri on the basis of optical microscopy. Where basic fuchsin and electron staining have been compared identical stratification patterns were obtained: in sculptured exines lacking observable stratification the exine appears to consist entirely of ektexine. Pollen with more than two exine layers has not been observed in our laboratory, thus an unequivocal example of a nexine 3 layer has not been demonstrated.

In describing the basal non-sculptured portion of the ektexine the authors have used the term, foot layer, though they recognize that with *Photinia* or maize exines this term may appear inadequate or awkward to some. The term does have an acceptable operational value, and the creation of a new term at this time does not seem advisable.

Recognition of exine subdivision on the basis of optical and electron stain reactions and coinciding fine structural patterns rather than arbitrarily on sculpturing is inescapable. In our observations the boundary between the ektexine and endexine has always been found to be without transition in mature exines. The precise boundary between the layers, and the number of occurrences of physical separation of the layers observed in our laboratory reinforce the stain and fine structural data. If further evidence of the ontogenetic individuality of exine strata as seen in *Parkinsonia aculeata* (Larson and Lewis) is obtained, recognition of the significance of stain differentiation will be mandatory whatever the layers are called.

That pollen grains similar in ornamentation and identical in the thickness of the non-sculptured portion of the exine may have basic differences in strata composition makes it imperative that taxonomic studies take cognizance of stratification. To be sure, the taxonomic value of pollen morphological studies will be increased if such observations become standard.

Sitte (1957) has suggested that differing exine resistance to corrosion is based on both chemistry and fine structure. The endexine patterns observed in *Ricinis* and *Saintpaulia* pollen tend to confirm this statement. A loosely lamellated and discontinuous endexine will present a greater surface area for chemical or biological activity and will subject to greater mechanical disruption than will a non-lamellate and continuous endexine.

Therefore, acetylated or otherwise chemically treated pollen may not always yield the most accurate data on exine stratification. This has been noted by Van Campo (1961) in studies of the aperture membrane."

r. Saad, 1963, p. 18–19, Fig. 570–577: "All . . . findings emphasize the existence of an intermediate layer between the exine and the intine [the medine]. This layer is usually lamellated and does not resist acetolysis. With fuchsin it does not or [only] faintly take the stain. Some authors like Erdtman and Afzelius consider it as the inner layer of the nexine (n2), others like Sitte and Bailey interpret it as the outer layer of the intine but Ehrlich interprets it as an independent layer.

"Working on pollen morphology and sporoderm stratification of *Linum* (Saad 1961) and Linaceae (Saad 1962), the author used acetolyzed as well as non-acetolyzed grains (only chlorinated and stained with an aqueous solution of ruthenium red). Thin sections in the fresh pollen grains stained with fuchsin were also studied.

The nexine has been found to consist of an outer deeply staining thin layer which corresponds to the foot layer of Faegri and an inner lightly staining one. In the colpate species of *Linum* a distinct intermediate layer, situated between the intine and the nexine, has been noticed. This layer has the following characters: 1. It does not resist acetolysis. -2. It swells during chlorination and bulges at the colpi forming plug-like structures. -3. It stains readily with ruthenium red but hardly with fuchsin. -4. It is usually thinner in the mesocolpia and thicker under the colpi (forming the main part of the colpus floor). -5. During sectioning rupturing often occurs between this layer and the intine. -6. It shows a faint lamellation . . ." [Saad, 1963, p. 20]. "The exine stains readily and deeply with basic fuchsin, does not take ruthenium red and resists acetolysis. It decreases in thickness over apertures forming thin aperture membranes . . . The intine . . . stains with ruthenium red but not with fuchsin . . ." [Saad, 1963, p. 28]. ". . . The destruction of medine by acetolysis and behavior towards stains indicates that this layer is not entirely composed of acetolysis-proof sporopollenine, as is the exine, but is encrusted with pectins or other compounds. In contrast to the cellulosic intine, the medine being non-cellulosic takes basic fuchsin." [Saad, 1963, p. 35–36]. "That the mesine of Rowley corresponds to the endonexine of Erdtman seems unacceptable since the endonexine is usually thin and highly refractile, whereas the mesine is a lamellated, thick, non-refracting layer. Undoubtedly the mesine corresponds to the present medine. As to the nomenclature of this intermediate layer the author is inclined to call it medine (Saad 1961 [*Grana Palynologica,* vol. 3, p. 109–125]). The word medine (from medius) reflects the meaning of a layer intermediate in position between the exine and the intine."

SPOROMORPH, n., pl. -s. cf. microspore, miospore, monospore, palynomorph, pollen, pollen grain, pollen granule, pollenospore, polospore, prepollen, small spore, spora, spora dispersa, spore, sporomorpha, sporomorphidium.

I. Erdtman, 1947, p. 107: "In studies of Quaternary deposits a pollen analyst is often confronted with pollen species which are more or less variable as to certain features, such as the number of apertures. Some grains may have three colpae, others four, five, or six, etc.; the grains could in other words be referred to different shape groups, or 'sporomorphs' (sporomorphae, sing. sporomorpha; abbreviation, spm, this term is linguistically of the same type as 'nothomorphae' suggested by Melville, 1939). A detailed diagnosis of any extant pollen species must, therefore, consider all available sporomorphs of the species in question."

II. Potonié, 1952, p. 143: *Tr.* Because the sporomorphs [Gestaltsgruppen] are not identical with the species of the natural system [of plants], Erdtman has introduced the term sporomorphae for the units of the artificial system. The *Sporites* cannot be subdivided in species but only in sporomorphae. In an artificial system the sporomorphae are treated formally in the same way as species.

SPOROMORPHA, n., pl. -ae. cf. sporomorph.

Erdtman, 1952, p. 469: "genetically fixed spore type of a plant species.* The term can be used in an abstract as well as in a concrete sense (as designation of spore types, or of individual spores belonging to these types)."

 * Genetically fixed pollen or spore types of a plant species may, if necessary, be referred to as 'sporomorphae' (s. str.). Genetically not fixed subtypes (modifications) of a sporomorpha as found, e.g., in *Dicliptera javanica, Nothofagus pumilio* (number of apertures 4, 5, 6, or 7; cf. p. 177), and *Thunbergia grandiflora* may be referred to as sporomorphae (s. lat.) or 'sporomorphidia.' The term sporomorpha and the term sporomorphidium can be used in an abstract as well as in a concrete sense (as designation of spore types, or of individual spores belonging to these types). In palynological investigations of fossil material (particularly Tertiary and older) it is often impossible to make a distinction between sporomorphae and sporomorphidia. Morphologically ± different 'sporomorphs' may sometimes represent only one species. On the other hand the representatives of a certain 'sporomorpha' may not at all be the product of one species only; in spite of their uniform appearance they may just as well be the product of several species or come from plants belonging to different genera or even to different families, Erdtman, 1947c.
sdb. Thomson and Pflug, 1953, p. 17.

SPOROMORPHIDIUM, n., pl. -ia. cf. sporomorph.

Erdtman, 1952, p. 469: "genetically not fixed spore subtypes (modifications) of a sporomorpha. The term can be used in an abstract as well as in a concrete sense (as designation of spore subtypes, or of individual spores belonging to the subtypes).

SPOROPHYLL, n., pl. -a. (Gr. phyllon, a leaf). cf. sporophyllum.

Jackson, 1928: "a leaf which bears spores;" "a leaf-like division of the thallus of an Alga bearing fruit, as in *Carpoclonium*."

SPOROPHYTE, n., pl. -s. (Gr. phyton, a plant).

Jackson, 1928: "in Ferns and Mosses, the plant in the life-cycle of alternation which produces spores."

SQUARE TETRAD, n., pl. -s. cf. aggregation, tetrad, tetragonal tetrad, spore.

STENOPALYNOUS, adj. (Gr. stenos, narrow, little). cf. figura, Groups of Sporomorphae, morphography.

Erdtman, 1952, p 469: "said of plant families, etc., characterized by ± slight variation in spore types

(their apertures, sporoderm stratification, etc.). Examples: Cruciferae and Gramineae . . . [Fig. 100–104]."
sdb. Pike, 1956, p. 51.

STEPHANAPERTURATE, adj. (Gr. stephanos, a crown). cf. aperture, colpate, cribellate, furrow, pore.

Erdtman, 1958, p. 4: "more than three apertures." [*see* GROUPS OF SPOROMORPHAE].

STEPHANOCOLPATE, adj. cf. colpate, furrow.

Faegri and Iversen, 1950, p. 128: "more than 3 furrows . . . all furrows meridional . . . without distinct pores or transversal furrows." [*see* GROUPS OF SPOROMORPHAE].

STEPHANOCOLPORATE, adj. cf. colpate, colporate, colpus, furrow, Groups of Sporomorphae, porate.

I. Faegri and Iversen, 1950, p. 128: "More than 3 furrows . . ." [*see* GROUPS OF SPOROMORPHAE].

II. Traverse, 1955, p. 95: "with five or more (four or more according to Iversen and Troels-Smith . . .), meridional furrows, each having an equatorial germ-pore or transverse furrow. Note that the writer uses as a prefix the number of furrows the form possesses." [*see* GROUPS OF SPOROMORPHAE].

STEPHANOPORATE, adj. cf. aperture, cribellate, porate, pore.

a. Faegri and Iversen, 1950, p. 128: "more than 3 pores . . . free . . . no furrows, restricted to the equatorial area." [*see* GROUPS OF SPOROMORPHAE and Fig. 70].

b. Traverse, 1955, p. 95: "n-Stephanoporate: having more than three pores, all on or very nearly on the equator. Note that the writer modifies the Iversen and Troels-Smith system somewhat to indicate by a prefix the number of pores."
sdb. Ingwersen, 1954, p. 41.
  Beug, 1961, p. 13.

STEPHANOTREME, adj. cf. aperture.

Erdtman and Straka, 1961, p. 65: "Spores with four to many zonally (monozonally or pleozonally) distributed apertures may be referred to as stephanotreme (monozono-stephanotreme or pleozono-stephanotreme; in the latter there are four or more than four apertures in each zone)."

ST-PATTERN, n. (Suprategillar-Tegillar Pattern). cf. LO-pattern.

Erdtman, 1952, p. 469: "pattern(s) due to S- (suprategillar) and T- (tegillar) elements."

STRAIGHT, adj.

Couper and Grebe, 1961, p. 2 and Fig. 172: [*see* SHAPE].

STRIA, n., pl. -ae. (L. a furrow).

a. Jackson, 1928, p. 1: "markings on the valves of Diatoms which present the appearance of lines;" "the spiral ridges of the oospore in Charads (Groves)."

b. Faegri and Iversen, 1950, p. 163: "Grooves between elongated sculpturing elements in striate grains."

c. Erdtman, 1952, p. 469: "narrow grooves ($\pm$ parallel; length at least twice the breadth), separated by ridges (lirae)."

sdb. Couper and Grebe, 1961, p. 8; [*see* ORNAMENTATION and Fig. 767].

STRIATE, adj. (L. striatus). cf. canaliculate, cicatricose, costate, ribbed, rugulo-striate, strio-rugulate.

a. Jackson, 1928: "marked with fine longitudinal parallel lines, as grooves or ridges."

b. Erdtman, 1947, p. 106: "striate; with streaks ('striae')."

c. Faegri and Iversen, 1950, p. 27: "Elements $\pm$ parallel. [*see* ORNAMENTATION and Fig. 704–706].

d. Kosanke, 1950, p. 11: [*see* Fig. 640].

e. Thomson and Pflug, 1953, p. 23: *Tr.* Elongated sculpture elements which are dominate in one direction.

f. Harris, 1955, p. 18: "Linear or 'fingerprint' patterns. The sculpture elements, though minute, exhibit parallelism . . ." [p. 20]: ". . . Striate, if excluded from group (II) . . . because of a vertical axis not less than 1 μ in diameter, would have to be considered with this group, with costate, rugulostriate, or strio-rugulate (*Thelypteris uliginosa*) as available alternative." [*see* ORNAMENTATION].

sdb. Iversen and Troels-Smith, 1950, p. 46; [*see* ORNAMENTATION].

Beug, 1961, p. 13.

Couper and Grebe, 1961, p. 8; [*see* ORNAMENTATION and Fig. 766–768].

STRIATION, n.

Jackson, 1928: "of cell-wall, markings believed to be due to the manner of formation in bands by the protoplasm."

STRIATO-RETICULATE, adj.

Erdtman, 1952, p. 469: "a sculptural pattern $\pm$ intermediate between striate and reticulate."

STRIATUS, adj. cf. striate.

STRIO-RUGULATE, adj. cf. striate.

STRUCTURA, STRUCTURE, n. cf. ornamentation, pattern, texture.

I. Jackson, 1928: "The peculiar organization of plants, with special modifications."

II. a. Potonié, 1934, p. 9: *Tr.* In the following, the structure of the exoexine is clearly differentiated from its sculpture. This is absolutely necessary for the description of certain specimens with which we are dealing. Nevertheless, it must be said that structure and sculpture are transi-

tional. One is often inclined to speak of structures in the case of pollen with less extended membranes and then recognizes how the structure, with strong extension of the exoexine, is broken down into sculpture-elements. Consequently one could define the exoexine in the sculptured pollen-membranes as that part which consists of the structure-elements. The structure-elements as we will see, may also still be connected laterally or they may consist of elevations which are completely independent from one another.

As far as possible, only the characteristics of cell wall expressed by certain designs in the optical sections of the exine are called *structures* . . . [Potonié, 1934, p. 10: *Tr.*]

Inner structure—The vertical view of the outer-layer of the air-sac of Pinaceae pollen gives the impression of a coarser and more irregular network to a system of lines (as compared to the remaining surface) where by the single lines are often quite inexactly limited and leave between them irregular areas or lumina. This picture becomes understandable if one applies it to the shrunken remains of the rodlet layer which cling to the inside wall of the air sac and which are pulled apart from each other by the expansion of the air sac. According to that, the coarse net-structure arises from the irregularities which are arranged on mesh lines on the inside of the outer wall of the air sac. This net-structure is the *inner sculpture* of the outer air sac wall which originates from the structure of the exine. [*see* ORNAMENTATION].

b. Erdtman, 1943, p. 52: "structure, texture; different patterns, in surface view usually more or less 'granular,' not produced by eventual sculpturing of the exine but by formative elements within the exine." [*see* ORNAMENTATION].

c. Faegri and Iversen, 1950, p. 18: "In accordance with Potonié (1934) we distinguish between the *structure* (*texture*) and the *sculpturing* of the exine. The term structure comprises all those characters which are due to the form and arrangement of the exine elements inside the tectum. It also comprises the form and arrangement of the individual elements in intectate types."

d. Iversen and Troels-Smith, 1950, p. 34–35: *Tr.* Pollen [Morphologic Definitions . . . .] *Structure of exine.*

ex = exina = exine: the exterior, very resistant, one- or two-layered membrane of a pollen grain.

matrix: the homogeneous fundamental substance of the exine.

gran = granula: sharply limited grains, rodlets and so on (structure elements) which are imbedded into or deposited on the homogeneous ground substance.

end = endexina = endexine: the inner homogeneous layer of two-layered exine.

ekt = ektexina = ektexine: the exterior continuous, or discontinuous layer of a two-layered exine which is constructed of granula.

tec = tectum: the exterior continuous skinlike part of an ektexine which more or less completely covers the endexine.

tec (perf) = tectum perforatum: tectum with holes.

col = columellae: the columnlike ektexine elements (granula) which support the tectum.

col (simpl) = columellae simplicea: simple columellae.

col (dig) = columellae digitatae: distally branched columellae.

col (conj) = columellae conjunctae: distally united groups of columellae. Also the granula of intectate pollen grains can be called columellae if they occur in combined structures (for instance in a reticulum).

cav = cavea: the cavity within the exine which originates by the separation of ektexine from endexine. [see Figs. 542, 624–626, and 690].

e. Pokrovskaya, et al., 1950: *Tr.* Structure, texture. The internal structure of the exine of a microspore, usually more or less granular and not affecting the outline of the pollen grain (e.g., the shield of pollen grains of *Picea*), in contrast to sculpture.

f. Cranwell, 1953, p. 17: "(= texture): pattern of granules endemic within the ektexine."

g. Harris, 1955, p. 27: "sculpture patterns are described in this manual on surface characters without any attempt to separate structure and sculpture."

h. Traverse, 1955, p. 94: "Internal organization of ektexine. This has two slightly different senses. If the ektexine is intectate, that is, if the ektexine is of separate elements not forming a continuous coat, the structure is the internal structure *and* external form of the elements of the ektexine. If the ektexine is tectate, the structure is the arrangement of the ektexinous elements in and beneath the tectum, as well as the internal structure of these elements themselves . . . It is evident that sculpture and structure can be quite different in a tectate pollen grain."

i. Krutzsch, 1959, p. 38–39: [see ORNAMENTATION].

sdb. Selling, 1947, p. 77; [see TEXTURE].

    Thomson and Pflug, 1953, p. 21–23; [see ORNAMENTATION].

    Potonié and Kremp, 1955, p. 13; [see ORNAMENTATION].

Beug, 1961, p. 13.

STRUCTURE ELEMENT, n., pl. -s. cf. ornamentation, sporodermis.

STYLOSPORE, n., pl. -s. (Gr. stylos, a column).

a. Jackson, 1928: "a spore borne on a filament."

b. Melchior and Werdermann, 1954, p. 14; [see SPORE (Exospores)].

SUB-, (L. under or below).

Jackson, 1928: "in compounds usually implies an approach to the condition designated, somewhat, or slightly."

SUBECHINATE, adj.

a. Wodehouse, 1928, p. 934: "as under Echinate, but with spines small and rounded on top. Example: *Ambrosia.*"

b. Wodehouse, 1935, p. 544: "provided with short and sometimes rounded spines, e.g., the grains of *Ambrosia elatior* . . . [Fig. 736]."

SUBECHINOLOPHATE, adj. cf. lophate.

I. Wodehouse, 1928, p. 934: "With a spiny surface partially thrown into ridges which are not sharply defined; intermediate between Echinate and Echinolophate. Examples: *Catananche caerulea, Vernonia noveboracensis.*"

II. Wodehouse, 1935, p. 544: "lophate, with crests bearing reduced spines, e.g., the grains of *Stokesia laevis* . . . [Fig. 761]."

SUBEQUATORIAL, adj. cf. equatorial, figura, pore, shape classes.

SUBISOPOLAR, adj. cf. polarity, shape classes.

I. Erdtman, 1947, p. 113: "hemispheres slightly different; example: tricolpate spore with colpae meeting in one pole." [see GROUPS OF SPOROMORPHAE].

II. Erdtman, 1952, p. 470: "in subisopolar spores there are certain ± slight differences between the distal and the proximal face (one may, e.g., be very convex, the other less convex, plane, or even concave). If there are other differences, e.g., in number and arrangement of apertures, the spores are heteropolar. Example: . . . [Fig. 534–537]." [see POLARITY].

sdb. Pike, 1956, p. 51.

SUBLAYER, n., pl. -s. cf. lamella, membrane.

SUBLOPHATE, adj. cf. lophate, ornamentation.

a. Wodehouse, 1928, p. 934: "With the surface partly thrown into ridges or crests which are imperfectly defined." [see SUBECHINOLOPHATE].

b. Wodehouse, 1935, p. 544: "With the surface thrown into ridges which are imperfectly defined, e.g., in the grains of *Catananche caerulea* . . . [Fig. 747]."

SUBOBLATE, adj. cf. shape classes.

a. Erdtman, 1943, p. 45: "P:E ratio 0.88–0.75 (7:8–6:8)." [see SHAPE CLASSES].

b. Erdtman, 1952, p. 470: "term used exclusively in describing radiosymmetric, isopolar spores with the ratio polar axis: equatorial diameter 0.75–0.80 (6:8–7:8)."
  sdb. Selling, 1947, p. 77.
  Pokrovskaya, et al., 1950.

SUBPILATE, adj. cf. pilate.
  Erdtman, 1952, p. 470: "with small, ± piloid sculptural elements."

SUBPROLATE, adj. cf. shape classes.
  a. Erdtman, 1943, p. 45: "P:E ratio 1.33–1.14 (8:6–8:7)." [see SHAPE CLASSES].
  b. Erdtman, 1952, p. 470: "term used exclusively in descriptions of radiosymmetric isopolar, spores where the ratio polar axis: equatorial diameter is 1.14–1.33 (8:7–8:6). Examples: . . . [Fig. 524–526]."
  sdb. Selling, 1947, p. 77.
  Pokrovskaya, et al., 1950.

SUBPSILATE, adj. cf. baculate, flecked, maculate, papillate, punctate, scabrate.
  a. Erdtman, 1948, p. 268: "As far as can be judged from an examination using highest magnification (2 mm apochromatic objective), the exine is subpsilate, i.e., almost smooth, without clearly defined sculpturing."
  b. Erdtman, 1952, p. 470: "almost smooth . . . ."

SUBRETICULATE, adj. cf. reticulate.
  Erdtman, 1952, p. 470: "with a very fine ± reticulate(?) pattern."

SUBSACCATE, adj. cf. cavate, pouched, rugulosaccate, vesiculate, winged.
  Erdtman, 1952, p. 470: "in subsaccate spores the sexine is loosened locally from the nexine (with bacula or ± baculoid elements as in the true sacci sticking to the under surface of the tegillum [or tectum]). Examples: . . . [Fig. 560–562]."

SUBSPHEROIDAL, adj. cf. shape classes, spheroidal.
  a. Faegri and Iversen, 1950, p. 163: "shape classes (q.v.), index 1.33–0.75."
  b. Erdtman, 1952, p. 470: "(Faegri and Iversen, 1950): suboblate, oblate spheroidal, prolate spheroidal, subprolate."
  c. Pike, 1956, p. 51: "The shape of a pollen grain when the ratio between the polar axis and equatorial diameter is between 0.75 and 1.33. 0.75–0.88 (suboblate); 1.00–0.88 (oblate-spheroidal); 1.14–1.00 (prolate spheroidal); 1.14–1.33 (subprolate)."
  sdb. Ingwersen, 1954, p. 41.

SUBTECTUM, n., pl. -a. cf. contact figure.
  Potonié and Kremp, 1955, p. 12: Tr. The lower part of the intratectum, that is of the interior of the tectum. [see Fig. 141–147].

SUBTRIANGULAR, adj.
  Norem, 1958, [see SHAPE CLASSES and Fig. 167j].

SUBZONOSULCATE, adj. cf. colpate, furrow.
  Erdtman, 1947, p. 113: ". . . as in zonosulcate, but with the sulcus-zone interrupted at one or more places." [see GROUPS OF SPOROMORPHAE].

SULCATE, adj. (L. sulcatus, furrowed). cf. anacolpate, colpate, dicolpate, furrow, monocolpate, trichotomocolpate.
  a. Jackson, 1928: "grooved or furrowed."
  b. Erdtman, 1945, p. 190–191: "As is well known, the furrow of a monocolpate pollen grain is not strictly homologous to the furrows of a tricolpate pollen. In monocolpate grains the furrow is borne on the distal part of the grain whilst in tricolpate grains the furrows are developed as meridional furrows crossing the equator of the grain at right angles. If it be considered that 'colpa' should not denote two not strictly homologous features, grains of the typical monocotyledonous monocolpate type may be termed 'monosulcate,' or 'trichotomosulcate,' provided the 'sulcus' presents a three-slit opening.
  "Terms such as 'polycolpate' ('nonacolpate,' etc.) may indicate pollen grains of two types, either grains with many equatorial (meridional) furrows, or grains with many furrows equally distributed over the general surface of the exine. Misinterpretations are avoided if the term 'colpae' is used to indicate furrows arranged more or less meridionally, and confined to the equatorial, more or less extended zone of the grain. Although originally applied by Potonié (1934) in another sense, the term 'ruga' may be used to denote furrows equally distributed over the surface of the grains. Of the three terms suggested 'ruga' is shortest, 'colpa' intermediate, and 'sulcus' longest. As to the features, which these terms are supposed to denote, the rugae in a polyrugate grain, as a rule, are shorter than the colpae in, e.g., a grain of the common tricolpate type. The colpae, in their turn, are as a rule comparatively shorter than the sulcus in a typical monocotyledonous monosulcate pollen grain."
  c. Selling, 1947, p. 77: "(pollen) with a sulcus or sulci. Prefixes: mono- and di-, denoting the number of ± straight sulci; also trichotomo- denoting one triradiate sulcus."
  sdb. Norem, p. 669.
  Beug, 1961, p. 10.

SULCOID GROOVE, n., pl. -s. cf. furrow, sulcus, tenuitas.

SULCULUS, n., pl. -i. cf. colpus, furrow, sulcus.
  Erdtman, 1952, p. 470: "± sulcoid apertures, parallel to the equator and usually situated between the latter and the distal pole. If apically united

the sulculi form a zone, or ring, parallel to the equator (as in the case in zonisulculate pollen grains). Example: . . . [Fig. 291–293, 295–300]."

SULCUS, n., pl. -i. cf. aperture, colpus, furrow, monocolpate, sulcate.

I. Jackson, 1928: "small grooves or fossulae in some Diatom valves;" "lamellae of certain Fungi (Lindley);" "= fossulae."

II. a. Potonié, 1934, p. 20–21: *Tr.* In the plan view of the germinal, the elongated rimula of *Tilia* extends far over the double circle enclosing the vesticulum . . . In *Tilia,* the rimula is quite clearly an open slit only within the double circle, functioning as an exit from the vestibulum. Outside the double circle, the rimula continues in both directions as only a small, narrow furrow, the *sulcus,* cutting into the exoexine . . . [Fig. 315]. The sulci, which point toward the poles and which are pointed at their ends, terminate quite far from the poles in *Tilia.* However, they are essentially nothing more than the long furrows of many pollen types differentiated on this basis, in which the furrows approach the poles, and in which there are transitions to those broad meridional areas which appear in fossil pollen grains, now evenly expanded, now sharply infolded. Consequently there is no boundary between simple sulci and small, infolded meridional areas. [*see* APERTURA].

b. Potonié, 1958, p. 91: *Tr.* Pollen grains which display only one "furrow" or fold generally parallel to the long axes and which [the furrow] usually function in germination.

The monocolpate germinal furrow (the colpus) is a more or less broad generally oblong sector of the pollen grain, which may show a ± thin exine; and cannot penetrate into the unswollen pollen grain. The two lengthwise edges of the sector in flattened fossil pollen grains can thereby be sharply folded. In contrast, in the Monoletes, the monolete tectum serves in germination. The tectum is (before its opening) a mostly sharp, protrubant small ridge-shaped wall. It is characterized by a vertex which passes through the proximal pole.

The monocolpate pollen grains are generally ± oblong. They exhibit different types of structure and sculpture of the exine, as well as a series of variations of "germinal furrows." [Potonié, 1958, p. 95: *Tr.*] Erdtman (1952, p. 12) places forms which have a sulcus appropriate to *his* definition in the *Monosulcites.* A sulcus touches or passes through the distal pole. On the other hand, germinal stripes which cross the equator meridionally are called *colpi* by Erdtman (1952, p. 13). In the concept of

Monocolpates this differentiation is neglected . . .

III. a. Erdtman, 1945, p. 190–191: [*see* SULCATE].

b. Erdtman, 1946, p. 70–71: "Typical 'monocotyledonous' pollen grains are provided with a longitudinal furrow borne on the distal part of the grain and crossing the polar axis at right angles . . . [Fig. 24–26, 289]. This kind of a furrow is termed a *sulcus.* By shortening and contraction a sulcus may ultimately develop into a porus. Exceptionally a sulcus may present a three- or four-slit opening. Such grains are termed trichotomosulcate (3-ts) and tetrachotomosulcate (4-ts) respectively . . ."

c. Erdtman, 1947, p. 105: "A germinal furrow confined to the distal part of the spore is termed a 'sulcus'."

d. Selling, 1947, p. 77: "a (straight or triradiate) furrow on the distal side of a pollen grain, one or—rarely—two on each."

e. Erdtman, 1952, p. 470: "(Erdtman, 1945 . . . ; the term has been used now and then during the last 150 years [cf., e.g., Potonié, 1934 . . .] to denote longitudinal apertures): aperture in the distal face of a spore (and usually with the distal pole in its center) with the ratio length:breadth >2. Examples: . . . [Fig. 249–263]."

f. Couper, 1958, p. 103: "The single furrow of some gymnospermous and monocotyledonous pollen grains. In contrast to laesura of monolete spores, the sulcus is on the distal surface, i.e., it faces outward in the tetrad."

IV. a. Faegri and Iversen, 1950, p. 163: "sulcus (Potonié, 1934) = short furrow, e.g., in *Tilia;*" "(Erdtman, 1954a) = distal furrow in *Monocotyledones.*"

b. Cranwell, 1953, p. 10: "Essentially, there are two types of [monocotyledonous pollen] grains: (A) those with bilateral symmetry, characterized by a single furrow or pore; and (B) those (usually spheroidal) with no recognizable aperture. The latter are regarded as derived from the aperturate type. Of the few variants known, *Alisma* is many pored: some have two apertures, but as Professor I. W. Bailey has recently stressed, none has three.

"Two of the three most typical pollen features, namely, distal position of the furrow and successive meiosis, are bound up with development and can be distinguished only in the early stages in the pollen mother cell, or if the grains remain permanently in tetrads.

"First, as to polarity: it was Goebel (1923, Organographie der Pflanzen; vol. 3, jena, p. 1538) who first demonstrated orientation in monocolpate grains by showing that the furrows of *Pinus Mugo* var. *Pumilio* (= *P.*

Pumilio) were turned outward and at right angles to the polar axis during tetrad formation. Newman (1928) showed that it was distal also in the Australian *Doryanthes,* but it was Pohl (1928) who, in describing tetrads of *Molineria* . . . [Fig. 243–244], advanced the theory that the furrow in the monocotyledons is homologous with that in the gymnosperms.

"Wunderlich (1936) and others have confirmed this orientation in many genera, while Wodehouse (1935), Swamy (1949), and others have recorded it for certain dicotyledons [Cranwell, 1953, p. 11]: as well. As a result, it has been assumed too readily that the furrow is distal in all monocolpate grains. Professor Bailey and his coworkers (1943, 1950), however, have proved that it may exceptionally be proximal, as in *Asimina* and *Amborella* Fritzsche (1837) had figured this position for the aperture in *Annona Cherimolia* (*A. tripetala*), and it was finally confirmed for Annonaceous tetrads by Dr. S. J. Golub, one of Professor Bailey's associates.

"A list of dicotyledonous families (all with Ranalian affinities), which may have monocolpate or derived types of pollen, will be found in Bailey and Nast (1943, 341). New Zealand genera in this category (cf. Cranwell, 1942), are as follows: *Ascarina, Beilschmiedia, Cassytha, Hedycarya, Laurelia, Litsea, Macropiper, Peperomia,* and *Pseudowintera* (cf. Table I). The only other native dicotyledons with even superficially comparable pollen grains are the acolpate *Aleurites,* the obscurely pored forms of *Drapetes,* and an occasional monoporate *Fuchsia* or *Epilobium.*

"Secondly, successive meiosis has been considered as essential a feature in the monocotyledons as simultaneous division in the dicotyledons. However, it is now clear from the work of Hofmeister, Guignard, Rosenberg, and various others (cf. Schnarf, 1929), that although most of the families are uniform in this respect, some, especially sedges, rushes, and orchids, are characterized by simultaneous division, while both types may occur in such homogeneous families as the Amaryllidaceae and Palmae, or even, as in *Aponogeton,* in the same pollen sac. On the other hand, successive division has also been traced in the Ranalian group (Polycarpicae) already mentioned (e.g., in *Cinnamomum*: Tackholm and Soderberg, 1917). The association of the single furrow with successive division in this group is of special interest. There is no agreement as to which type of division may be the more advanced.

"Thirdly, the furrow of the monocolpate grain never encloses a pore; and finally, the furrow margins are rarely thickened, the exine pattern often extending with little modification over the furrow membrane. Furrows of this type are found only occasionally in the tricolpate and derived dicotyledonous types, but they do occur fairly freely in the Polycarpicae. Wodehouse (1936, 495), in discussing the grains of *Ranunculus,* for instance, considers this poreless furrow 'of profound phylogenetic significance.' The tendency to elimination of the furrow is also discussed by him (1935, 1937). In the native flora, *Hectorella* (Caryophyllaceae: aberrant), the Cruciferae, *Teucrium* (. . . in the Verbenaceae), and a few other genera in the Ranunculaceae in particular, share this type of furrow."

c. Cranwell, 1953, p. 17: "originally used for grooves in a grain (e.g., Lindley, 1848) but used tentatively by Erdtman and adopted by Selling (1947) for a straight or triradiate furrow on the distal side of a pollen grain. Disulcate: two on one side, or zonate . . ."

SULCUS SIMPLEX, n.
Potonié, 1934, p. 24: *Tr.* On the other hand, we also have here the route which could have led to certain pollen types with simple sulcus (S. simplex) not yet discussed. The simple sulcus, i.e., the ordinary furrow, shows no differentiation in its equatorial regions; it lacks any special apparatus, and in some cases could have been a transitional stage leading to the meridional area. [*see* APERTURE, COLPUS].

SUPER-, (L. above).
Jackson, 1928: "often modified into supra."

SUPRARETICULATE, adj. cf. reticulate, tectate-reticulate.
Erdtman, 1952, p 470: "Suprareticulate pollen grains have a reticulate pattern due to suprategillar elements. Figs. 6–8 in Faegri and Iversen (1950, p. 26) [*see* Fig. 698–700] are said to illustrate a suprareticulate ('tectate-reticulate') pollen grain. It is, however, difficult and sometimes impossible to decide—by means of an ordinary microscope—whether a pollen grain is provided with a reticulum-bearing tegillum (i.e., 'tectate-reticulate' according to Iversen and Troels-Smith) or simply reticulate ('intectate-reticulate' according to Iversen and Troels-Smith) with isolated bacula rising from the bottom of the lumina as in *Catopheria chiapensis* . . . [Fig. 754–756], *Cobaea* spp. and other plants."

SUPRATEGILLAR, adj.
Erdtman, 1952, p. 471: "Suprategillar patterns are due to sculptural elements rising above the general surface of a tegillum."

SUTURA, SUTURE, n., pl. -s. (L. sutura, suere, to sew). cf. aperture, contact figure, tetrad mark.
  a. Jackson, 1928: "a junction or seam of union; a line of opening or dehiscence."
  b. Potonié and Kremp, 1955, p. 11: *Tr.* . . . dehiscence furrow or seam . . . [*see* Y-MARK].
  c. Kosanke, 1950, p. 14: [*see* Fig. 109–110].
  d. Couper and Grebe, 1961, p. 3: "the line of dehiscence of a tetrad scar . . . [Fig. 115–121]. Compare with *laesura* which, however, includes the *labrum* if present."

SWARM SPORE, n., pl. -s. cf. zoospore.
  Jackson, 1928: "a motile naked protoplasmic body, a zoospore."

SWIMMING-APPARATUS, n. cf. contact figure, floats, floating body, massa.
  Jackson, 1928: "in *Azolla,* three apical episporic spongy masses of tissue, surrounding a central conical body with an array of fine filaments (Campbell)."

SYMMETRY, n. (Gr. symmetria). cf. figura, morphography, shape classes.
  I. Jackson, 1928: "capable of division into similar halves;" "used of topography when it shows uniform changes (elements)."
  II. Thomson and Pflug, 1953, p. 18–19 and Fig. 114: *Tr.* The symmetry of the sporomorphae is determined by their arrangement in the tetrad. The specific arrangement of tetrads within a larger botanical unit is often the same or similar. Therefore, for our morphological system the establishment of the symmetry elements of the grain is of great importance.

  The symmetry of many sporomorphae shows two different poles, which, if their arrangement within the tetrad is known, are called the "proximal" and "distal" pole; the former is turned toward the tetrad whereas the latter is turned away from it. The axis connecting the poles is called the polar axis. It is perpendicularly cut in half by the equatorial plane, whose symmetry axes are called equatorial axes. The point of intersection of polar axis and equatorial plane is called the "centrum." The terms interior and exterior signify centrifugal and centripetal, also behind signifies centripetal and before centrifugal.
  III. Krutzsch, 1959, p. 35: *Tr.* Symmetry: microspores possess precise symmetry ratios (Thomson and Pflug, 1953, p. 18); only a few are built wholly asymmetrically. The symmetry ratios generally permit a division into two hemispheres, which according to their positions in the mother cell are designated as "proximal" and "distal": the terms proximal and distal can be rigorously used if the situation of the grain of fossil material is effectively demonstrated.

Appropriately one distinguishes a "proximal" and "distal" pole; a line joining the two is the polar axes, often with several polar-symmetry planes. Perpendicular to these are one or more equatorial symmetry axes and sometimes an equatorial symmetry plane. The equatorial contour . . . is sometimes of taxonomic significance. In the bilaterally constructed monolete spores beyond the equatorial contour besides the dehiscence contour . . . is a taxonomically useful character.

SYMMETRY AXIS, n., pl. -es. cf. axis, equatorial axis, figura, symmetry.
  Thomson and Pflug, 1953, p. 18: *Tr.* The axis connecting the poles is called the polar axis. It is perpendicularly cut in half by the equatorial plane, whose symmetry axes are called equatorial axes.

SYMMETRY ELEMENT, n., pl. -s. cf. figura, shape classes, symmetry.

SYMPILATE, adj. cf. pilate.
  Erdtman, 1952, p. 471: "in sympilate grains the capita of the pila abut upon each other (and coalesce to a certain extent?)."

SYN- (Gr. sym, with). Adhesion or growing together. Used in compounds.

SYNCOLPATE, adj. cf. colpate, furrow.
  Erdtman, 1952, p. 471: "(Faegri and Iversen, 1950, p.p.): with colpi anastomosing at the poles. Examples: . . . [Figs. 34–35, 445–451]."
  sdb. Faegri and Iversen, 1950, p. 128; [*see* GROUPS OF SPOROMORPHAE and Fig. 53–58].
    Pike, 1956, p. 51.
    Beug, 1961, p. 13.

SYNDEMICOLPATE, adj. cf. colpate, furrow.
  Erdtman, 1952, p. 471: "said of pollen grains in which the demicolpi of the proximal face meet at the proximal pole and those of the distal face at the distal pole. Example: . . . [Fig. 470–472]."

SYNEXINA MIHI, n. cf. synexine.
  Kupriyanova, 1956: *Tr.* The membrane of the mother cells. [*see* SPORODERMIS].

SYNEXINE, n. (Gr. syn-, with). cf. general exine, sporodermis, synexina mihi.
  Kupriyanova, 1956: *Tr.* It is impossible not to mention here about another layer covering not only individual pollen grains but the whole tetrad, and which can be observed in pollen tetrads of Juncaceae, Cyperaceae, Thurniaceae, Orchidaceae, Ericaceae, and in certain other families. The layer mentioned is nothing but the membrane of the maternal cells. To distinguish it from the exine of the singular pollen grain it can be called general exine or synexine (synexina mihi). The synexine is well stained by basic fuchsin and it can be referred to the group of ektexine layers. [*see* SPORODERMIS].

T

TASICOLPATE, adj. (Gr. tasis, straining). cf. colpate, furrow.
Wodehouse, 1935, p. 544: "bearing furrows in some systematic arrangement, apparently resulting from stresses acting over the surface of the grain; distinct from the furrow of a monocolpate grain which arises from the collapse of the grain on its unsupported side."

TASITHYNIC, adj. (Gr. tasic, a straining; ithuneto, in a straight line). cf. figura, furrow, Groups of Sporomorphae, heterotasithynic, isotasithynic, morphography.
Wodehouse, 1935, p. 544: "due to lateral stresses the stresses that arise from shrinking, as in plaster, mud, or pollen-grain surfaces."

TAXONOMY, n., pl. -ies. (Gr. taxis, order; nomos, law). cf. anatomy, Groups of Sporomorphae, morphography, morphology.
Jackson, 1928: "classification."

TECTATE, TECTATUS, adj. (L. tectus, covered, hidden). cf. columellate, infrapunctate, intragranulate, planitegillate, psilotegillate, punctitegillate, tegillate.
a. Faegri and Iversen, 1950, p. 17: "Many *Liliiflorae* possess a reticulum with very narrow meshes. As long as the diameter of the holes (*lumina* Potonié, 1934) is more or less equal to the breadth of the walls (*muri* Potonié, 1.c.), the grains may still be considered intectate. If the holes are distinctly smaller, we may consider the grain to possess a *perforate tectum* (*Silene maritima*) and consequently classify it as tectate."
b. Harris, 1955, p. 27: "means that the wall is at least two-layered and that the innermost layer (outside the intine) is completely veiled by the layer next outside it, except in the case of the perforate or the fenestrate tectum. In the perforate tectum the diameter of the holes is less than the width of the walls which separate them. The fenestrate tectum is exemplified in the pollen grains of the *Liguliflorae* where the tectum is broken by a limited number of rather large openings which are arranged in a symmetrical pattern (Faegri and Iversen, 1950, p. 17)."
sdb. Cranwell, 1953, p. 17; [*see* TECTUM].
Beug, 1961, p. 13.

TECTATE-PERFORATE, adj. cf. tectum perforatum.
Faegri and Iversen, 1950, p. 17: "tectum with holes."
[*see* ORNAMENTATION and Figs. 624, 693–697].

TECTATE-RETICULATE, adj. cf. reticulate, suprareticulate.
Faegri and Iversen, 1950, p. 26: [*see* ORNAMENTATION and Fig. 698–700].

TECTOID, adj.
Harris, 1955, p. 27: ". . . is equivalent to the statement that semi-mature spores have been seen which were quite smooth and that mature spores of the same species have been seen which possessed a complete outer coat which, however, there is reason to believe may not always be preserved in peat or older deposits. The reason for belief that the outer coat may not be preserved may vary in different cases—the outer coat may be structurally very weak, or it may be known from the examination of peats that this layer tends to be shed. The term conveys information about the layering of the wall which may be difficult to observe directly in mature spores, without recourse to sectioning [*see* 'TECTATE' and 'PERINE']."

TECTUM, n., pl. -a. (L. a roof). cf. contact figure, ectosexine, exolamella, list, tegillum, trilete mark.
I. a. Faegri and Iversen, 1950, p. 17: "if the ektexine forms another membrane outside the endexine (*tectum*), the grains are called *tectate*, notwithstanding whether the tectum is separated from the endexine by a cavity or not." [*see* ORNAMENTATION and Figs. 622–626, 690, 693–700].
b. Ingwersen, 1954, p. 41: "external membranous part of ektexine more or less covering the endexine; also used here when ektexine elements form a massive layer with the same effect."
c. Harris, 1955, p. 27: "the surface formed by the fusion of the elements of the ektexine or of their outer ends . . . [*see* 'TECTATE']."

165

sdb. Iversen and Troels-Smith, 1950, p. 35; [*see* STRUCTURE].

Cranwell, 1953, p. 17.

Kupriyanova, 1956; [*see* SPORODERMIS].

Beug, 1961, p. 13.

II. Erdtman, 1952, p. 19: "Among flowering plants one of the basic elements of the sexine appears to consist of small drumstick-shaped rods (pila, . . . [Fig. 648]), projecting at right angles from the nexine surface. Each pilum has a head (caput) supported by a rod-like neck (baculum). The capita form the upper part of the sexine (the ectosexine), the bacula the lower, basal part of it (the endosexine). If the capita amalgamate . . . [Fig. 648] or—hypothetically—a layer of some sort should be formed on the top of the pila, coalescing with or enveloping the capita whilst leaving the bacula free, a tegillum (small roof) is formed. (The term tectum is used in descriptions of thick, ± stratified tegilloid layers . . .)"

III. a. Potonié and Klaus, 1954, p. 525: *Tr.* In 1950, Faegri and Iversen described as tectum that part of the exine, which was termed as "Exolamella" by R. Potonié already in 1934. We do not use the term tectum in the sense that it was introduced by Faegri and Iversen recently, but in the sense of R. Potonié, in order to make explicit the nomenclature situation.

b. Potonié and Kremp, 1955, p. 11: *Tr.* The three rays or tecta (of the tetrad mark) originate at the apex . . . The tecta are characteristically formed, more or less high, sometimes very low, roof or wall-like pleats or ridges of the exine . . . [*see* Y-MARK and Fig. 144–148].

TECTUM PERFORATUM, n., pl. -a.

Iversen and Troels-Smith, 1950, p. 35: *Tr.* tectum with holes. [*see* STRUCTURE].

sdb. Faegri and Iversen, 1950, p. 17; [*see* ORNAMENTATION].

Beug, 1960, p. 13.

TEGILLATE, adj. cf. infrapunctate, intragranulate, planitegillate, psilotegillate, punctitegillate, suprategillar, tectate, unditegillate.

TEGILLUM, n., pl. -a. (L. tegere, to cover). cf. ectosexine, exolamella,` tectum.

Erdtman, 1952, p. 471: "an ectosexinous, ± homogeneous layer usually distinctly separated from the nexine by a baculate zone (endosexine)." [*see* TECTUM].

TELEUTOSPORE, n., pl. -s. (Gr. teleuta, an end). cf. teliospore.

a. Jackson, 1928: "a resting bilocular spore of Uredineae, on germination producing a promycelium."

b. Melchior and Werdermann, 1954, p. 15. [*see* SPORE].

TELIOSPORE, n., pl. -s. (Gr. telos, completion). cf. teleutospore.

Jackson, 1928: "= teleutospore."

TENUATE, adj. (L. tenuis, thin). cf. aperture, colpate, furrow, tenuitas.

Erdtman, 1952, p. 471: "with one or several tenuitates. Example: . . . [Fig. 264–265]."

sdb. Norem, 1958, p. 668. [*see* Fig. 265–1].

TENUI-EXINOUS, adj.

Erdtman, 1952, p. 471: "said of spores with exine which is distinctly thin in relation to the size of the spores. Example: . . . [Fig. 598–599]."

TENUIMARGINATE, adj. cf. aperture, crassimarginate, pore.

Erdtman, 1952, p. 471: "with thin margins."

TENUINEXINOUS, adj.

Erdtman, 1952, p. 471: "with thin nexine; thickness of nexine less than half that of sexine."

TENUISCLERINOUS, adj.

Erdtman, 1952, p. 471: "said of spores with sclerine distinctly thin in relation to the size of the spores."

TENUISEXINOUS, adj.

Erdtman, 1952, p. 471: "with thin sexine; thickness of sexine less than half that of nexine. Examples: . . . [Figs. 534–537, 593–596, 600–601]."

TENUITAS, n., pl. -tates. (L. tenuis, thin). cf. furrow, sulcoid groove, sulcus.

a. Potonié, 1934, p. 21: *Tr.* Very instructive in understanding many of the previously discussed patterns of germinal apparatus is the fossil which has been described as *Pollenites megadolium digitatus* . . . [Fig. 322, 323, 332]. In this case, a long, narrow meridional ruga is present, which because of a modest infold and its narrowness, is designated as a sulcus. [Potonié, 1934, p. 22: *Tr.*] In the middle of the sulcus there is a strongly differentiated germinal point or exitus. How this differentiated condition arises is shown in the type figured as *Pollenites sinuatus* . . . [Fig. 330, 334–335]. Here in the exitus, the meridional ruga is crossed by a very distinctly formed equatorial ruga. In *P. megadolium sinuatus* this equatorial ruga exists as a slightly inward-bent depression which is crossed by the sulcus and is thus divided into two halves. It seems the region of this depression has a thinner exine; still it could not be determined whether the exine here consists only of intexine or not. If it should be confirmed that we are dealing with a less thick wall division, we would be able to speak of a *tenuitas* . . . In more enlarged specimens, as in fossil *Pollenites megadolium* . . . [Fig. 331], the tenuitas is arched toward the outside, so that the exitus is occupied by a hill over which the sulcus crosses in the form of a furrow. In recent relatives of the pollen under discussion, the hill expands and smooths out considerably by swelling, so that it is no longer divided by the sulcus (for a similar situation see . . . [Fig. 325–328]). Let us now further consider the fossils in which the tenuitas is not protrudent, but is indented. Here the indentation can deepen more and more in such a manner that the upper and lower edges of the tenuitas approach each other, more and more covering the hollow forming below them . . . [Fig. 322, 323]. The *exitus digitatus* is developed. [see APERTURA, TENUITUS, and EXITUS].

b. Erdtman, 1952, p. 471: "thin, ± aperturoid exine area, not as distinctly delimited as 'true' apertures. . . . [Fig. 264–265] shows a pollen grain with three tenuitates."

c. Potonié and Kremp, 1955, p. 12: *Tr.* Not to be mistaken for the tecta of the Monoletes are the similarly elongated colpi of the Monocolpates . . . [Fig. 246]. Whereas in the case of tecta, we are dealing with wall-like, elongate protrusions of more or less unweakened exine, the colpi of the monocolpates are either more or less elongate primary or secondary weakenings of the exine-thickness (*tenuitas*) or only furrows or "seams," but not walls or tecta. In fossils, the tenuitas is usually indented or folded in . . . [Fig. 246].

d. Erdtman, 1957, p. 3–4: "The aperture (of saccate pollen grains) is distal and should often, perhaps more appropriately, be referred to as tenuitas (i.e., a thin aperturoid area functioning as an aperture and gradually merging into the surrounding exine). It has earlier, as a rule, been described as a sulcus, or a sulcoid groove."

TENUITEGILLATE, adj.

Erdtman, 1952, p. 471: "with thin tegillum; thickness (depth) of tegillum not more than half the length of the supporting sexinous elements."

TENUITY, n. cf. tenuitas.

TETRACHOTOMOSULCATE, adj. (Gr. tetraxa, four-fold; tomos, a cutting). cf. colpate, furrow.

a. Erdtman, 1947, p. 113: "with a four-slit aperture." [see GROUPS OF SPOROMORPHAE].

b. Erdtman, 1952, p. 471: "with a four-slit sulcoid distal aperture."

TETRACOLPORATE, adj. (Gr. tetras, four.). cf. colpate, colporate, colpus, furrow, Groups of Sporomorphae, porate.

Traverse, 1955, p. 95: "Having four meridional furrows, each having an equatorial germ-pore or transverse furrow. Iversen and Troels-Smith call this stephanocolporate. Since this is such a common variant of tricolporate forms it does not seem advisable to set it off as so distinct from tricolporate."

TETRAD, n., pl. -s. (Gr. tetradion, a set or group of four). cf. aggregation, cross tetrad, Groups of Sporomorphae, linear tetrad, pollen tetrad, pollen tetrahedron, pollina occlusa, pseudomonad, rhombohedral tetrad, rhomboidal tetrad, spore, square tetrad, tetragonal tetrad, tetrahedral tetrad, tetras, tetrasporaceous, tetraspore, tetrasporic, tetrasporine, tetrasporous.

a. Jackson, 1928: "a body formed of four cells, as in the formation of pollen in the pollen-mother-cells;" "= TETRASPORE;" "in heterotypic nuclear division, one division of a chromosome into four parts, forming a set."

b. Potonié, 1934, p. 8: *Tr.* In some genera, the four daughter cells that come from a mother cell remain together as a *pollen tetrad* (Ericaceae). [Potonié, 1934, p. 13: *Tr.*] The predominance of just three germinal places is due to the fact that pollen cells usually lie in groups of fours in the pollen-mother-cell, forming a *tetrad*. They are usually arrayed as a tetrahedral tetrad . . . [Fig. 2]. [see APERTURA].

c. Erdtman, 1943, p. 52: "Tetras: tetrad; a union of four pollen grains or spores formed by one mother-cell. If the grains are arranged in one plane, the tetrads may be tetragonal . . . [see Fig. 19], or rhomboidal . . . [Fig. 22]. In the first case all four cells meet at the center of the tetrad, in the second case only two of them meet. If the grains are arranged in two planes, the tetrads are tetrahedral . . . [Fig. 20] or, in exceptional cases, hexahedral. In the last case, the grains are about twice as long as broad and arranged crosswise in two stories. This case and transitions to tetragonal tetrads . . . [Fig. 21] may be found, e.g., in *Picea*."

d. Faegri and Iversen, 1950, p. 15: "The pollen grain is formed in the male apparatus of the flower, that is, in the anther. The interior of the anther consists of a sporogeneous tissue from which originate the so-called pollen mother cells. With few exceptions each of these gives rise to 4 pollen grains that are in most plants ultimately free from one another. In some genera they remain together, forming tetrads, or other rarer types (dyads, polyads) of composite grains (cf. list in Erdtman, 1945b, p. 286). Even if the grains are eventually isolated from each other, the tetrad stage constitutes an important step in their development, and the surface pattern shows a distinct relation to the orientation of the grain in the tetrad (Wodehouse, 1935, p. 159).

e. Pokrovskaya, et al., 1950: *Tr.* Combination of four microspores originating from one mother cell.

If microspores are arranged in one plane, the tetrads may be square or rhomboidal. In the first case all four cells contact one another in the center of the tetrad, in the second case, only two of them are in contact. If the microspores are arranged in two planes, then the tetrads are tetrahedral (pollen of most of the Dicotyledoneae) or in exceptional cases hexagonal (most of the Monocotyledonae).

f. Cranwell, 1953, p. 17: "a unit of 4 grains formed by one mother cell, *or,* a union of 4 grains which function independently."

g. Traverse, 1955, p. 95: "Having the pollen grains joined in fours, presumably the four products of a single pollen mother cell. In addition to the characters of the whole tetrad, each grain of the tetrad is described as if it were separate. Tetrads found in this investigation were all *tetrahedral*— that is, they are arranged as if each grain were in one apex of a pyramidal tetrahedron."

sdb. Wodehouse, 1935, p. 159.
Erdtman, 1945, p. 287: [*see* Fig. 14–18].
Selling, 1947, p. 77.

TETRAD MARK, n., pl. -s. cf. contact figure, tetrad.
Couper and Grebe, 1961: [*see* CONTACT FIGURE and TETRAD and Fig. 115–119].

TETRAD SCAR, n., pl. -s. cf. contact figure, tetrad.

TETRADYMOUS, adj. (Fr. tetradymos, four fold).
Jackson, 1928: "having four cells or cases;" "when the lamellae of an Agaric are arranged so that alternate lamellae are shorter than the intermediates, and one complete lamella terminates a set of four pairs, short and long."

TETRAGONAL, adj.
Norem, 1958, [*see* SHAPE CLASSES and Fig. 167h].

TETRAGONAL TETRAD, n., pl. -s. (Gr. tetra, four; gonia, corner, angle). cf. aggregation, spore, tetrad.
Potonié, 1934, p. 13: *Tr.* Each of the four balls joined in a tetrahedron, touches the three other ones at three points . . . [Fig. 2]. These three *contact points* form an isosceles triangle (that lies in the proximal polar hemisphere). In systematically higher pollen, germinal apparatus begin at these contact points and change position from the proximal polar hemisphere to the equator.

In pteridophyte spores, just as in pollen, the tetrad is not always arranged as a tetrahedron in all cases. The central points of the four cells may also be arranged in a plane within the mother cell . . . [Fig. 1] so that the cells are enclosed, for example, in a rectangular box. Wodehouse (1929) therefore speaks of *tetragonal tetrad.* This tetrad is characteristic of monocotyledons. In this situation each pollen cell touches only two of the grains be-

longing to the same tetrad. [*see* APERTURA, TETRAD, and Figs. 3–5].

TETRAHEDRAL TETRAD, n., pl. -s.
a. Potonié, 1934, p. 13: *Tr.* The predominance of just three germinal places is due to the fact that pollen cells usually lie in groups of four within the pollen mother cell, forming a *tetrad;* they are usually arrayed as a *tetrahedral tetrad* . . . [Fig. 2]. [*see* APERTURA, Potonié and Kremp, (1954), and TETRAD].

b. Kupriyanova, 1948, p. 75: *Tr.* . . . in which the pollen is closed within a membrane of the maternal pollen cell and is arranged as a tetrahedral tetrad. [*see* GROUPS OF SPOROMORPHAE].

sdb. Kozo-Polyanskiy, 1945, p. 237.

TETRAS, n. cf. aggregation, spore, tetrad.

TETRASPORACEOUS, adj. cf. aggregation, spore, tetrad.
Jackson, 1928: "connected with the production of TETRASPORES;" bearing tetraspores."

TETRASPORE, n., pl. -s. aggregation, spore, tetrad.
Jackson, 1928: "a spore formed by division of the spore-mother-cell into four parts."
sdb. Melchior and Werdermann, 1954, p. 16; [*see* SPORE].

TETRASPORIC, adj. cf. tetrasporaceous.

TETRASPORINE, adj. cf. tetrasporaceous.

TETRASPOROUS, adj. cf. tetrasporaceous.

TETRATREME, adj. cf. aperture.
Erdtman and Straka, 1961, p. 65: "(4-treme) spores: with four apertures." [*see* GROUPS OF SPOROMORPHAE].

TEXTURA, n. cf. texture.
Kupriyanova, 1948, p. 70: *Tr.* . . . is created by partial separation of the layers of the exina . . .

TEXTURE, n., pl. -s. (L. textura). cf. ornamentation, pattern, structure.
I. Jackson, 1928: "applied by Starback to hyphal structures in Discomycetes . . ."
II. a. Erdtman, 1943, p. 52: "Structura: structure, texture: different patterns in surface view usually more or less 'granular,' not produced by eventual sculpturing of the exine but by formative elements within the exine." [*see* STRUCTURE].

b. Selling, 1947, p. 77: "a pattern not produced by sculpture."

c. Kupriyanova, 1948, p. 70: *Tr.* The sculpture (ornamentation) of the exine must be differentiated from the texture (textura). The latter is created by partial separation of the layers of the exine; in this way forming a slight roughness on the surface of the exine. For example, as the pollen grains of grass. [*see* ORNAMENTATION, STRUCTURE, Faegri and Iversen (1950)].

d. Cranwell, 1953, p. 17: [*see* STRUCTURE].

e. Traverse, 1955, p. 94: "A noncommital term used by the author when he was unable to ascertain the exact nature of sculpturing, usually because of poor preservation, although the exine was obviously sculptured."

THORN-LIKE SCULPTURE ELEMENT, n. cf. cone, conus, echinus, spina, spine.
Kupriyanova, 1948, p. 70: *Tr.* A type of ornamentation. [*see* ORNAMENTATION].

TIP OF THE COLPUS, n. cf. colpus, furrow.

TOP VESICLE, n., pl. -s. (L. vesicula, dim. of vesica, a bladder, blister). cf. bulla apicalis, contact figure, dehiscence cone, gula, nozzle.

TORUS, n., pl. -i. (L. a bed). cf. contact figure, kyrtome.
I. Jackson, 1928: "the receptacle of a flower, that portion of the axis on which the parts of the flower are inserted; when elongated it becomes the GONOPHORE and GYNOPHORE."
II. Thomson and Pflug, 1953, p. 26: *Tr.* The "torus" is significant for the *"gleichenoid"* spores which occur during the late Mesozoic in a great variety of forms. These *"gleichenoid"* spores still occur in the oldest Tertiary and must here be considered as important index fossils. The torus is a kind of swollen protrusion or invagination of exospore lamellae. This protrusion or invagination is restricted to a narrow or broad region which runs parallel to the Y-mark. The torus can embrace the entire exospore wall or it can be restricted to only the inner lamellae, while the outer exospore lamellae smoothly cover the swelling. These features have not yet been investigated in detail.
III. Potonié and Kremp, 1955, p. 13: [*see* KYRTOME and Fig. 122].
IV. Krutzsch, 1959, p. 38: *Tr.* Dissociations or other differentiations are found in late Mesozoic and early Tertiary spores and furthermore in spore-like objects (= pollen) where frequently they parallel the tetrad rays on both sides. These are the *tori,* as shown in . . . [Fig. 123–128].
The tori can be differentiated:
a) the predominantly narrow band-tori;
b) fold-tori (with transitions to "simple pleats" and to band-tori);
c) field-tori (planelike broad and more or less long), e.g., the *arealis* group;
  c_1) plane field-tori, short
  c_2) full field-tori, long (circumfluent)
d) full-fold [Voll-Falten] or plane field-tori; they reach to the corners and are more or less closely connected with each neighboring part; they comply habitually with some plicae in pollen; partially with pores or with pore structures in the corner circuits (tp = torus-pores).

e) "lane"-tori ["Bahnen"-(Cursi) tori]; so far noticed only distally; local, straightened [gerichtete] solution areas.
f) "pseudotori" and "pseudotoroid claws;" both are only fold-features without special structural peculiarities opposite the surrounding spore wall, probably mainly of secondary nature.
The tori can be shaped in many different ways. They may be situated distally or proximally, or may occur on both hemispheres at the same time. Thus a great abundance of types are worked out. The area between torus and apex is also called the pole- or apex-field; also it is called area by R. Potonié (1934).

T-PATTERN, n., pl. -s. (tegillate-pattern). cf. LO-pattern.
Erdtman, 1952, p. 471: "designations used only in descriptions of tegillate (or tectate) spores; connote patterns etc. produced by details of the tegillum (tectum) itself."

TRANSCOLPATE, adj. (L. trans, across, over). cf. colpate, furrow.
Erdtman, 1952, p. 471: "with colpi more elongated equatorially than meridionally. Example: . . . [Fig. 426]."

TRANSLUCENT, adj. chagrenate, laevigate.

TRANSVERSAL FURROW, n., pl. -s. cf. transverse furrow.

TRANSVERSE, adj. (L. transversus, past part. of transvertere, to turn or direct across). cf. figura.
a. Jackson, 1928: "across, right and left as to bract and axis, collateral;" "Lindley gives 'broader than long' as the definition of *transversus.*"
b. Traverse, 1955, p. 91: "A part lying either on the equator or on the surface parallel to the equator."

TRANSVERSE FOLD, n., pl. -s. cf. colpus, equatorial ruga, furrow.

TRANSVERSE FURROW, n., pl. -s. cf. colpus, furrow, transversal furrow.

TRANSVERSE MEDIAN OF THE COLPUS, n. cf. medianum colpi.

TRAPEZIFORM, adj. cf. shape classes.
a. Jackson, 1928: "an unsymmetrical four-sided figure, as a trapezium, almost the same as rhomboid."
b. Zaklinskaya, 1957: [*see* Fig. 227:f].

TREMA, n., pl. tremata cf. aperture.
Erdtman and Straka, 1961, p. 65: "Aperture." [*see* GROUPS OF SPOROMORPHAE].

TRI-, (L. tres, three). A prefix meaning three or triple.

TRIACONTARUGATE, adj. (L. triaconta, thirty, and rugate). cf. rugate.

Erdtman, 1952, p. 471: "(cf. Wodehouse, 1935, p. 179): with 30 rugae (corresponding polyhedron: pentagonal dodecahedron)."

TRIAD, n. (L. trias). cf. aggregation, spore, tetrad.
Selling, 1947, p. 77: "a unit of three grains (spores) formed by one mother-cell."

TRIANGULAR, adj. (L. triangulus). cf. shape, shape classes.
a. Jackson, 1928: "with three angles."
b. Zaklinskaya, 1957: for outline of disaccate pollen grains [see Fig. 227:g].
c. Couper and Grebe, 1961: [see SHAPE].

TRIANGULUS, adj. (L. triangular). cf. shape classes.
Pokrovskaya, et al., 1950: Tr. Used to classify the amb of certain microspores. [see CIRCUMSCRIPTIO].

TRICHOTOMOCOLPATE, TRICHOTOMOCOL-PATUS, adj. cf. colpate, furrow, trichotomo-fissurate, trichotomosulcate, trichotomous.
Erdtman, 1945, p. 190: "possessing a germinal furrow or aperture with a three-slit opening."

TRICHOTOMO-FISSURATE, adj. cf. colpate, furrow, trichotomocolpate, trichotomosulcate, trichotomous.
Kupriyanova, 1948: Tr. The pollen grains with an opening that has three slits... [see GROUPS OF SPOROMORPHAE].

TRICHOTOMOSULCATE, adj. cf. colpate, furrow, trichotomocolpate, trichotomo-fissurate, trichotomous, trisulcate.
a. Erdtman, 1947, p. 113: "with a three-slit aperture." [see GROUPS OF SPOROMORPHAE].
b. Selling, 1947, p. 9: "(= with one triradiate sulcus). . . . rays of the sulcus not bordered by any pronounced thickenings such as found at the tetrad scars of the trilete pteridophyte spores." Selling, 1947, p. 78: "(pollen) with a triradiate sulcus (on the distal side of the grain)."
c. Erdtman, 1952, p. 471: "Erdtman, (1945): with a three-slit sulcoid distal aperture. Example: . . . [Fig. 247]."
sdb. Beug, 1961, p. 13.

TRICHOTOMOUS, adj. (Gr. cutting, in a three-fold manner).

TRICOLPATE, TRICOLPATUS, adj. cf. colpate, furrow, trimesocolpate.
a. Wodehouse, 1928, p. 934: "Possessing three longitudinal furrows or expansion folds, generally enclosing germinal pores. Example: Tanacentum camphoratum."
b. Wodehouse, 1935, p. 545: "possessing three meridionally arranged germinal furrows."
c. Selling, 1947, p. 9: "(= with three colpae without pores or transverse furrows)."

sdb. Erdtman, 1943, p. 52.
Faegri and Iversen, 1950, p. 128; [see GROUPS OF SPOROMORPHAE and Fig. 60].
Pokrovskaya, et al., 1950.
Traverse, 1955, p. 95.
Beug, 1961, p. 13.

TRICOLPODIPORATE, adj. cf. colpate, colpodiporate, colporate, colpus, furrow, Groups of Sporomorphae, porate.
Selling, 1947, p. 49: "with three colpi, each enclosing three pores or three transverse furrows, e.g., Myoporum." [see COLPODIPORATE].

TRICOLPORATE, adj. colpate, colporate, colpus, furrow, Groups of Sporomorphae, porate, trimesocolporate.
a. Faegri and Iversen, 1950, p. 128: "Three furrows, meridional arranged with pores or transversal furrows (sometimes missing in one or two furrows; see . . . [Fig. 64])."
b. Traverse, 1955, p. 95: "Having three longitudinal furrows, all arranged on meridians, each furrow having an equatorial transverse furrow or a germ-pore."
c. Ingwersen, 1954, p. 41: "pollen type with 3 colpi and 3 pores, the latter centrally in the colpi."
sdb. Erdtman, 1945, p. 190 [see COLPORATE].

TRIFOLIUM, n., pl. -ia. (L. trefoil). cf. triletus, triradial.
Potonié, 1956, p. 53: Tr. While in the Lagenotriletes a gula originates by a somewhat conelike structure of the over-developed tecta, in the Capulitriletes a trifoil (trifolium) is formed by the over-development of the planklike tecta. The tecta in this case do not have a widely broadened base. The diameters do not diminish from base to apex, so that between the three blades three sharp deep nooks are found in which the contact areas are unreduced. [see CONTACT FIGURE and Fig. 136].

TRILATIPORATE, adj. [see LATIPORATE].

TRILETE, TRILETUS, adj. cf. annulotrilete, contact figure, triradial, vesicotrilete.
a. Erdtman, 1943, p. 52: "trilete; possessing a triradiate tetrad scar."
b. Selling, 1947, p. 78: "spore with one triradiate tetrad scar (on the proximal side of the spore). cf. Trichotomosulcate pollen."
sdb. Erdtman, 1947, p. 113; [see GROUPS OF SPOROMORPHAE].
Pokrovskaya, et al., 1950.
Beug, 1961, p. 13.

TRILETE APERTURE, n., pl. -s. cf. contact figure, trilete mark.

TRILETE MARK, TRILETE MARKING, n., pl. -s. cf. contact figure, dehiscence mark, tetrad scar, triradiate, Y-mark, Y-radii.
a. Kosanke, 1950, p. 14: [see Fig. 109–110].

b. Potonié and Kremp, 1955, p. 10; [*see* Y-MARK and Fig. 141–148].

TRILETOID, adj. cf. contact figure.
Erdtman, 1947, p. 113: "with a three-slit aperture, not as distinct as a triradiate scar." [*see* GROUPS OF SPOROMORPHAE].

TRILOBATE, TRILOBATUS, adj. (L. trilobus). cf. shape classes.
a. Jackson, 1928: "three-lobed."
b. Norem, 1958, p. 675: "clover-leaf shaped with more or less sharp angles between lobes." [*see* SHAPE CLASSES and Fig. 167 p].

TRIMESOCOLPATE, adj. cf. tricolpate. [*see* MESOPORATE].

TRIMESOCOLPORATE, adj. cf. tricolporate. [*see* MESOPORATE].

TRIMESOPORATE, adj. cf. triporate. [*see* MESOPORATE].

TRIPORATE, adj. cf. aperture, Groups of Sporomorphae, mesoporate, porate, pore.
a. Kupriyanova, 1948: *Tr.* Pollen grains with three pores. [*see* GROUPS OF SPOROMORPHAE].
b. Faegri and Iversen, 1950, p. 128: "3 pores, restricted to equatorial area, no furrows present." [*see* GROUPS OF SPOROMORPHAE and Fig. 69].
c. Traverse, 1955, p. 95: "Having three germ-pores in the exine, all on or approximately on the equator."
sdb. Ingwersen, 1954, p. 41.
    Beug, 1961, p. 13.

TRIQUETE, adj. (L. triquetrus, three-pointed).
Norem, 1958, p. 675: "equilateral triangle with concave sides and rounded apices." [*see* SHAPE CLASSES and Fig. 167o].

TRIRADIAL, adj. (tri- + L. radius, ray). cf. contact figure, trilete, triletus.

TRIRADIATE CREST, n., pl. -s. cf. contact figure, crista triradiata.

TRIRADIATE FIGURE, n., pl. -s. cf. contact figure.

TRIRADIATE MARKING, n., pl. -s. cf. contact figure.

TRIRADIATE RIDGE, n., pl. -s. cf. contact figure.

TRISACCATE, adj. with three air sacs. [*see* Fig. 197 and 203].

TRISCHISTOCLASIC, adj. (Gr. tri, three; schisto, branching or portal; clasic, a breaking). cf. colpate, furrow.
Wodehouse, 1935, p. 545: "triradiate cracking, the system in which the furrows of the pollen grains of the higher dicotyledons tend to form, a system similar to the cracking of drying mud or shrinking plaster, as if produced by equilateral stresses."

TRISULCATE, adj. cf. trichotomosulcate.
Norem, 1958, p. 669: "Tri(chotomo)sulcate . . .

Note: The syllables within the parentheses may be dropped . . . [Fig. 248–1 and 2]—Trisulcate."

TRITREME, adj. cf. aperture.
Erdtman and Straka, 1961, p. 65: "(3-treme) spores: with three apertures." [*see* GROUPS OF SPOROMORPHAE].

TUBER, n., pl. -s. (L. a tumor). cf. elevation, excrescence, mammilla, process, projection, protuberance, protrusion, tubercle, tuberculum.
I. Jackson, 1928: "a thickened and short subterranean branch, beset with buds or 'eyes'."
II. Erdtman, 1947, p. 106: "projections . . . lower, more rounded (than verrucae) and very densely placed." [*see* TUBEROSE].

TUBERCLE, n., pl. -s. (L. tuberculum). cf. tuber.
I. Jackson, 1928: "a little tuber;" "a wart-like apothecium in *Verrucaria*;" "any similar excrescence, as on roots, ascribed to the action of symbiotic organisms;" "a tuberous root, as of the *Dahlia* (Crozier);" "the bulbil of charads."
II. Kupriyanova, 1948, p. 71: *Tr.* The flattened spine appearance can be called tubercles (tuberculata). [*see* ORNAMENTATION and Fig. 659].

TUBERCLED, adj. cf. tuberculate.
Jackson, 1928: "covered with warty excrescences, as the seeds of *Silene*."

TUBERCULAR, adj. cf. tuberculate.
Jackson, 1928: "having tubercles or like a tubercle."

TUBERCULATA, adj. (L. tuberculatus). cf. tuberculate.
I. Kupriyanova, 1948, p. 70: *Tr.* The flattened spine appearance can be called tubercles (tuberculata). [*see* ORNAMENTATION and Fig. 659].
II. Zaklinskaya, 1957: for tuberculate ornamentation of disaccate pollen grains see Fig. 223.

TUBERCULATE, adj. cf. mammillate, tubercled, tubercular, tuberculose, tuberculous, tuberiferous, tuberose, tuberosus.
Jackson, 1928: "beset with knobby projections or excrescences."

TUBERCULOSE, TUBERCULOUS, adj. cf. tuberculate.
Jackson, 1928: "consisting of or tubercules."

TUBERCULUM, n., pl. -a. (L. a little tuber). cf. tuber.

TUBERIFEROUS, adj. cf. tuberculate.
Jackson, 1928: "tuber-bearing."

TUBEROSE, adj. (L. tuberosus). cf. tuberculate.
a. Jackson, 1928: "producing tubers;" "resembling a tuber."
b. Erdtman, 1947, p. 106: "cf. Verrucosus: projections (tubera) lower, more rounded, and very densely placed."
c. Faegri and Iversen, 1950, p. 163: "(Erdtman, 1947): with projections lower than in verrucate,

more rounded, and very densely placed. (Potonié, 1934): 'höckerig'."

TUBULUS, n., pl. -i. (L. a small pipe).
I. Jackson, 1928: "the pores or hymeneal tubes of some Hymenomycetous Fungi as *Polyporus;*" "in Pyrenomycetes, the prolonged apex of perithecium pierced by a canal, the same as NECK."
II. Erdtman, 1952, p. 472: "fine channels running through the nexine. Tubuli have been observed in *Conospermum incurvum* (Proteaceae) and several nyctaginaceous plants (illustrations in Sv. bot. Tidskr. 46 (2), 1952)."

TUMESCENCE, n., pl. -s. (L. swelling up). cf. aperture, dissence.
Thomson and Pflug, 1953, p. 36: *Tr.* If the ektexine in cross-section thickens as a wedge starting from the center of the side, we speak of a "tumescence" (e.g., in the *Triatriopollenites rurensis*-type).

TUMESCENT, adj.
Jackson, 1928: "somewhat tumid."

TUMID, adj. (L. tumidus, swollen).
Jackson, 1928: "inflated, swollen."

TYPUS BILATERALIS SPORANUM FORMATIONIS. cf. tetrad, tetragonal, monolete.
Pokrovskaya, et al., 1950: *Tr.* Bilateral type of the formation of spores in tetrads proceeds in this manner: that only two sides of each spore are in contact with the neighboring spores, similar to lobes of an orange.

TYPUS RADIALIS SPORANUM FORMATIONIS. cf. tetrad, tetrahedral, trilete.
Pokrovskaya, et al., 1950: *Tr.* The radial type of the formation of spores. The spores form in a tetrad, in such a manner that three sides of each spore are in contact with the neighboring spores.

U

ULCERATE, adj. (L. ulcus, a sore, ulcer). cf. ana-porate, aperture, Groups of Sporomorphae, porate, pore, ulcus.

Erdtman, 1952, p. 472: "with ulcus."

ULCUS, n., pl. -i. cf. aperture, diaporus, endoporus, exitus, exoporus, germinal apparatus, germ pore, os, pore, pseudopore.

Erdtman, 1952, p. 472: "a single, ±pore-like aperture, usually situated at the distal pole or ± near it. Examples: . . . [Fig. 101–102, 271–282]."

UNADORNED, adj. cf. laevigate, psilate, smooth.

UNDITEGILLATE, adj. (L. unda, a wave). cf. tegillate.

Erdtman, 1952, p. 472: "with a tegillum with an undulating upper surface. The undulations are due to variations in the thickness of the tegillum itself or—if the thickness is the same throughout—to variations in the length of the supporting bacula."

UNDULATE, adj. (L. undulatus, wavy). cf. waved.

I. Jackson, 1928: "wavy."

II. Harris, 1955, p. 20: "Waved, more or less iso-diametric (elevations). Example: *Gleichenia circinata.*" [*see* ORNAMENTATION].

UREDINIOSPORE, n., pl. -s. cf. uredospore.

Jackson, 1928: "(Arthur) = uredospore."

UREDOSPORE, n., pl. -s. (L. uredo, a blight, blast, burning itch). cf. urediniospore.

a. Jackson, 1928: "a spore formed by acrogenous abjunction from a sterigma, germinating immediately and producing a mycelium which bears other uredospores alone, or with teleutospores."

b. Melchior and Werdermann, 1954, p. 15; [*see* SPORE].

V

VALLAE, VALLUM, n. pl. valli, (L. an earthen wall or rampart set with palisades). cf. lira, list, ridge, rugula, stripe, wall.

Faegri and Iversen, 1950, p. 163: "ridges in striate and rugulate sculpturing types." [*see* ORNAMENTATION and Fig. 688–689].

sdb. Iversen and Troels-Smith, 1950, p. 35; [*see* ORNAMENTATION].

VALVA, VALVE, n., pl. -ae, -s. (L. leaf of a folding door). cf. auricula, ear.

I. Jackson, 1928: "Valve, valva: a piece into which a capsule naturally separates at maturity;" "the segment of a calyx meeting in vernation without overlapping;" "in Diatoms, each half of the silicified membrane in side view;" "the lid of an ascidium (Crozier);" "the flowering glume of grasses (Stapf);" "a partially detached flap of an anther; *Valvae Seminum* = COTYLEDONS."

II. Potonié and Kremp, 1955, p. 15: *Tr.* The swelling in the equatorial region of spores (Zonales) may be more pronounced at the "corners" of the equatorial section of the spore than on the rest of the equator. Then we speak of *valvae*. [*see* EQUATORIAL REGION and Fig. 149].

sdb. Potonié and Kremp, 1956, p. 86; [*see* ZONA].

VALVEANUS, adj.

Jackson, 1928: "when a partition arises from the expansion of the inner substance of a valve."

VELUM, n., pl. -a. (L. an awning). cf. aerostatic umbrella, marginal frill.

I. Jackson, 1928: "a special envelope in Agarics within which the growth of the sporophore takes place;" "by Persoon applied to the CORTINA;" "the membranous indusium in *Isoëtes* (A. Braun)."

II. a. Thomson and Pflug, 1953, p. 27–28: *Tr.* The species of the zone-bearing *Tsuga* pollen may be mentioned here. The zone-development in these gymnosperms forms differently; it forms by elongation of the radial elements of the rodlet layer. In this way, an "aerostatic umbrella" is formed which is designated as a "marginal-frill" (velum).

The "*Diversifolia*-type" (Rudolph) has the velum completely developed; in the "*Canadensis*-type" the marginal frill seems to be vestigial regressed.

b. Potonié and Kremp, 1955, p. 19: *Tr.* Another development [of the saccus] is the velum; it is, so to speak, as a single limbus over the entire sac of the Monosaccites (see *Tsugapollenites igniculus*. Potonié, 1951). [*see* LIMBUS].

c. Couper and Grebe, 1961, p. 5 and Fig. 201: "This term should be applied only to a thin *exoexinal* layer, characteristically thrown into small, closely spaced folds or convolutions and which is extended globally."

VENTRAL, adj. (L. ventralis, pertaining to the belly). cf. distal, distalis, exterial.

a. Jackson, 1928: "the anterior or inner face of a carpel, opposed to dorsal;" "relating to the VENTER."

b. Wodehouse, 1935, p. 545: "The side of a grain turned outward in its tetrad. In monopored or monocolpate grains it is the side upon which the pore or furrow is borne. In other grains the dorsal and ventral sides are generally not distinguishable from each other after the grains have separated from their tetrads."

VENTRAL ROOT OF THE BLADDERS.

Wodehouse, 1935: [*see* SACCATE and Fig. 205–206].

VERMICULAR, adj. (L. vermicularis). cf. vermiculate.

VERMICULATE, adj. (L. vermiculatus, vermiculus, a little worm). cf. fossulate, grooved, vermicular, vermiculus.

a. Jackson, 1928: "worm-shaped, thickened and bent in places, as the root of *Polygonum Bistorta*, Linn."

b. Kosanke, 1950, p. 11: *see* Fig. 640:G.

c. Harris, 1955, p. 19: "With grooves like worm tracks. The example is from Kosanke (1950) where the term is used (cf. fossulate)." [*see* ORNAMENTATION].

sdb. Couper and Grebe, 1961, p. 8; [*see* ORNAMENTATION and Fig. 686].

VERRUCA, n., pl. -ae. (L. a wart). cf. wart.
a. Jackson, 1928: "a wart or elevation sometimes of a glandular nature;" "a sessile apothecium, as in *Verrucaria;*" "the perithecium of some Fungi."
b. Potonié, 1934, p. 11: *Tr.* The grana are usually more or less round to globular, the verrucae are more irregular in outline.
c. Faegri and Iversen, 1950, p. 163: [*see* Fig. 623–626].
d. Iversen and Troels-Smith, 1950, p. 35: *Tr.* = warts: the largest diameter is as large as the height or larger; sculpture elements are neither pointed or constricted. [*see* ORNAMENTATION].
e. Erdtman, 1952, p. 472: "wartlike sculptinous projections (basal diameter as a rule longer than any other tangential diameter)."
f. Ingwersen, 1954, p. 41: "more or less isodiametric sculpturing elements which in at least one dimension are greater than or equal to 1 μ and whose greatest diameter is greater than or equal to the height. The elements are neither distally pointed nor constricted at the base. The ratio of greatest to smallest diameter is less than 2."
g. Potonié and Kremp, 1955, p. 14: *Tr.* The verrucae have a more irregular outline and, more important, on one and the same spore they vary somewhat in shape and size; in plan view, one can notice smaller verrucae next to somewhat larger ones, round next to less round, lobulate next to bean-shaped, etc. Consequently, when the elements stand closely together, they form a more irregular mosaic than the grana . . . [Fig. 741]. Furthermore, the verrucae are usually not as high or not much higher than their width. Usually their greatest width is no more significant than their least; however, a ratio of 2:1 may occur. The surface of the verrucae can be rounded, flat, or irregular. [*see* ORNAMENTATION].
sdb. Erdtman, 1943, p. 52.
 Kupriyanova, 1948, p. 70; [*see* ORNAMENTATION and Fig. 660].
 Beug, 1961, p. 13.
 Couper and Grebe, 1961, p. 6; *see* Fig. 678.

VERRUCATE, VERRUCATUS, adj. cf verrucatus, verrucose, verrucosus, wartlike, warty.
a. Erdtman, 1943, p. 52: "There is a gradual transition from grana to the warts of verrucate exines." [*see* ORNAMENTATION and Fig. 639].
b. Harris, 1955, p. 19: "Broad, or, if more isodiametric, larger than granulate (*Lygodium articulatum*)." [*see* ORNAMENTATION].
c. Traverse, 1955, p. 94: "Sculpturing of projections not specially constricted or modified in any way, at least one dimension of which exceeds 1 μ."

d. Couper, 1958, p. 102: "Granulate—more or less isodiametric; not more than 1/20th of the equatorial diameter of the spore if the latter is over 20 μ. Verrucate—more or less isodiametric, larger than granules." [*see* ORNAMENTATION].
sdb. Kupriyanova, 1948; [*see* ORNAMENTATION].
 Ingwersen, 1954, p. 41.
 Beug, 1961, p. 13.

VERRUCOSE, adj. (L. verrucosus, full of warts). cf. verrucate.
Erdtman, 1947, p. 106: "with wart-like projections ('verrucae'); the basal diameter of a projection of this kind is as a rule longer than any other tangential diameter of the projection . . ."
sdb. Kosanke, 1950, p. 11; [*see* ORNAMENTATION and Fig. 640:I].
 Pokrovskaya, et al., 1950.

VERTEX, n., pl. -es. (L. that which revolves about itself). cf. contact figure, crestline, list, murus, ridge.
I. Jackson, 1928: "the apex of an organ;" "± the pileus of Agarics."
II. Potonié and Kremp, 1955, p. 11: *Tr.* The ridge of the tecta. [*see* Y-MARK and Fig. 141–148].

VESICLE, n., pl. -s. (L. vesicula, a little bladder). cf. saccate, saccus.
Jackson, 1928: "a small bladder or cavity;" "Grew's term for CELL."

VESICOTRILETE, adj. (L.). cf. vesiculate.
Norem, 1958, p. 668: [*see* Fig. 196–1].

VESICULA AERIFERA, n., pl. -ae. (L. vesicula, a little bladder). cf. air sac, air sack, bladder, cavea, sac, saccus, sack, wing.
Pokrovskaya, et al., 1950: *Tr.* Air sac, swelling. Swellings on the body of microspores of the families *Pinaceae* and *Podocarpaceae*.

VESICULAR, adj.
Jackson, 1928: "composed of vessels."

VESICULATE, adj. (L. vesiculatus). cf. bisaccate, disaccate, monosaccate, perisaccate, saccal, saccatus, subsaccate, vesicotrilete, vesiculosus, winged.
I. a. Cranwell, 1953, p. 17: "winged gymnosperm grains."
 b. Traverse, 1955, p. 95: "Having parts of the exine modified to bladders, which stand apart from the main body of the grain. These grains, cf. *Pinus,* are also referred to as monocolpate, because of the single furrow, but vesiculate seems better, at least for *Pinus,* where the furrows are usually very difficult to observe."
sdb. Faegri and Iversen, 1950, p. 128.
 Norem, 1958, p. 668.
II. Harris, 1955, p. 27: "with air sacs. Used in this manual for rounded excrescences as in the spores of *Pyrrosia serpens* (compare SACCATE)." [*see* PERISPORE and Fig. 564].

VESICULOSUS, adj. cf. vesiculate.
Pokrovskaya, et al., 1950: *Tr.* Furnished with air sacs.

VESTIBULATE, adj. aperture, atrium, pore, vestibulum.
Ingwersen, 1954, p. 41: "type of pore with vestibulum." [*see* Vestibulum].

VESTIBULE, n., pl. -s.
Jackson, 1928: "a chamber above the stoma formed by the depression of the guard-cells and growth of cells round them, as in Cycas."

VESTIBULUM, n., pl. -a. (L. a forecourt). cf. porus vestibuli.
  I. a. Potonié, 1934, p. 19: *Tr.* The *porus vestibuli* is formed thus: that the exoexineous margin of the fovea is compressed over its intexininous floor . . . [Fig. 391]. The margin of the depression then looks like a porus (known as porus vestibuli . . . [Fig. 391]), which leads into one of the small intexine-enclosed ante-chambers, the *vestibulum*.
  Vestibulum apparatus—A more thorough examination of germinal apparatus, e.g., *Corylus*, . . . shows that in this case the germinal apparatus belongs in part to the strongly differentiated type of vestibulum-apparatus (e.g., in *Corylus* sp. from sapropel Lake Ahlbeck sediment). Here the exoexine is clearly separated from the intexine in all parts of the membrane by a distinct separation line. Possibly the intexine detaches itself slightly from the exoexine at the germinal apparatus . . . [Fig. 392], but this is not clear. Thus too, a distinct porus vestibuli would grow out of the fovea, pierce the outer exine, and drive into a small, more or less lense-shaped vestibule. In any case, this impression can be conjured-up many times. However, in certain optical cross sections, it can also occur in an apparatus as shown in . . . [Fig. 393]. The ante-chamber, the *vestibulum* in *Corylus,* is in many instances still very small but it can in other cases be distinctly perceived. An opening which connects the vestibulum and cell chamber of the pollen grain is not present. However, the intexine is probably here many times thinner than in its remaining portions. Hence, the porus leads only from the outside into the vestibulum. . . . [Fig. 394] shows the germinal apparatus of *Corylus* sp. in a vertical view of the exine surface. The porus is evident as a small point. At a distance, two parallel circles surround the porus. The diameter of the inner circle is about three times as big as that of the porus. The outer circle is not always clearly defined; it encircles the first at just a little distance. Dokturowsky and Kudrjaschow (1924) give an explanation of vestibulum apparatus by drawings without text. [*see* Apertura, Annulus, Aquatorialruga, Arcus, Exitus Digitatus, Ruga Compressa].
  b. Erdtman, 1943, p. 52: "a small chamber under an aspidate pore."
  c. Pokrovskaya, et al., 1950: *Tr.* A pore chamber.
  II. a. Faegri and Iversen, 1950, p. 163: "cavity forming the pore and being separated from the interior of the grain by a low rim or by a separation between different layers of the exine, e.g., *Betula* or *Circaea*. Originally used by Potonié (1934) with a slightly different definition." [*see* Furrow and Fig. 357].
  b. Thomson and Pflug, 1953, p. 33: *Tr.* If the endexine continues independently after splitting and encloses an endoporus which is not more than three times as large as the exoporus, the structure is called a "vestibulum." [*see* Apertura and Fig. 381].
  c. Ingwersen, 1954, p. 41: "more or less sharply delimited small cavity within the pore; vestibulate: type of pore with vestibulum."
  sdb. Pike, 1956, p. 51.
  III. Hyde, 1954, p. 256: "Erdtman (1943, p. 52) applied the term *vestibulum* (which had been used by Potonié, 1934, in the sense of a circular surface cavity lined with intexine) to a 'small chamber under an aspidate pore.' He went on: 'Aspidate germ pores protrude as rounded domes. *The protrusions are due to a thickening of the intine underlying the region of the pore.*' (The sentence here italicized was a quotation from Wodehouse (1935, p. 363), whose studies were all made on non-fossilized pollen.) It might perhaps have been inferred, therefore, that Erdtman was using *vestibulum* as synonymous with Zander's *Keimhof,* though the context provides no evidences for or against this. In his later treatise (1952) Erdtman omits the term *vestibulum* altogether. Faegri and Iversen's definition (1950, p. 163) of *vestibulum,* viz., 'a cavity forming the pore and separated from the interior of the grain by a low rim, or by a separation between different layers of the exine,' is clearly intended to exclude the intine, and the same applies to Iversen and Troels-Smith's (1950), viz., 'a small anteroom [Vorraum] within the pore which arises in such a way that the exine at the pore edge as seen in optical section has a split-in-two appearance'." [*see* Oncus].

VESTIBULUM APPARATUS, n. cf. porus vestibuli.

VESTIBULUM PORI, n.
Iversen and Troels-Smith, 1950, p. 33: *Tr.* Small "vestibulum" within the pore which originates by the exine having a two-pronged aspect at the edge of the pore in optical cross section. [*see* PORE].

VESTIGIAL SPINE, n., pl. -s. (L. vestigium, footprint, sign). cf. conus, echinus, ornamentation.
Wodehouse, 1935, p. 545: "those of less prominence than of subechinate grains. Example: *Xanthium* . . . [Fig. 711]."

W

WALL, n., pl. -s. (L. vallum, a wall set with palisades). cf. lira, list, murus, ridge, vallum, vertex.

WART, n., pl. -s. cf. verruca.
Jackson, 1928: "a hard or firm excrescence."

WARTY SCULPTURE, n., pl. -s. cf. verrucate.
Kupriyanova, 1948, p. 70: *Tr*. Warty sculpture, that is an exine with warts, the basal diameter of which is greater than its height. [*see* ORNAMENTATION and Fig. 660].

WAVED, adj. cf. undulate.
Jackson, 1928: "undulate, or sinuate."

WEDGE, n., pl. -s. cf. clava, club.

WIDTH, n., pl. -s. (Measurement of spores and pollen—in disaccate grains *see* SACCATE). cf. breadth, dimension, figura.

WING, n., pl. -s. cf. air sac, air sack, bladder, cavea, sac, saccus, sack.
a. Jackson, 1928: "= ALA, any membranous expansion attached to an organ;" "a lateral petal of a papilionaceous corolla."
b. Wodehouse, 1935, p. 545: "the bladdery projection flanking or surrounding, frill-like, the germinal furrow of the grains of some Abietineae and Podocarpineae. It is generally greatly distended and attached by its ventral roots along the furrow rim and by its dorsal roots just ventrad of the marginal ridge." [*see* Fig. 205–206].
sdb. Harris, 1941, p. 50; [*see* DIMENSION].

WINGED, adj. cf. cavate, pouched, rugulo-saccate, saccate, subsaccate, vesiculate.
I. Jackson, 1928: "alate."
II. Harris, 1955, p. 27: "an alternative to 'possessing a perispore,' used in this manual only when the general contour of the spore is modified by the presence of this layer and implying that the perispore, irrespective of projections which may be said to form the sculpture pattern is almost separated from the exine by a space (cavate)." [*see* PERISPORIUM].

WRINKLED, adj. cf. corrugate, costate, ribbed, rugose, rugous, rugulate.
a. Jackson, 1928: "rugose, creased."
b. Kupriyanova, 1948, p. 71: *Tr*. A type of ornamentation. [*see* ORNAMENTATION].

Y

Y-MARK, n., pl. -s. cf. aperture, contact figure, dehiscence mark, Y-splitmark.

I. Potonié, 1934, p. 12: *Tr.* In contrast to pollen, these spores usually show a triradiate mark, the *Y-mark,* which is formed by *dehiscence lists* and *dehiscence furrows* . . . [Fig. 107–108], and which aids in germination. We find this arrangement in bryophytes too. Comparatively, the uni-radiate dehiscence ridge is a rare type among Pteridophytes. [*see* APERTURE, CONTACT AREA and Y-SPLITMARK].

II. Thomson and Pflug, 1953, p. 26: *Tr.* The Y-mark is a tri-radiate, linear, colpus-like solution of one or more exosporlamellae which does not encompass the complete exospore. However, the unbroken lamellae are easily torn mechanically. [*see* SOLUTION AREA].

III. Potonié and Kremp, 1955, p. 10: *Tr.* Y-mark, *Tecta, Vertex, Suture.* The dehiscence mark (laesura, Erdtman, 1952) in spores forming from a tetrahedral-tetrad appears as a *tetrad mark, trilete mark,* or *Y-mark*; in spores separated from the tetragonal-tetrad it appears as a *monolete mark.* Contact areas can remain in the angles of the Y-mark and on both sides of the monolete mark. The triradiate, that is the trilete or Y-mark, is also often found in the fossil state with closed sutures (q.v.). The three rays or tecta originate at the apex . . . [Fig. 132–133]. [Potonié and Kremp, 1955, p. 11: *Tr.*] In plan view, they form angles of approximately the same size. Then the apex is the highest point. The tecta are characteristically formed, more or less high, sometimes very low, roof or wall-like pleats or ridges of the exine. This can be best seen in cross sections of the spores . . . As seen in [those] cross sections, the crestline, that is the ridge or *vertex* of the tecta, is in many cases closed. This can also be found in the plan view of macerated spores [e.g., . . . *Laevigatisporites glabratus* (Zerndt)]. Often one does not notice anything about the *suture,* "dedehiscence furrow" or seam. Only when the suture opens [Fig. 129] can it be seen, in the plane view as well as in the cross section of the exine. What one would often regard as the suture in the macerated spore is the intratectum . . . [Fig. 141–148].

Intratectum, Labra—If one focuses the microscope on the uppermost part of the tectum, that is on the vertex in microspores compressed along the equator, an image often appears as in . . . [Fig. 143]; the suture is not, or hardly recognizable. If one focuses lower, one sees an image as in . . . [Fig. 147] which is the *intratectum* . . . [Fig. 142], that is the more or less narrow interior of the tectum. In . . . [Fig. 142], the intratectum is surrounded, in cross section, by the walls of the tectum, that is by the *labra,* lips, "swellings," or "dehiscence-ridges." The name "lips" would actually be proper only when the suture has opened . . . [Fig. 148]; however, it is also applied with certain justification to the walls of the tectum that are still connected. [Potonié and Kremp, 1955, p. 12: *Tr.*] The side walls, the so-called lips, can appear to be thicker than the remaining exine of the spore. They can appear thicker than they actually are, especially if the tectum, because of its state of preservation, no longer stands more or less perpendicular so that the optical cross section is inclined . . . [Fig. 141].

Subtectum—In focusing still lower a view as in . . . [Fig. 144] may appear. The intratectum may under the circumstance gradually widen; it is widest in the *subtectum* . . . [Fig. 142]. The labra, which originally could also have widened, have disappeared.

Contravertex—Finally (not always) a view as in . . . [Fig. 146] appears; here the subtectum has its greatest dimension. Sometimes it has an irregular more or less distinctly flexuose outline; in the center of its rays, a new Y-ray appears brightly, *the contravertex* . . . [Fig. 146:C], i.e., the nestling of the distal exine opposite the Y-mark in the subtectum.

Consequently, quite different images (also

photos) can be obtained from one and the same Y-mark.

Fissure—The problem becomes even more complicated when the suture opens. When this happens a simple fissure may appear, in which case the edges of the separated walls of the tectum and also of the apex show nothing in particular. But sometimes the parts of the exine surrounding the fissure show, after splitting, very characteristic types of transformation, of rolling up, etc. This behavior cannot be described just by the word "state of preservation" in the usual sense. It seems to be a fact that any given form species exhibits in any certain state of preservation its own characteristics. It may be sometimes of importance for the determination of a form species whether or not the suture opens more or less easily to form a fissure. . . . *Tuberculatisporites difficilis* (Wicher).

. . . [Fig. 148] shows the open apex and the open tecta of a type which is otherwise not very significant. After the suture has opened, the rather high walls of the tecta—the labra—are separated. Corresponding outer lines of Fig. 148 were originally joined with one another to form the crestline (the vertex) of the Y-mark. Because the apex was originally somewhat higher than the tecta, three spoonlike lobes of the exine now stand in the place of the apex; this is a phenomenon which appears only in certain form species; one is reminded of *Lagenicula* . . . [Fig. 131].

Y-RADIUS, n., pl. -i. (L. radius, staff, spoke, rod). cf. contact figure, Y-mark, Y-splitmark.

Y-SPLITMARK, n., pl. -s. cf. aperture, contact figure, Y-mark.

Thomson and Pflug, 1953, p. 27: *Tr.* A further differentiation of the Y-ridge is the splitting of the ridge ends into two branches (Y-splitmark). [*see* Fig. 113–114].

Z

ZONA, n., pl. -s. (L. zona, a belt, zone). cf. arista, cingulum, collar, crassitude, crassitudo, equatorial collar, equatorial region, flange, frassa, ring, zone.
  a. Thomson and Pflug, 1953, p. 27: *Tr.* If the exosporelamellae are split so wide in the equatorial region that the interspace created is wider than the thickest wall, then it is called a zone (Zona). Often the zone encircles the equator as a rim which is usually air filled and which probably improves the spore's suspension qualities as an aerostatic tube and is probably analogous to the air sacks of the conifers e.g., *Cingulatispore levispeciosus.*
  b. Potonié and Kremp, 1955, p. 15: *Tr.* If the equator has a wide membranous flange which is not wedge-shaped in cross section, we are dealing with a *zona* or *frassa* which is also called equatorial collar or simply collar (Kragen) (as in *Cirratriradites*). [*see* EQUATORIAL REGION and Fig. 155].
  c. Potonié and Kremp, 1956, p. 86 and Fig. 149–158: *Tr.* Zona or frassa is a broad, closed, ± membranous rim or collar situated on the equator. The *corona* is an equatorial wreath which is ± dissolved in the fimbria, i.e., in hairs, into branching and anastomosing hairs, fringes, or the like. The *cingulum* (= Arista) is a smaller ± massive equatorial ring or hoop and ± wedged cross sectioned. The *crassitudo* arising ± in the polar hemisphere undergoes a gradual thickening of the exine toward the equator, often without an equatorial sharpening. The *auriculae* are thickenings or lobes of the three triangular tips of the equatorial section; they can appear as ± stronger pillows of the exine (*valvae*), or as little ears (auriculae, s. str.) which are situated on the triangular tips. (The formation of the pillowlike valvae as well as auriculae and kyrtomes happened partly by virtue of the fact that certain layers and parts of the exine thicken) perhaps also in part by virtue of the fact that localized separations occur between the exine layers. The air sacs of the Saccites, called sacci, could be developed if these separations increase in size so

that large gas-filled spaces arise between the separating exine layers.)
  d. Couper and Grebe, 1961, p. 4: "a structural feature which appears as an equatorial extension of the exoexinal layer . . . [Fig. 188–189]. It is similar in some respects to a saccus . . . but lacks the exoexinal structural elements of the latter. Spores with a zona are described as *zonate.*"

ZONACOLPATE, adj.
  Norem, 1958, p. 672: "Syncolpate—with furrows fused into . . . rings . . . [Fig. 442–1]."

ZONAPERTURATE, adj. cf. colpate, colporate, furrow, Groups of Sporomorphae.
  Erdtman, 1952, p. 472: "with apertures forming one ± continuous zone (cf. 'zonisulculate') or several zones parallel to each other. cf. also 'zonorate'."

ZONATE, adj. (L. zonatus). cf. annulate, annulotrilete, zona.
  I. Jackson, 1928: "marked circularly, as the leaves of *Pelargonium zonale,* L'Herit."
  II. a. Wodehouse, 1935, p. 545: "provided with one or more furrows, each encircling the grain as a lesser circle, wholly in one hemisphere and usually parallel to the equator."
  sdb. Faegri and Iversen, 1950, p. 163.
  III. b. Cranwell, 1953, p. 11: "Continuous or Twinned Furrows. Aberrant or imperfectly understood types include those with merged (fused) furrows, usually separable into two elements, *zonate* if running around the equator, and *dicolpate* if running meridionally." [*see* DICOLPATE].

ZONE, n., pl. -s. cf. zona.
  Jackson, 1928: "of temperature, with its influence on distribution;" "a belt of more or less uniform vegetation;" "the connection between two valves of a Diatom;" "the hoop or girdle."

ZONED FURROW, n., pl. -s. cf. colpus, furrow, zonosulcatus.

ZONICOLPATE, adj. cf. colpate, furrow, stephanocolpate.
  Erdtman, 1958, p. 137: [*see* GROUPS OF SPOROMORPHAE and Fig. 105].

ZONIPORATE, adj. cf. colpate, colporate, porate.
Erdtman, 1958, p. 137: [*see* GROUPS OF SPOROMOR-PHAE and Fig. 105].

ZONISULCATE, adj. cf. colpate, furrow.
Erdtman, 1952, p. 470: "more or less sulcoid apertures, parallel to the equator and usually situated between the latter and the distal pole. If apically united the sulculi form a zone, or ring parallel to the equator (as in the case of zonisulculate pollen grains." [*see* SULCULUS].

ZONORATE, adj. cf. colpate, colporate, Groups of Sporomorphae, porate.
Erdtman, 1952, p. 472: "with anastomosing ora forming a continuous equatorial oral zone. Example: . . . [Fig. 487–488]."

ZONOSULCATE, adj. cf. colpate, furrow.
Erdtman, 1947, p. 113: "with one continuous sulcus-zone." [*see* GROUPS OF SPOROMORPHAE].
sdb. Kupriyanova, 1948; [*see* GROUPS OF SPORO-MORPHAE].

ZONOTREME, adj. cf. aperture.
Erdtman and Straka, 1961, p. 66: "(monozono-treme) spores: usually pleotreme spores with the centers of the apertures situated at the middle line of a zone coinciding with or parallel to the equatorial belt." [*see* GROUPS OF SPOROMORPHAE].

ZOOSPORE, n., pl. -s. (Gr. zoion, an animal; spore, seed).
a. Jackson, 1928: "a free-moving spore, an asexual reproductive cell with cilia, sometimes a plano-gamete."
b. Melchior and Werdermann, 1954, p. 15; [*see* SPORE].

ZYGOSPORE, n., pl. -s. (Gr. zygon, a yoke).
Jackson, 1928: "a body produced by the coalescence of two similar gametes."

ZYGOTE, n., pl. -s. (Gr. zygotos, yoked).
Jackson, 1928: "a body produced by fertilization or conjugation of two gametes;" "by Bateson extended to denote the individual which develops by somatic divisions from the cell resulting from the gametic union."

References

"Anonymous." 1945. Studies on Pollen Analysis. *Nature,* v. 155, p. 264.

————. 1958. Towards Terminological unification in pollen and spore morphology. *Grana Palynologica (N.S.),* v. 1, no. 3, p. 1–5, Stockholm.

Beug, H. J. 1961. Leitfaden der Pollenbestimmung für Mitteleuropa and angrenzende Gebiete. G. Fischer-Verlag, 63 p., 8 pl., 17 text-figs., Stuttgart.

Bower, F. O. 1923. The ferns (Filicales). Treated comparatively with a view of their natural classification. Vol. 1, Analytical examination of the criteria of comparsion. 359 p., 309 text-figs., Univ. Press, Cambridge.

————. 1935. Primitive land plants. Macmillan, 658 p., London.

Couper, R. A. 1958. British Mesozoic microspores and pollen grains, a systematic and stratigraphic study. *Palaeontographica,* Abt. B, v. 103, no. 4–6, p. 75–179, 11 text-figs., Stuttgart.

———— and H. Grebe. 1961. A recommended terminology and descriptive method for spores: Compte Rendu, III. Réunion de la Commission Internationale de Microflore du Paléozoique, Krefeld, May 11–13, 1961, Report of Group No. 16, 15 p., 32 figs., distributed by Cerchar de France, Verneuil-en-Halafte, Oct. 3, 1961 (multilith print).

Cranwell, L. M. 1942. New Zealand pollen studies. *Rec. Auckland Inst. and Museum,* v. 2, no. 6, p. 280–308, Auckland.

————. 1953. New Zealand pollen studies: The Monocotyledons. *Bull. Auckland Institute and Museum,* no. 3, 91 p., 8 pl., Harvard Univ. Press, Cambridge, Mass.

De Jekhowsky, B. 1958. Méthodes d'utilisation stratigraphique des microfossiles organiques dans les problèmes pétroliers. *Révue, Inst. Français du Pétrole et Annales des Combustibles Liquids,* v. 13, no. 10, p. 1391–1418, 3 pl., Paris.

Dijkstra, S. J. and P. H. van Viersen Trip. 1946. Eine monographische Bearbeitung der karbonischen Megasporen mit besonderer Berücksichtigung von Südlimburg (Niederlande). *Mead. Geol. Stichting,* Ser. C–III–1, no. 1, 101 p., 16 pl., Maastricht.

Dokturowsky, W. and W. Kudrjaschow. 1924. Schlüssel zur Bestimmung der Baumpollen im Torf. *Geol. Archiv.,* v. 3, p. 180–183, Königsberg and Dahlem.

Eames, A. J. 1936. Morphology of vascular plants (Psilophytales to Filicales). McGraw-Hill, 433 p., 215 text-figs., New York.

Erdtman, G. 1943. An introduction to pollen analysis. New Ser. Pl. Sci. Books, F. Verdoorn, v. 12, 239 p., 15 text-figs., 28 pl., Waltham, Mass.

————. 1945. Pollen morphology and plant taxonomy. III. *Morina* L. With an addition on pollen-morphological terminology. *Svensk Bot. Tidskrift,* v. 39, no. 2, p. 187–191, figs. 1–9, Uppsala.

————. 1945. Pollen morphology and plant taxonomy. V. On the occurrence of tetrads and dyads. *Svensk Bot. Tidskrift,* v. 39, no. 3, p. 286–297, 2 figs., Uppsala.

————. 1946. Pollen morphology and plant taxonomy. VI. On pollen and spore formulae. *Svensk Bot. Tidskrift,* v. 40, no. 1, p .70–76, Uppsala.

————. 1947. Suggestions for the classification of fossil and recent pollen grains and spores. *Svensk Bot. Tidskrift,* v. 41, no. 1, p. 104–114, Uppsala.

————. 1948. Did dicotyledonous plants exist in early Jurassic times? *Geol. Foreningens Forhandlingar,* v. 70, no. 2, p. 265–271, Stockholm. (Reprinted as *Grana Palynologica,* Part 1, 1948).

————. 1951. On the *"Tricolporites protrudens* problem." *Svensk Bot. Tidskrift,* v. 45, p. 355–357.

————. 1952. Pollen morphology and plant taxonomy. Angiosperms. Almquist and Wiksell, 539 p., 261 text-figs., Stockholm.

————. 1957. Pollen and spore morphology. Plant taxonomy. Gymnospermae, Pteridophyta, Bryophyta (Illustrations). (An introduction to Palynology. II). Almquist and Wiksell, 151 p., text-figs., Stockholm.

————. 1958. On terminology in pollen and spore morphology. *Uppsala Universitets Arsskrift,* v. 6, p. 137–138.

————. 1958. Towards terminological unification in pollen and spore morphology. *Grana Palynologica (N.S.),* v. 3, p. 3–5, Stockholm.

————. 1966. Sporoderm morphology and morphogenesis, a collocation of data and suppositions. *Grana Palynologica,* v. 6, no. 3, p. 318–323, Stockholm.

———— and H. Straka. 1961. Cormophyte spore classification. *Geol. Foreningens Forhandlingar,* v. 83, no. 1, p. 65–78, Stockholm.

———— and Vishnu-Mittre. 1956. On terminology in pollen and spore morphology. *Palaeobotanist,* v. 5, no. 2, p. 109–111, Lucknow. (Reprinted in *Grana Palynologica* (N.S.), v. 1, no. 3, p. 6–9, 1958.)

Faegri, K. and J. Iversen. 1950. Text-book of modern pollen analysis. Munksgaard, 168 p., 17 text-figs., Copenhagen.

Fischer, H. 1890. Beiträge zur vergleichenden Morphologie der Pollenkörner. 72 p., pl. 1–3, Berlin.

References

Fischer, H. 1892. Beiträge zur Morphologie der Farnsporen; Jahresberichte der Schlesischen Gesellschaft vaterländischer Kultur. Breslau.

Fitting, H. 1900. Bau und Entwicklungsgeschichte der Makrosporen von *Isoëtes* und *Selaginella* usw. *Bot. Zeitung,* v. 58, p. 107.

Fritzsche. 1832. Beiträge zur Kenntnis des Pollen. Berlin.

————. 1837. Über den Pollen. *Mém. Sav. Étrang Acad. Sci. St. Petersburg,* v. 3, p. 649–672, pl. 1–13.

Funkhouser, J. W. 1959. A survey of non-spore-pollen palynology (Abstract). *9th Internat. Bot. Congress, 1959, Proc.,* v. 2, p. 126, Montreal.

Grayson, J. F. 1956. The conversion of calcite to fluorite. *Micropaleontology,* v. 2, no. 1, p. 71–78, text-figs. 1–g. tab. 1, New York.

Guennel, G. K. 1952. Fossil spores of the Alleghenian coals in Indiana. *Indiana Dept. Conserv., Geol. Surv., Rept. of Prog.,* no. 4, 40 p., 9 text-figs., 4 pl., Bloomington.

Harris, W. F. 1955. A manual of the spores of New Zealand Pteridophyta. A discussion of spore morphology and dispersal with reference to the identification of the spores in surface samples and as microfossils. *New Zealand, Dept. Scientific and Industrial Research, Bull.* 116, 186 p., 4 text-figs., Wellington.

Hoeg, O. A., M. N. Bose, and S. Manum. 1955. On double walls in fossil megaspores. With description of *Duosporites congoensis* n. gen., n. sp. *Nytt Magasin for Botanikk,* v. 4, p. 101–108, 1 text-fig., 2 pl., Oslo.

Hyde, H. A. 1944. Pollen analysis and the museums. *Museums Jour.,* v. 44, p. 145–149, London.

————. 1954. Oncus, a new term in pollen morphology. *The New Phytologist,* v. 54, no. 2, p. 255–256, 1 text-fig., Cambridge.

Ingwersen, P. 1954. Some microfossils from Danish late-Tertiary lignites. *Danmarks Geol. Undersögelse,* 2. Raekke, no. 80, p. 31–63, pl. 3–6, 1 text-fig., Copenhagen.

Iversen, J. and J. Troels-Smith. 1950. Pollenmorfologiske definitioner og typer (Pollenmorphologische Definitionen und Typen). *Danmarks Geol. Undersögelse,* 4. Raekke, v. 3, no. 8, 52 p., 16 pl., Copenhagen.

Jackson, B. D. 1928. A glossary of botanic terms with their derivation and accent. Duckworth and Co., 481 p., London. (2nd edition, reprinted June 1949.)

Kosanke, R. M. 1950. Pennsylvanian spores of Illinois and their use in correlation. *Illinois State Geol. Survey Bull.,* v. 74, 128 p., 18 pl., 7 text-figs., Urbana.

Kozo-Polyanskiy, B. M. 1945. Novye uspekhi polinistiki i problema zvolyutsii vysshikh rasteniy [Modern advances in pollinistics and the evolution of higher plants]. *Uspekhi Sovr. Biologii,* v. 19, no. 2, p. 236–247, (in Russian).

Kremp, G. O. W. 1959. Can expanding palynology escape taxonomic chaos? *Int. Bot. Congress, Montreal,* 1959, v. 2, Abstracts, p. 206–207.

Krutzsch, W. 1954a. Bemerkungen zur Benennung und Klassifikation fossiler (insbesondere tertiärer) Pollen und Sporen. *Geologie,* v. 3, no. 3, p. 258–311, Berlin.

————. 1954b. Möglichkeiten zur Benennung und Bezeichnung von fossilen Pollen und Sporen. *Geologie,* v. 3, no. 5, p. 649–654, Berlin.

————. 1959. Mikropaläontologische (sporenpaläontologische) Untersuchungen in der Braunkohle des Geiseltales. *Geologie,* Jahrgang 8, Beiheft 21–22, p. 1–425, 49 pl., Berlin.

Kupriyanova, L. A. 1948. Morfologiya pyl'tsy odnodol'nykh rasteniy (Materialy k filogenii klassa) [Pollen morphology of the monocotyledons (Material for the phylogeny of the class)]. *Trudy Botan. Inst. AN SSSR, ser. 1, flora i sistematica vysshikh rasteniy,* no. 7, p. 163–262, (in Russian), Leningrad.

————. 1958. K voprosu o stroenii obolochki pyl'tsevykh zeren [The structure of the membrane of the pollen grains]. *Botan. Zhurnal,* v. 41, no. 8, p. 1212–1216, (in Russian).

Larson, D. A., J. J. Skvarla, and C. W. Lewis, Jr. 1962. An electron microscope study of exine stratification and fine structure. (1). *Pollen et Spores* v. 4, no. 2, p. 233–246, Paris.

Leitgeb, H. 1883. Über Bau and Entwicklung einiger Sporen. *Ber. Deutsch. Bot. Ges.,* Berlin.

Loose, F. 1934. Sporenformen aus dem Flöz Bismarck des Ruhrgebietes. *Arbeiten, Institut f. Paläobotanik u. Petrographie d. Brennsteine,* no. 4, p. 127–164, 2 text-figs., pl. 7, Preuss. Geol. L. A. Berlin.

Mädler, K. 1954. *Azolla* aus dem Quartär und Tertiär sowie ihre Bedeutung für die Taxonomie älterer Sporen. *Geol. Jahrbuch,* v. 70, p. 143–158, 1 fig., 1 pl., Hannover.

Mangin, M. L. 1889. Observations sur la membrane du grain de pollen. *Bull. Soc. Bot. France,* v. 36, comptes rendus, p. 274–283, Paris.

Melchior, H. and E. Werderman. 1954. A. Engler's Syllabus der Pflanzen-familien mit besonderer Berücksichtigung der Nutzpflanzen nebst einer Übersicht über die Florenreiche und Florengebiete der Erde, I. Band, Allgemeiner Teil, Bakterien bis Gymnospermen. Bornträger, 366 p., 137 text-figs., Berlin.

Merriam-Webster. 1956. Webster's New Collegiate Dictionary. 1174 p., text-figs., Springfield.

Mohl, Hugo von. 1835. Sur la structure et les formes des grains de pollen. *Ann. Sci. Nat., Bot.,* ser. 2, v. 3, p. 148–180; 220–236; 304–346, pl. 9–11, Bern.

Needham, M. 1750. Nouvelles observations microscopiques. 525 pp., 8 pl., Paris.

Norem, W. L. 1958. Keys for the classification of fossil spores and pollen. *Jour. Paleontology,* v. 32, no. 4, July, p. 666–676, 40 text-figs.

References

Oldfield, F. 1959. The pollen morphology of some of the west European Ericales. Preliminary description and a tentative key to their identification. *Pollen et Spores,* Mus. Nat d'Historie Naturelle, v. 1, no. 1, p. 19–48, 1 tab., 1 text-fig., 2 pl., Paris.

Pike, K. 1956. Pollen morphology of Myrtaceae from the south-west Pacific area. *Australian Jour. Botany,* v. 4, no. 1, p. 13–53.

Pokrovskaya, I. M. (edit.) 1950. Pyl'tsevoy analiz [Pollen analysis]. Authors: A. I. Gladkova, V. P. Grichuk, E. D. Zaklinskaya, and L. A. Kupriyanova. *Gosgeolisdat,,* 571 p., 55 text-figs., 17 pl. (in Russian), Moscow. (French translation by E. Boltenhagen, 1958, *Annales, Service d'Inform. Géol. du B. R. G. G. M.,* no. 24, 434 p., Paris).

Potonié, H. 1912. Grundlinien der Pflanzen-Morphologie im Licte der Paläontologie. Zweite, stark erweiterte Auflage des Heftes: "Ein Blick in die Geschichte der botanischen Morphologie und die Perikaulon-Theorie." G. Fischer, 259 p., Jena.

Potonié, R. 1931. Pollenformen der miocänen Braunkohle. 2. Mitt. Sitzungsberichte, Gesellschaft naturforschender Freunde, Berlin, 1931, p. 24–26, 2 pl., Berlin.

———. 1934. Zur Mikrobotanik der Kohlen und ihrer Verwandten. 1. Zur Morphologie der fossilen Pollen und Sporen. *Arbeiten Inst. Paläeobot. u. Petrographie Brennsteine,* no. 4, p. 5–24, 44 text-figs., Preuss. Geol. L. A., Berlin.

———. 1934. Zur Mikrobotanik der Kohlen und ihrer Verwandten. 2. Zur Mikrobotanik des eocänen Humodils des Geiseltals. *Arbeiten Inst. Paläeobot. u. Petrographie Brennsteine,* no. 4, p. 25–125, Preuss. Geol. L. A., Berlin.

———. 1952. Zur Morphologie und morphologischen Nomenklatur der *Sporites* H. Potonié 1893. *Palaeontologische Zeitschrift,* v. 25, no. 314, p. 143–154, Stuttgart.

———. 1956. Synopsis der Gattungen der Sporae dispersae, I. Teil: Sporites. *Beih. Geol. Jahrbuch,* v. 23, 103 p., 11 pl., Hannover.

———. 1958. Synopsis der Gattungen der Sporae dispersae, II. Teil: Sporites (Nachträge), Saccites, Aletes, Praecolpates, Polyplicates, Monocolpates. *Beih. Geol. Jahrbuch,* v. 31, 114 p., 11 pl., Hannover.

——— and W. Klaus. 1954. Einige Sporengattungen des alpinen Salzgebirges. *Geol. Jahrbuch,* v. 68, p. 517–546, 11 text-figs., pl. 10, Hannover.

——— and G. Kremp. 1954. Die Gattungen der paläozoischen Sporae dispersae und ihre Stratigraphie. *Geol. Jahrbuch,* v. 69, p. 111–194, 5 text-figs., 17 pl., Hannover.

——— and ———. 1955, 1956. Die Sporae dispersae des Ruhrkabons, ihre Morphographie und Stratigraphie mit Ausblicken auf Arten anderer Gebiete und Zeitabschnitte. *Palaeontographica, Abt. B;* 1955, Teil I, v. 98, Liefg. 1–3, p. 1–156, 37 text-figs., pl. 1–16; 1956, Teil II, v. 99, Liefg. 4–6, p. 85–191, text-figs. 38–87, pl. 17–22; 1956, Teil III, v. 100, Liefg. 4–6, p. 65–121, 3 tab., Stuttgart.

Purkinje, J. E. 1830. De cellulis antherarum fibrosis nec non de granorum pollinarium formis. *Commentatio phytotomica.* Pressburg.

Reinsch, P. R., 1884. Micro-Paleophytologia Formationis Carboniferae, E. T. J. Erlangae, v. I, 79 p., 16 pl; vol. II, 14 p., pl. 17–35, Leipzig.

Rowley, J. R. 1957. The fine structure of the pollen wall in the Commelinaceae Reichenb. Thesis, Univ. Minnesota Library; University microfilms, Pub. No. 22, 743; Dissertation Abstracts 17 (9) p. 1881–1882.

———. 1959. The fine structure of the pollen wall in the Commelinaceae. *Grana Palynologica,* v. 2, no. 1, p. 3–31.

Saad, S. I. 1961. Pollen morphology and sporoderm stratification of *Linum. Grana Palynologica,* v. 3, no. 1, p. 109–125.

———. 1962. Palynological studies in the Linaceae. *Pollen et Spores,* v. 4, no. 1, p. 65–82.

———. 1963. Sporoderm stratification: the "Medine," a distinct third layer in the pollen wall. *Pollen et Spores,* v. 5, no. 1, p. 17–29.

Schopf, J. M. 1938. Spores from the Herrin (No. 6) coal bed in Illinois. *State Geol. Survey Illinois, Report of Investigations No. 50,* 55 p., 8 pl., Urbana.

Sears, P. B. 1930. Common fossil pollen of the Erie Basin. *Bot. Gaz. 89,* p. 95–106.

Selling, O. H. 1947. Studies in Hawaiian pollen statistics, Part II. The pollen of the Hawaiian Phanerogams. *Spec. Bishop Mus.,* no. 38, Gothenburg.

Sladkov, A. N. 1962. On spore and pollen morphology of recent plants in the USSR. In *For the First International Conference on Palynology (Tucson, USA), Reports of Soviet Palynologists* (extended theses), Academy of Sciences of the USSR, Inst. of Geology, p. 5–13, Moscow.

Smith, G. M. 1950. The fresh-water algae of the United States. McGraw-Hill, 719 p., 559 text-figs., New York.

Stanley, E. A. and G. O. W. Kremp. 1959. Some observations on the geniculus in the pollen of *Quercus prinoides. Micropaleontology,* v. 5, no. 3, p. 351–354, 1 text-fig., New York.

Steéves, M. W. and E. S. Barghoorn. 1959. The pollen of *Ephedra. Jour. of the Arnold Arboretum,* v. 40, no. 3, p. 221–255, 2 tab., 2 charts, 1 text-fig., pl. 1–4, Harvard Univ.

Takhtadzhyan, A. L. and A. A. Yatsenko-Khmelevskiy, 1945. Palynologia caucasia, I, Op'it standartizatsii palinologicheskoy terminologii [Palynologia caucasia, I, An essay on the standardisation of te palynolocic terminology]. *Izv. AN Armyansk. SSR, Est. Nauki,* no. 5–6, p. 31–46, (in Russian).

References

Thiergart, F. 1938. Die Pollenflora der niederlausitzer Braunkohle, besondes im Profil der Grube Marga bei Senftenberg. *Jahrbuch, Preuss. Geol. Landesanstalt,* 1937, v. 58, p. 282–351, pl. 22–30, Berlin.

Thomson, P .W. and H. Pflug. 1953. Pollen und Sporen des mitteleuropäischen Tertiärs. *Palaeontographica,* Abt. B, v. 94, Liefg. 1–4, p. 1–138, 4 tab., 20 text-figs., pl. 1–15, Stuttgart.

Traverse, A. 1955. Pollen analysis of the Brandon lignite of Vermont. *Bureau of Mines, U.S. Dept. Interior, Reports of Investigations 5151,* 107 p., 13 pl.

Tschudy, R. H. 1961. Palynomorphs as indicators of facies environments in Upper Cretaceous and Lower Tertiary strata, Colorado and Wyoming. *Wyoming Geol. Assoc. Guidebook, 16. Annual Field Conference, 1961,* p. 53–59.

Ueno, J. 1958. Some palynological observations of Pinaceae. *Jour. Inst. Polyt.,* Osaka City Univ., D9, p. 169–188.

Van Campo-Duplan, M. 1954. Considerations generales sur les caractères des pollens et des spores et sur leur diagnose. *Bull. de la Soc. Bot. Français,* v. 101, no. 5–6, p. 250–281, Paris.

Wicher, C. A. 1934. Sporenformen der Flammkohle des Ruhrgebietes. *Arb. Inst. Paläobot. u. Petrographie der Brennsteine,* v. 4, p. 165–212, Berlin.

Winslow, M. R. 1959. Upper Mississippian and Pennsylvanian megaspores and other plant microfossils from Illinois. *Illinois State Geol. Survey Bull. 86,* 101 p., 8 text-figs., 1 tab., 16 pl., Urbana.

Wodehouse, R. P. 1928. The phylogenetic value of pollen-grain characters. *Ann. of Bot.,* v. 42, p. 891–934, figs. 1–2, pl. 20–21, London.

————. 1933. Tertiary pollen. II. The oil shales of the Eocene Green River formation. *Bull. Torr. Bot. Club,* v. 60, p. 479–524, 56 text-figs.

————. 1935. Pollen grains. Their structure, identification and significance in science and medicine. McGraw-Hill, 514 p., 6 tab., 123 text-figs., New York.

Zaklinskaya, E. D. 1957. Stratigraficheskoe enachenie pyl'tsy golosemennykh kaynozoyskikh otlozheniy Pavlodarskogo Priirtysh'ya i Severnogo Priaral'ya [The stratigraphic significance of the gymnospermous pollen grains of the Cenozoic deposits of the Irtysh Basin and of the northern Aral Basin]. *Trudy Geol. Inst. AN SSSR,* no. 6, 184 p., 32 text-figs., 17 pl. (in Russian), Moscow.

Plate 1

Figure 1–5. TETRADS. From Potonié, 1934, figs. 11–13, 16, 18.

1. Tetragonal tetrad
2. Tetrahedral tetrad
3. Rhombohedral tetrad
4. Tetrahedral tetrad (Ericaceae tetrad)
5. Tetragonal tetrad of a monolete spore
   a. Side view
   b. Single grain
   c. Surface view of tetrad

Figure 6–13. GROUPS OF SPOROMORPHAE. Principal types of pollen grains. From Faegri and Iversen, 1950, pl. V.

6. Polyad, e.g. *Acacia*
   a. Plan view
   b. Side view
7, 10–12. Tetrads
   7a,b. Square tetrad
   10a,b. Tetrahedral tetrad
   11. Cross tetrad
   12a,b. Rhomboidal tetrad
8. Dyad, e.g. *Scheuchzeria*
9. Inaperturate (Monad)
13a,b. Vesiculate

Figure 14–18. DIFFERENT TYPES OF TETRADS. From Erdtman, 1945, fig. 1.

14. Tetrahedral tetrad
15. Cross tetrad
16. Square tetrad
17. Rhomboidal tetrad
18. Linear tetrad

Figure 19–23. TETRADS (Scheme). If situated in the plane of the figure the polar axes of the single grains are indicated by full lines, otherwise by broken lines or a dot. From Erdtman, 1943, fig. 6.

19. Tetragonal tetrad in lateral and surface view
20. Tetrahedral tetrad in surface and lateral view
21. Transitional form between tetragonal and hexahedral tetrad
22. Rhomboidal tetrad in lateral and surface view
23. Hexahedral tetrad in two different positions

Figure 24–33. FORMULA OF SPORES AND POLLEN SYMBOLS. From Erdtman, 1946, fig. 1. Further explanations in text.

24. Monosulcate, plano-convex pollen grain
25. Monosulcate, concavo-convex pollen grain
26. Monosulcate, biconvex grain
27. Monolete, concavo-convex spinose spore
28. Nonaperturate pollen grain
29. Dodekacolpate, peroblate pollen grain
30. Tricolporate, prolate-perprolate pollen grain
31. Hexaporate grain
32. Polyporate grain
33. Polyrugate grain

Figure 34–37. EXAMPLES OF PLANTS WITH DIMORPHIC POLLEN. From Erdtman, 1952, fig. 197 B and D.

34. Primulaceae, *Primula farinosa* f. *longistyla*, 7.6 × 10.2 μ
35. Primulaceae, *Primula farinosa* f. *brevistyla*, 6.8 × 14.2 μ
36. Primulaceae, *Primula veris* f. *brevistyla*, 30 × 25 μ
37. Primulaceae, *Primula veris* f. *longistyla*, 14 × 16 μ

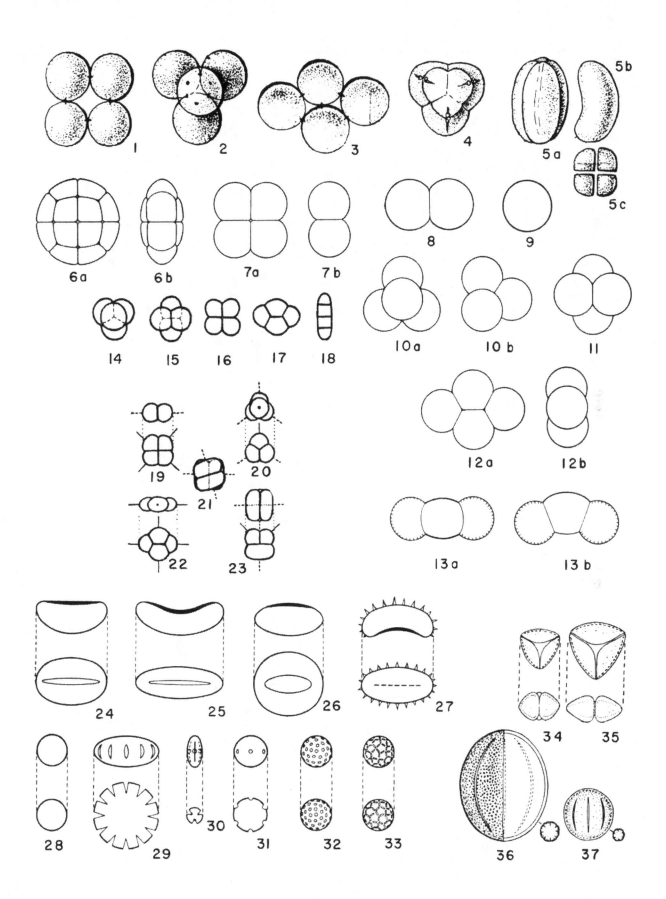

189

Plate 2

Figure 38–46. GROUPS OF SPOROMORPHAE. Pollen and spore types. The pollen grains and spores are drawn in four different positions. The drawings are arranged in eleven columns. Po, polar view; Eq, equatorial view; Pr, proximal part; Di, distal part; Lat, lateral view. From Erdtman, 1943, text-fig. 3.

38. Tetrahedral tetrad
39. *Carex* pollen
40. Trilete spore
41. Tricolpate pollen (*Trapa* pollen with triradiate scar)
42. Tricolpate pollen
43. Three-pored pollen, e.g. *Engelhardtia*
44. Pollen with three-slit opening (e.g. *Johnsonia*)
45. Monocolpate pollen grains
    a. Longitudinal positions
    b. Transverse positions
46. Monolete spores
    a. Longitudinal positions
    b. Transverse positions

Figure 47. Dyad. Scheuchzeriaceae, *Scheuchzeria palustris,* 44 × 30μ. From Erdtman, 1943, fig. 29.

Figure 48. MONOLETE SPORE. Polypodiaceae, *Dryopteris thelypteris,* transverse position, lateral view, 29 × 55μ. From Erdtman, 1943, fig. 482.

Figure 49. MONOCOLPATE POLLEN GRAIN. Nymphaceae, *Nuphar luteum,* oblique lateral view, distal part down, polar axis about 28μ. From Erdtman, 1943, fig. 257.

Figure 50. TETRAD SCAR. Hydrocaryaceae, *Trapa natans,* polar view (proximal pole) of young pollen grain, 58μ. From Erdtman, 1943, fig. 223.

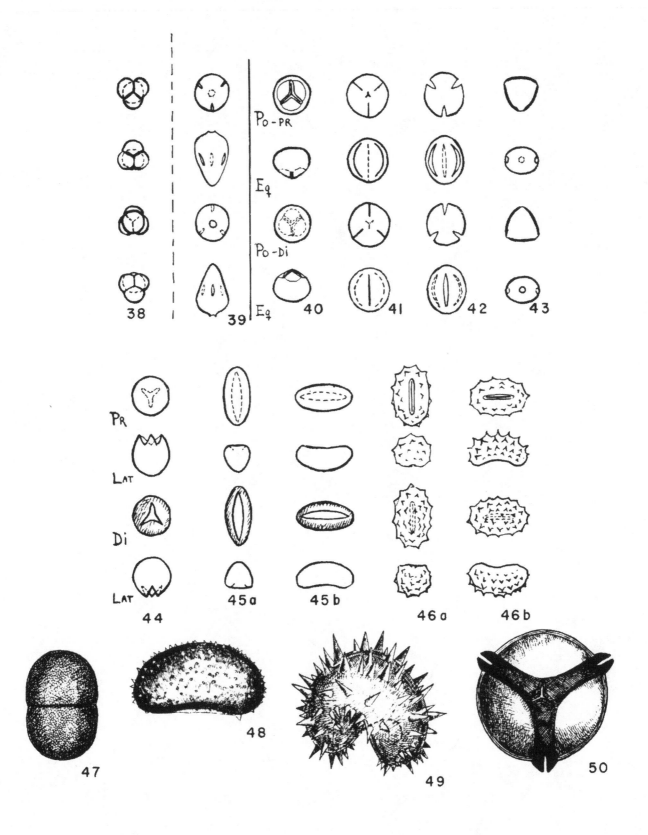

Plate 3

Figure 51–74. GROUPS OF SPOROMORPHAE. Principal types of pollen grains. From Faegri and Iversen, 1950, pl. VI-VIII.

51. Monoporate
52. Monocolpate
53–58. Syncolpate
59. Dicolpate
60. Tricolpate
61. Stephanocolpate
62–63. Pericolpate

64. Tricolporate
65. Stephanocolporate
66–67. Pericolporate
68. Diporate
69. Triporate
70. Stephanoporate
71. Periporate
72. Fenestrate
73. Heterocolpate
74. Extraporate

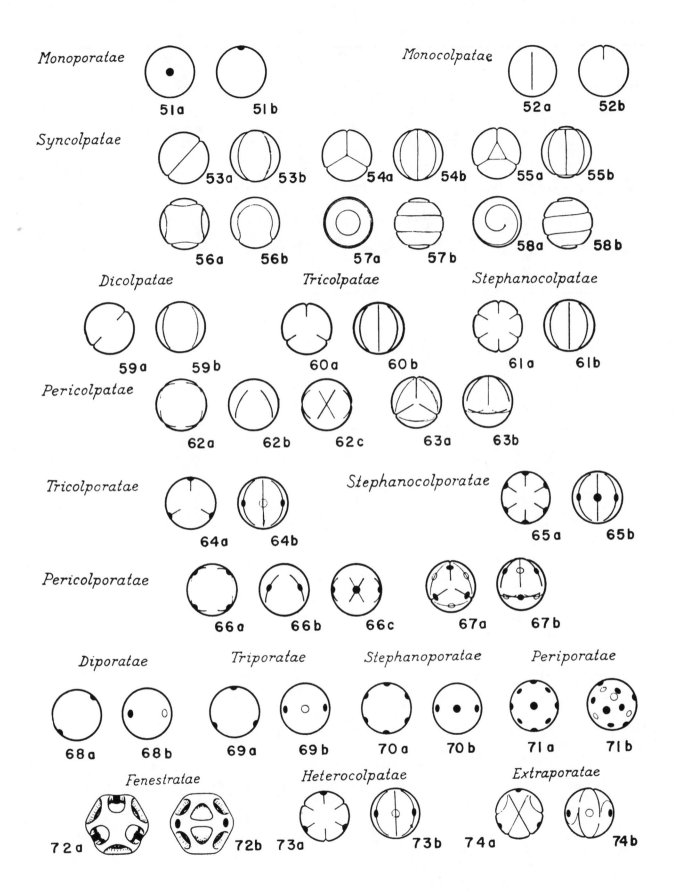

Monoporatae 51a 51b

Monocolpatae 52a 52b

Syncolpatae 53a 53b 54a 54b 55a 55b

56a 56b 57a 57b 58a 58b

Dicolpatae 59a 59b

Tricolpatae 60a 60b

Stephanocolpatae 61a 61b

Pericolpatae 62a 62b 62c 63a 63b

Tricolporatae 64a 64b

Stephanocolporatae 65a 65b

Pericolporatae 66a 66b 66c 67a 67b

Diporatae 68a 68b

Triporatae 69a 69b

Stephanoporatae 70a 70b

Periporatae 71a 71b

Fenestratae 72a 72b

Heterocolpatae 73a 73b

Extraporatae 74a 74b

193

Plate 4

Figure 75. POSITION AND SHAPE OF APER-
TURES. From Erdtman, 1952, fig. 1.

   I. Apertures, polar
  II. Apertures, nonpolar
  A. Proximalipolar
  B. Distalipolar
  C. Equatorial
  D. Global
     1. laesura; 2. hilum; 3. sulcus; 4. ulcus; 5. colpi;
     6. pori; 7. rugae; 8. foramina. di. po. = distal
     pole; equ. = equator; pr.po. = proximal pole.

Figure 76–99. PALYNOGRAMS (orientation and
position, etc.). From Erdtman, 1952, fig. 3. For further detail see text.

76–81. Triaperturate pollen grains (equatorial view)
82–87. Triaperturate grains (polar view)
88–93. Tetra-aperturate grains (polar view)
94. Tetrahedral pollen tetrad
95. Monosulcate pollen grain (distalipolar view)
96. Monosulcate grain (equatorial view, longitudinal position)
97. Monolete spore (proximalipolar view)
98. Monolete spore (equatorial view, longitudinal position)
99. Trilete spore (proximalipolar view)

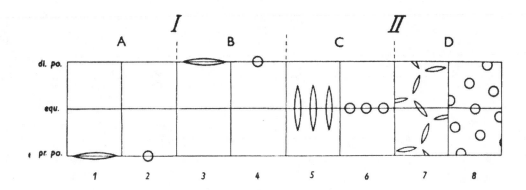

75

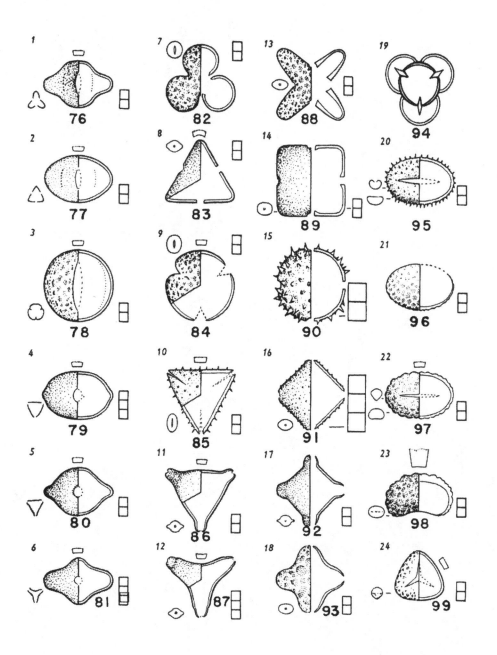

Plate 5

Figure 100–104. STENOPALYNOUS. From Erdtman, 1952, figs. 73 and 114.

100. Cruciferae, *Biscutella auriculata*, 39 × 33μ
101. Gramineae, *Zea mays,* ca. 120 × 85μ
102. Gramineae, *Dendrocalamus strictus*, ca. 48 × 45μ

103. Cruciferae, *Matthiola tricuspidata*, 26μ
104. Gramineae, *Triticale (Secale cereale × Triticum aestivum)* cult.

Figure 105. GROUPS OF SPOROMORPHAE. Basic aperture terminology. From Erdtman, 1958, p. 137.

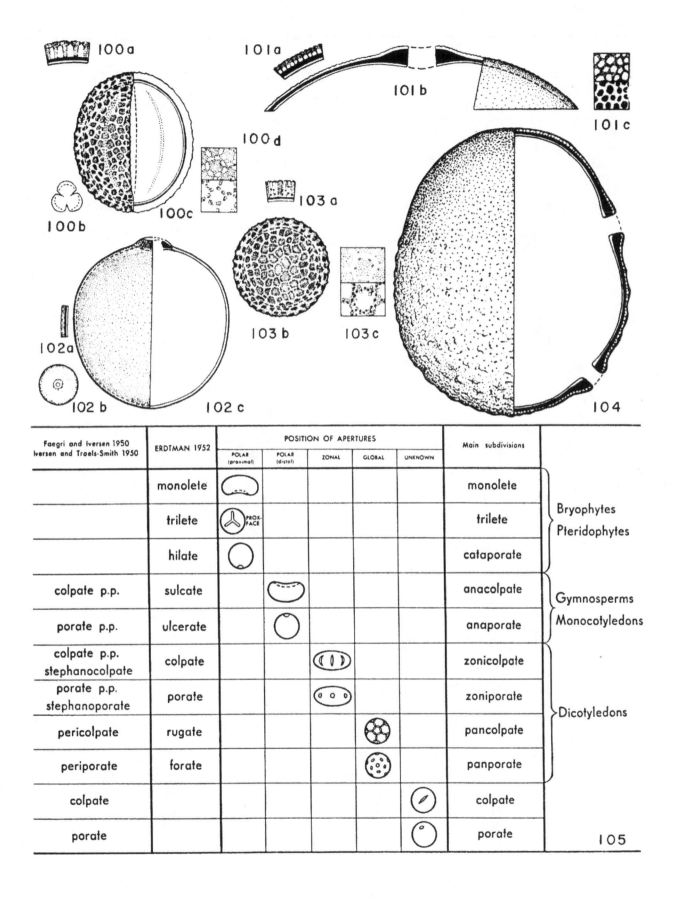

100a · 100b · 100c · 100d

101a · 101b · 101c

103a · 103b · 103c

102a · 102b · 102c

104

Faegri and Iversen 1950 Iversen and Troels-Smith 1950	ERDTMAN 1952	POSITION OF APERTURES					Main subdivisions
		POLAR (proximal)	POLAR (distal)	ZONAL	GLOBAL	UNKNOWN	
	monolete	⬭					monolete
	trilete	⬭ PROX-FACE					trilete
	hilate	⬭					cataporate
colpate p.p.	sulcate		⬭				anacolpate
porate p.p.	ulcerate		⬭				anaporate
colpate p.p. stephanocolpate	colpate			⬭			zonicolpate
porate p.p. stephanoporate	porate			⬭			zoniporate
pericolpate	rugate				⬭		pancolpate
periporate	forate				⬭		panporate
colpate						⬭	colpate
porate						⬭	porate

Bryophytes
Pteridophytes

Gymnosperms
Monocotyledons

Dicotyledons

105

Plate 6

Figure 106. SPORE. Orientation of radially symmetrical spores. Frim Winslow, 1959, fig. 1.
a. Spores in tetrad association
b. Proximal view of a single spore showing trilete suture and contact areas
c. Lateral view of single spore

Figure 107–108. Y-MARK AND CONTACT AREAS of Carboniferous spores. From Potonié, 1934, figs. 17 and 15.

Figure 109–110. SPORES. From Kosanke, 1950, figs. 3 and 4.

109. Diagrammatic drawing of a radially symmetrical spore illustrating:
 A. Flange and striations
 B. Ray of trilete mark
 C. Original position of suture or commissure
 D. Contact area
 E. Lip
 F. Arcuate (?) ridge

110. Diagrammatic drawing of bilateral spore illustrating:
 A. Original position of monolete mark
 B. Lip

Figure 111. SPORE. Orientation of bilaterally symmetrical spores. From Winslow, 1959, fig. 2.
a. Spores in tetrad association
b. Proximal view of single spore showing monolete suture and contact areas
c. Transverse lateral view of single spore

Figure 112. CONTACT AREA. *Laevigatisporites* sp., 50x. From Potonié and Kremp, 1955, fig. 7.

Figure 113. Y-SPLIT MARK. H, main ray; s, forked part; a, angle of split. From Thomson and Pflug, 1953, fig. 4a.

Figure 114. MONOLETE SPORE. Equatorial view. S, dehiscence (single ray split mark); SE, plane of symmetry, a second plane of symmetry is situated at the plane of the figure. From Thomson and Pflug, 1953, fig. 5a.

Figure 115–121. TETRAD MARK AND ASSOCIATED FEATURES. From Couper and Grebe, 1961, figs. 7–10.

115a,b. Laesura, contact area, curvatura
116, 117. Laesura, with margo and commissure
 116. Plan view
 117. Cross section
118, 119. Laesura, with tectum and commissure
 118. Plan view
 119. Cross section
120, 121. Reduced curvatura

Figure 122. KYRTOME. *Ahrensisporites* sp., ca. 50μ. From Potonié and Kremp, 1956, textfig. 41.

Figure 123–128. TORUS. From Krutzsch, 1959, fig. 11.

123. Narrow band-tori and fold-tori
124. Short, plane field-tori
125. Circumfluent field-tori
126a. Full-fold [Voll-Falten] field-tori
 b. Full-fold field-tori with pore, structures in the corner circuits (tp, torus pores)
127. "Lane"-tori
128. "Pseudotori" and "pseudotoroid claws"

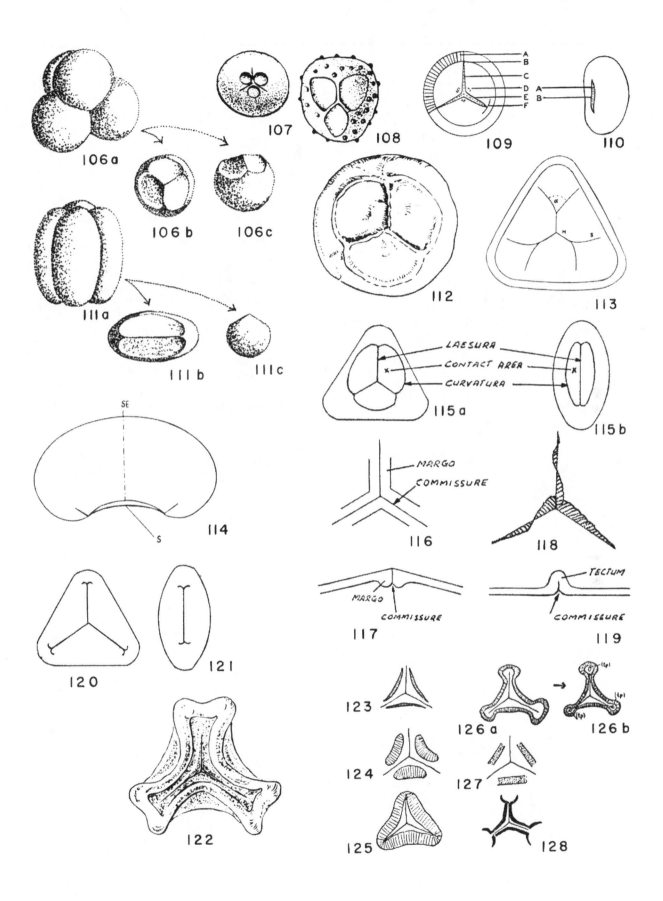

199

Plate 7

Figure 129. SUTURE. *Triletisporites* sp., ca. 400x. From Potonié and Kremp, 1955, fig. 23.

Figure 130. MASSA. From Couper and Grebe, 1961, fig. 11.

Figure 131. GULA. *Lagenicula crassiaculeata* Zerndt, ca. 60μ. From Potonié and Kremp, 1955, fig. 35.

Figure 132–133. APEX. *Leiotriletes* sp., ca. 570μ. From Potonié and Kremp, 1955, fig. 3.

Figure 134–135. MASSA. From Potonié, 1956, figs. 109, 110.

134. *Cystosporites* sp., ca. 1100μ.
135. *Azolla* sp., ca. 400μ.

Figure 136. TRIFOLIUM. *Capulisporites* ca. 330μ. From Potonié, 1956, p. 53, fig. 65.

Figure 137–140. APERTURE, CONTACT AREA. From Wicher, 1934, pl. 8, figs. 23, 27, 20, 19.

137. *Sporites diffusopilosus* Wicher, polar view, 1140μ.
138. *Sporites hirsutus* Loose, 780μ.

139. *Sporites subfuscus* Wicher, 1350μ.
140. *Sporites valens* Wicher, 2380μ.

Figure 141–148. STRUCTURE OF THE Y-MARK. For detail see Y-MARK. From Potonié and Kremp, 1955, fig. 1. ap, apex or vertex; l, labra; it, intratectum; st, subtectum; c, contravertex.

141–142. Sketches of vertical section of the exine
143, 144, 146, 147. Various horizontal sections through the Y-mark, as appearing in the optical section
143. Y-mark, optical section I of fig. 142
144. Y-mark, optical section III of fig. 142
146. Y-mark, optical section IV of fig. 142
147. Y-mark, optical section II of fig. 142
145. Spore of *Lepidostrobus foliaceus* with floating body (after Scott).
148. Split open apex of a laevigate miospore; before splitting the outer edges of the labra were connected to a vertex; because of the height of the apex spoon-like lobes were formed.

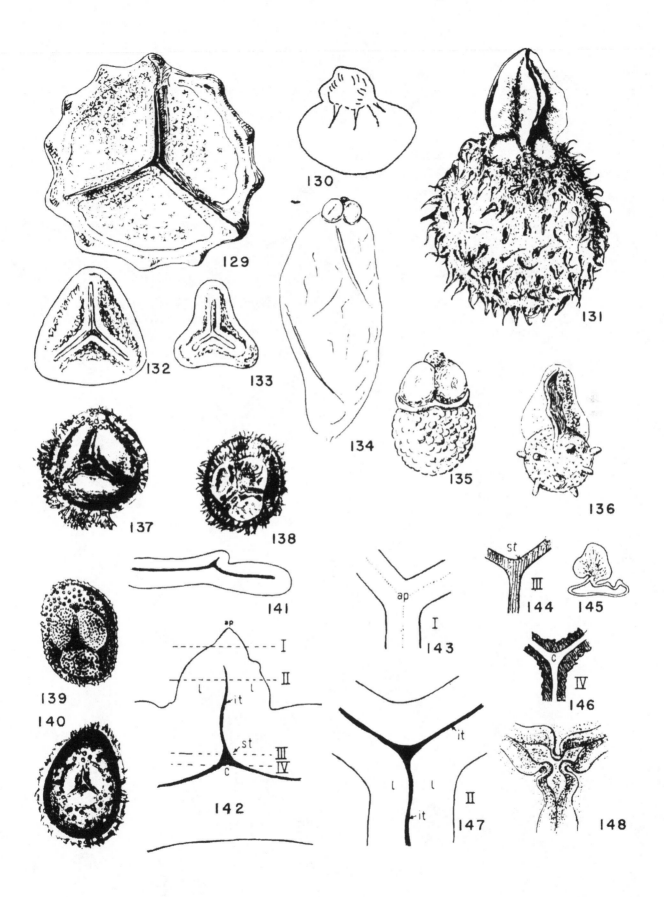

Plate 8

Figure 149–159. EQUATORIAL REGION. From Potonié and Kremp, 1956, figs. 38–40, 42, 43, 45, 55–59.

149. Valvae. *Triquitrites* sp., ca. 50μ

150, 151. Auriculae

    150. *Tripartites* sp., ca. 50μ

    151. *Valvisisporites* sp., ca. 1000μ

152–154. Cingulum

    152. *Lycospora* sp., ca. 30μ

    153. *Cadiospora* sp., ca. 110μ

    154. *Rotaspora* sp., ca. 40μ

155. Zona or Frassa. *Cirratriradites* sp., ca. 70μ

156–159. Corona

    156. *Triangulatisporites* sp., ca. 600μ

    157. *Superbisporites* sp., ca. 2000μ

    158. *Rotatisporites rotatus* (Bartlett), ca. 1000μ

159. Corona of free fimbriae. *Radiatisporites radiatus* (Zerndt), ca. 1030μ

Figure 159–1. ANNULATE. From Norem, 1958, text-fig. 3.

Figure 159–2. ANNULOTRILETE. From Norem, 1958, text-fig. 4.

Figure 159–3. AURICULATE. From Norem, 1958, text-fig. 5.

Figure 160. RADIAL AND INTERRADIAL POSITION. From Couper and Grebe, 1961, fig. 1.

Figure 161. SCUTULUM. *Dulhuntyispora dulhuntyi* Potonié. From Potonié, 1956, pl. 4, fig. 38.

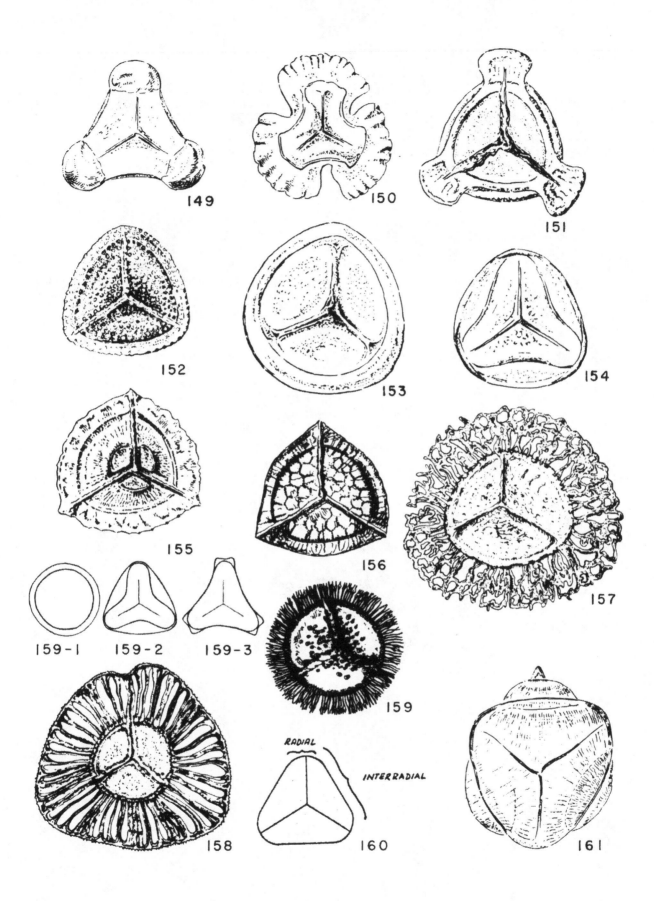

149

150

151

152

153

154

155

156

157

159-1   159-2   159-3

159

158

RADIAL

INTERRADIAL

160

161

Plate 9

Figure 162–167. CONTOUR. From Thomson and Pflug, 1953, p. 19, textfig. 1.

162. Concave triangular
163. Convex triangular
164. Distorted-concave triangular
165. Straight triangular
166. Circular
167. Tricolpate

Figure 167a–167p. SHAPE CLASSES. From Norem, 1958, text-figs. 24–36, 38–40.

167a. Ellipsoidal
167b. Globoid
167c. Peroblate
167d. Oblate
167e. Spheroidal
167f. Prolate
167g. Perprolate
167h. Tetragonal
167i. Deltoid
167j. Subtriangular
167k. Reniform
167l. Lenticular
167m. Crescent
167n. Spatulate
167o. Triquete
167p. Trilobate

Figure 168–183. SHAPE. Descriptive terms used in describing the shape of spores. From Couper and Grebe, 1961, figs. 2–5.

168–170. Equatorial contour as seen in polar view
    168. Circular;   169. Triangular;   170. Oval
171–173. Sides seen in polar view
    171. Convex;   172. Straight;   173. Concave
174–176. Angles seen in polar view
    174. Sharply rounded; 175. Rounded; 176. Flat

177–180. Shape in equatorial view, proximal profile
    177. Flat; 178. Convex; 179. Pointed; 180. Concave
181–183. Shape in equatorial view, distal profile
    181. Flat; 182. Convex; 183. Pointed

Figure 184–194. Structural features of spores. From Couper and Grebe, 1961, fig. 12, 13, 16–21.

184–189. Surrounding features.
    184. CINGULUM
        a. Polar view
        b. Equatorial view
    185–187. DISSECTIONS. Polar view
        185. Inner margin dissected into small rounded lumina
        186. Dissections in the form of elongated lumina, extending from inner margin to over half the width of zona (or cingulum)
        187. Dissections in the form of rounded and/or elongated lumina, not extending to inner margin
    188–189. ZONA.
        188. Polar view
        189. Equatorial view
190–194. Non-surrounding features.
    190. CORONA.
    191–194. CRASSITUDE.
        191. Radial crassitude
        192. Inter-radial crassitude
        193. Proximal crassitude
        194. Distal crassitude

Figure 195. DIMENSIONS MEASURED ON TRI-LETE SPORES. (For monolete spores see Fig. 248.) P, polar diameter; E, equatorial diameter. From Couper, 1958, fig. 4.

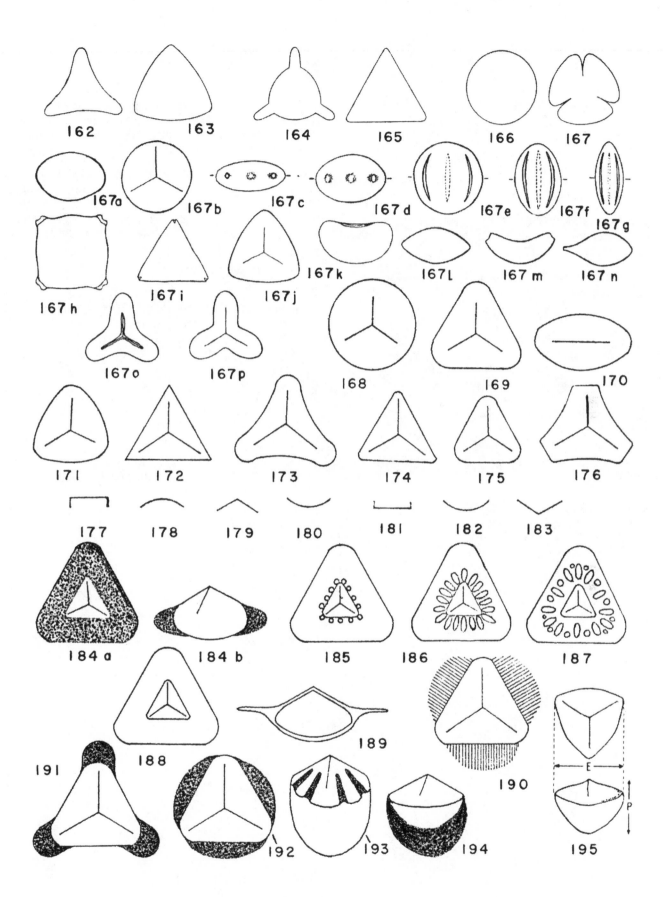

Plate 10

Figure 196. SACCUS (MONOSACCATE). From Couper and Grebe, 1961, fig. 15.

a. Polar view
b. Section in equatorial view

Figure 196–1. VESICOTRILETE. From Norem, 1958, text-fig. 2.

Figure 197. TRISACCATE. Schematic outline of a pollen grain of the Podocarpaceae. From Zaklinskaja, 1957, fig. 18 (6).

Figure 198–200. LIMBUS. From Potonié and Kremp, 1955, textfigs. 71, 72, and 75.

198. *Microsporites* sp., ca. 350µ.
    a. Proximal view
    b. Meridional section
199. *Nuskoisporites* sp., 130µ.
    a. Oblique view of the proximal side
    b. Oblique view of the distal side
    c. Meridional section
200. *Endosporites* sp., ca. 150µ.
    a. Proximal view
    b. Meridional section

Figure 201. VELUM. From Couper and Grebe, 1961, fig. 14.

Figure 202. ROUNDED SACCUS (Compare with LIMBUS). *Florinites* sp., 75µ. Meridional section. The distal germinal area is at the bottom of the figure. From Potonié and Kremp, 1956, fig. 75.

Figure 203. TRISACCATE POLLEN GRAIN. *Alatisporites pustulatus* Ibrahim, ca. 86µ. From Loose, 1934, pl. 7, fig. 4.

Figure 204. SACCATE. Morphologic elements of importance on pollen with air sacs. a, body; b, air sacs; c, furrow; d, cap; e, ridge. From Pokrovskaja, *et al.,* 1950, fig. 16.

Figure 205–206. SACCATE. Pollen grains of *Pinus scopulorum,* diagrammatic transverse optical sections: ex, exine; int, intine; cp, cap; mr, marginal ridge; fur, furrow; bl, bladders; dr, dorsal root of the bladders; vr, ventral root of the bladders. From Wodehouse, 1935, fig. 78.

205. Dry and contracted
206. Moist and expanded

Figure 207. DIMENSIONS MEASURED ON WINGED POLLEN GRAINS. LB, length of body; LA, length of air sac (bladder); BB, breadth of body; BA, breadth of air sac (bladder). From Couper, 1958, fig. 41.

Figure 208–210. BASIC DIMENSIONS MEASURED ON WINGED POLLEN GRAINS OF THE PINACEAE.

a-a, total length of the grain
b-b, height of the body
c-c, width (diameter) of the body
d-d, length of the body
e-e, height of the air sacs
f-f, width (diameter) of the air sacs
g-g, height of the ridge
After Zaklinskaya, 1957, fig. 20.
208. Side view
209. Polar view with the body on top
210. Polar view with the body at the bottom

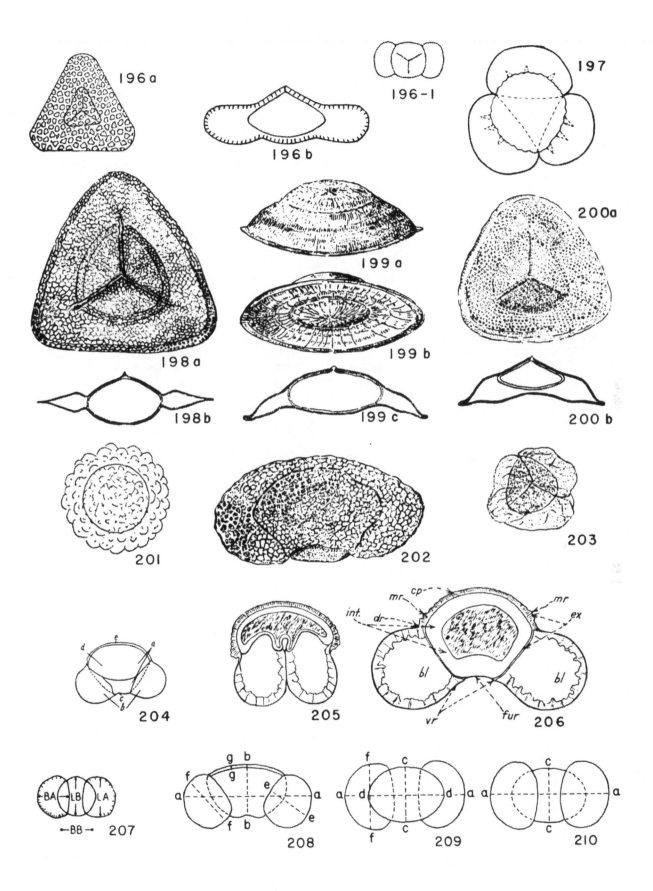

Plate 11

Figure 211. SACCATE.

A. *Abies nephrolepis,* 650x
   a. Lateral longitudinal view
   b. Lateral transverse view
   c. Polar view (distal face)
B. *Abies magnifica,* lateral longitudinal view, 650x
C. *Abies mariesii,* saccus pattern (about 2500x). 1: distal pole, 4: proximal pole. Corpus: height 1–4, breadth 5–5, depth 3–3. Sacci: height 9–8, breadth 10–10, depth 11–11. Total grain: height 4–12 (A:a,b), breadth 13–13, depth 3–5–3 (A: b), 3–2–1–2–3 (A:c). Eleven of the inserted photomicrographs show the pattern of the outer surface of the sacci at different foci (the uppermost photomicrograph in each saccus shows the pattern at high adjustment of the microscope, the lowermost the same at low adjustment). Two of the photomicrographs between numerals 11 and 8 in A: a and B show a part of the saccus in optical cross section. The remaining four photomicrographs exhibit the pattern of the corpus. Those in A:b are phase contrast pictures. From Erdtman, 1957, fig. 2.

Figure 212. LIGULA. Taxodiaceae, *Pollenites polyformosus* Thiergart, ca. 25μ. Redrawn from Thiergart, 1938, pl. 23, fig. 5.

Figure 213. POLLEN WITH FISSURA (e.g. *Taxodium,* ca. 20μ). From Potonié, 1934, fig. 36.

Figure 214–215. BASIC DIMENSIONS MEASURED ON WINGED POLLEN GRAINS OF THE PODOCARPACEAE.

a-a, total length of the grain
b-b, height of the body
c-c, width (diameter) of the body
d-d, length of the body
e-e, height of the air sacs
f-f, width (diameter) of the air sacs
h-h, distance between sacs at the point of their attachment to the body
After Zaklinskaja, 1957, fig. 17.
214. Polar view
215. Side view

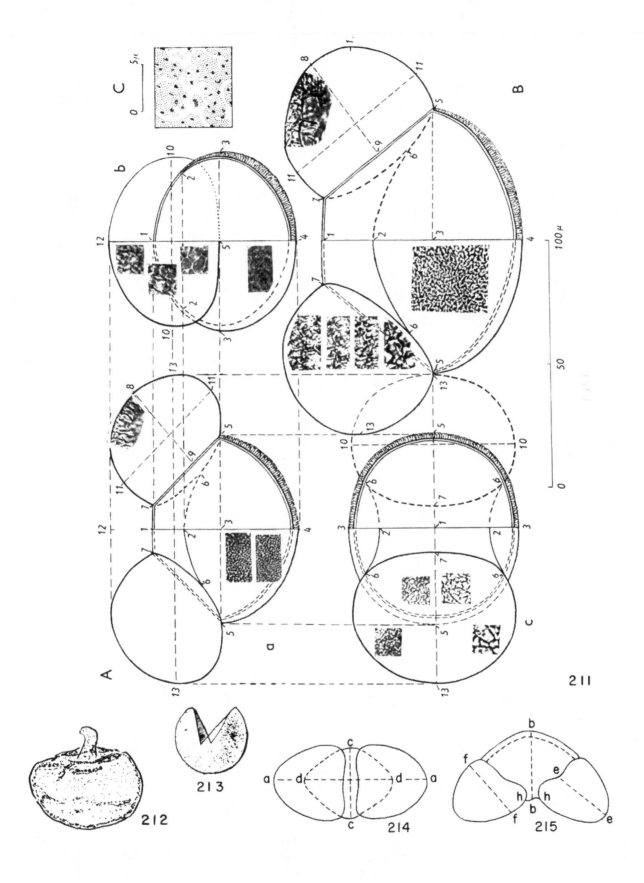

211

213

212

214

215

Plate 12

Figure 216–218. SACCATE. Schematic representation of the sac attachment. From Pokrovskaya, *et al.,* 1950, fig. 23.

216. Sacs attached to the lateral sides of the body, angle almost 90 degrees
217. Sacs somewhat displaced toward the ventral side of the body, angle less than 90 degrees
218. Sacs attached to the ventral side of the body, angle sharp, small

Figure 219–221. CAP. Schematic representation of various species of pollen of the Pinaceae (side view). Zaklinskaya, 1957, fig. 25.

219. Pollen grains, the body covered by a distinct cap
220. Pollen grains with cap margins indistinct
221. Pollen grains with cap margins indistinct

Figure 222–225. ORNAMENTATION OF SACCATE POLLEN GRAINS. From Zaklinskaya, 1957, figs. 24, 26, 19.

222. Schematic representation of the various reticulum types of the sacs and of the pollen of Pinaceae
a. reticulum distinct, irregularly meshed; b. reticulum distinct, regularly meshed, with large meshes; c. reticulum distinct, regularly meshed, with small meshes; d. reticulum distinct, with large, roundish meshes; e. reticulum distinct, with small, roundish meshes; f. reticulum distinct, with irregular, elongated meshes; g. reticulum indistinct, interrupted, simple; h. reticulum indistinct, combined (double, superimposed)
223. Cap sculpturing of pinaceous pollen grains

a. surface large-tuberculate; b. surface chagrenate-corrugate; c. surface small-tuberculate; d. surface small-tuberculate; e. surface small-pitted; f. surface distinctly or indistinctly punctate

224. Reticulum structure of the sacs of various fossil pollen types of the Podocarpaceae
a. distinct reticulum with unequal, irregularly distributed or elongated meshes of various sizes and outlines; b. distinct reticulum with irregular, usually large meshes, which are elongated along the line running from the point of attachment of the sac to its periphery; c. reticulum large-meshed, more or less distinct or interrupted, with heavy relief (apparently) lines; d. reticulum small-meshed, interrupted, with irregular and varied meshes; these meshes are sometimes not distinct because the muri, surrounding the meshes, are interrupted
225. Sculpture of the body surface of various pollen types of the Podocarpaceae
a. chagrenate-corrugate; b. tuberculate; c. small-tuberculate; d. small-pitted

Figure 226. SACCATE. *Dacrydium taxoides,* 1000x. For explanation of numerals see Fig. 211. From Erdtman, 1957, fig. 27.

a. Upper figure, pollen grain in lateral, longitudinal view, surface and section. Lower figure, proximal face, surface (left) and optical section (right)
b. Outline of pollen grain in lateral, transverse view (250x, outer contour of the corpus marked by the broken line)

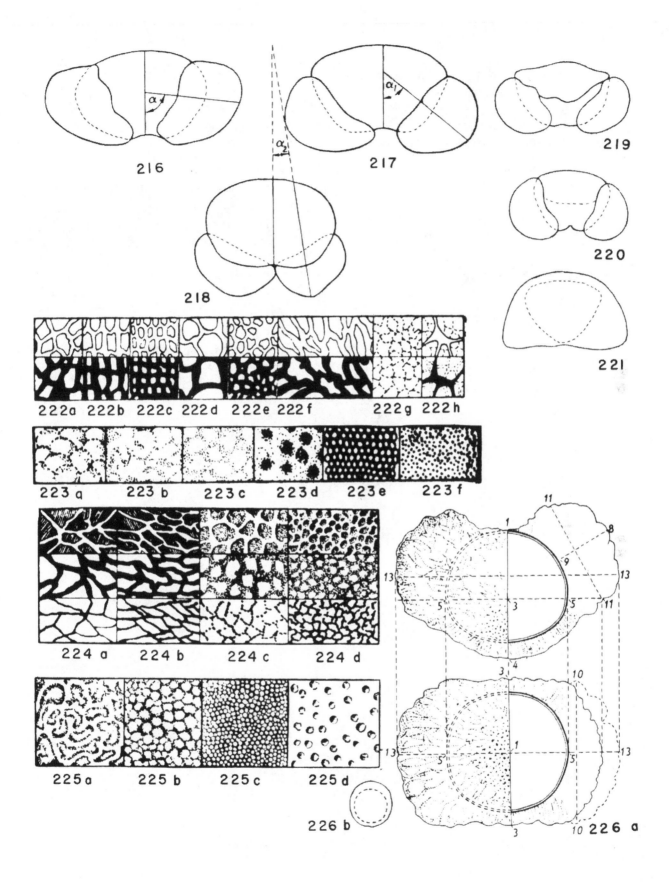

216

217

218

219

220

221

222a 222b 222c 222d 222e 222f 222g 222h

223 a 223 b 223 c 223 d 223 e 223 f

224 a 224 b 224 c 224 d

225 a 225 b 225 c 225 d

226 b

226 a

Plate 13

Figure 227. SHAPE OF BODY OF DISACCATE POLLEN GRAINS. Schematic representation of the body shape of various species of *Pinus*.

a, roundish outline; b, trapeziform-roundish outline; c, rhomboidal-roundish outline; d, angular-elliptical outline; e, elliptical; f, trapeziform outline; g, triangular outline; h, angular outline. Redrawn from Zaklinskaya, 1957, fig. 21.

Figure 228–230. CAP. Schematic representation of various pollen grains of the Pinaceae. From Zaklinskaya, 1957, fig. 28.

228. Body outline even
    a. Side view
    b. Polar view
229. Body outline undulate only in the region of sac attachment
    a. Side view
    b. Polar view
230. Body outline undulate
    a. Side view
    b. Polar view

Figure 231–235. RIDGE OUTLINE. Types of outlines of the dorsal part of the body of pinaceous pollen. From Zaklinskaya, 1957, fig. 27.

231. Outline even
232. Outline small-undulate
233. Outline large-undulate
234. Outline dentate-undulate
235. Outline slightly undulate

Figure 236–241. SHAPE. Schematic representation of the outline of the sacs in various disaccate types and the manner of their attachment to the body. From Zaklinskaya, 1957, fig. 22.

236. Spheroidal
    a. Side view   b. Polar view
237. Angular-spheroidal
    a. Side view   b. Polar view
238. Ellipsoidal
    a. Side view   b. Polar view
239–240. Hemispherical
    a. Side view   b. Polar view
241. Conical
    a. Side view   b. Polar view

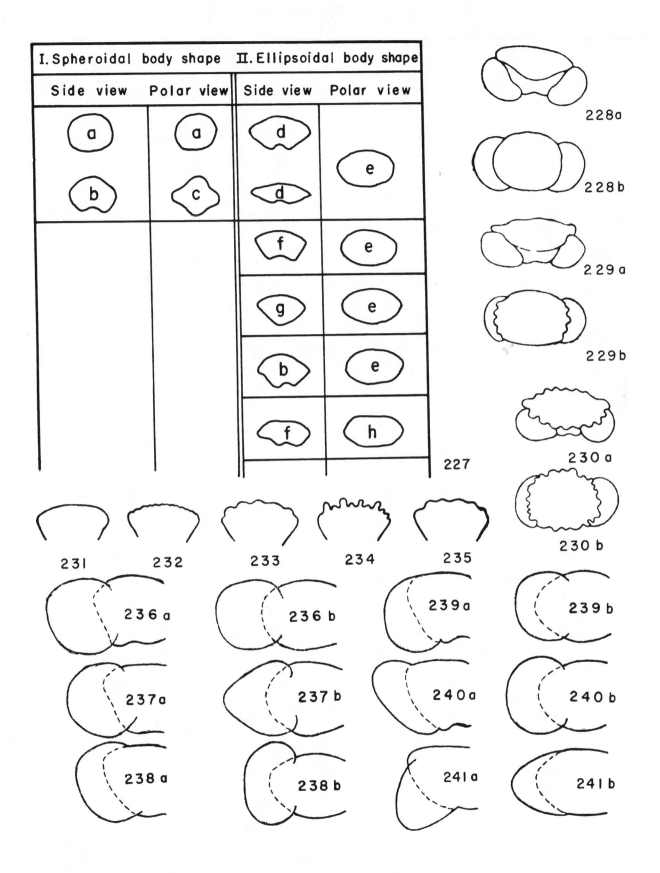

I. Spheroidal body shape		II. Ellipsoidal body shape	
Side view	Polar view	Side view	Polar view
a	a	d	
b	c	d	e
		f	e
		g	e
		b	e
		f	h

227

228a

228b

229a

229b

230a

230b

231   232   233   234   235

236a   236b   239a   239b

237a   237b   240a   240b

238a   238b   241a   241b

Plate 14

Figure 242. GROUPS OF SPOROMORPHAE. Types of pollen grains of the monocotyledonous plants.
1. Pollen grains with a three-slit opening (Pollina trichotomo fissurata)
2. Pollen grains with one furrow (Pollina mono-sulcata)
3. Pollen grains with two furrows (Pollina bisulcata)
4. Pollen grains with a spiral furrow (Pollina spirali-sulcata)
5. Pollen grains with a zoned furrow (Pollina zonosulcata)
6. Pollen grains with one simple pore (Pollina monoporate, porus simplex)
7. Pollen grains with one pore (Pollina mono-porata, porus annularis)
8. Pollen grains with two pores (Pollina diporata)
9. Pollen grains with three pores (Pollina triporata)
10. Pollen grains with many pores (Pollina poly-porata)
11. Pollen grains with pores different in form (pollina diverse porata)
12. Pollen grains without apertures (Pollina non-aperturata)
13. Pollen grains without exine (Pollina nuda)
14. Enclosed pollen grains (Pollina occlusa)
From Kupriyanova, 1948, fig. 3.

Figure 243–244. SULCUS. Young *Molineria* tetrads: Furrows occur on distal ("Basis") side. From Cranwell, 1953, fig. 1, after Pohl, 1928.

243. Lateral view
244. End view

Figure 245. MONOCOLPATE. *Ginkgo biloba* (Ginkgoales) ventral view, 28μ long. From Wodehouse, 1935, pl. 2, fig. 6.

Figure 246. TENUITAS. *Entylissa* sp., ca. 40μ. From Potonié and Kremp, 1956, fig. 88.

Figure 247. TRICHOTOMOSULCATE. Palmae, *Acanthorhiza mocinni*, 29μ, distal side. From Erdtman, 1952, fig. 177c.

Figure 248. DIMENSIONS MEASURED ON MONOLETE SPORES AND MONOSULCATE POLLEN GRAINS. B, breadth; L, length; D, depth. After Couper, 1958, fig. 41.

Figure 248–(1 and 2). TRISULCATE. Equatorial and polar view. From Norem, 1958, text-fig. 11a,b.

Figure 249–263. SULCUS. From Erdtman, 1952, figs. 177D, B, A, 139.

249–255. Palmae, *Jessenia bataua*, 45μ
  249. Equatorial view, transverse position
  250. Equatorial view, longitudinal position
  251–255. Distal views
256–257. Palmae, *Lepidocaryum gracile*, 37μ
  256. Equatorial view, transverse position
  257. Distal view
258–260. Palmae, *Wallichia oblongifolia*, 27μ
  258. Distal view
  259. Equatorial view, longitudinal position
  260. Equatorial view, transverse position
261–263. Liliaceae, *Lilium bulbiferum*, 95μ
  261. Left side: half pollen grain (polar view, proximal face)
  Right side: half pollen grain (polar view, distal face)
  Right margin: optical section
  262. Equatorial view, longitudinal position
  263. Equatorial view, transverse position

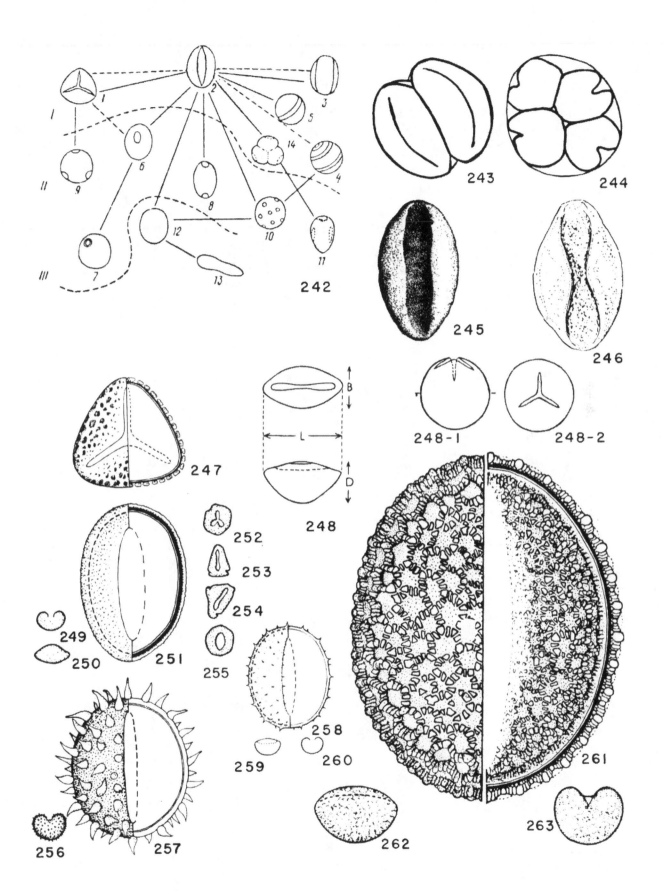

242

243

244

245

246

247

248

248-1

248-2

249

250

251

252

253

254

255

256

257

258

259

260

261

262

263

215

Plate 15

Figure 264–265. TENUATE. Ruppiaceae, *Ruppia maritima*. Grains heteropolar, bilateral, faintly arcuate, 3-tenuate (thin ends, thin area at convex pole, longest axis about 70μ). From Erdtman, 1952, fig. 225.

264. Polar view
265. Side view

Figure 265–1. TENUATE. From Norem, 1958, text-fig. 8.

Figure 266–267. DICOLPATE. Monimiaceae. Furrow outlines are dotted on undersurface of grains. From Cranwell, 1953, textfigs. 2 and 3 (after Money, Bailey, and Swamy, 1950, figs. 15 and 16, respectively).

266. *Laurelia novae-zelandiae*
267. *Dotyphora sassafras*

Figure 268–270. OPERCULUM. Gramineae. From Wodehouse, 1935, pl. 5, figs. 2, 3; pl. 6, fig. 1.

268. *Phleum pratense,* ventral view, 35μ in diameter
269. Germ pore and surrounding exine of a pollen grain of *Avena fatua,* diameter of the aperture 7.5μ
270. *Castalia odorata,* side view, expanded, 30μ

Figure 271–282. ULCUS. From Erdtman, 1952, figs. 179, 220B, 148B.

271–272. Pandanaceae, *Sararanga sinuosa,* 18μ
273. Pandanaceae, *Pandanus odoratissimus* var. *suvaensis,* 21μ
274. Pandanaceae, *Freycinetia arborea,* 21μ
275–276. Restionaceae, *Staberhoa cernua,* 43μ
Section through ulcus

277–281. Pandanaceae, *Pandanus eydouxia,* 33μ
282. Magnoliaceae, *Drimys winteri,* ca. 50μ. Top figure: polar view (di, distal pole); g, tetrad; h, uppermost pollen grain in g (proximalipolar view); i, equatorial view (section, unexpanded grain; p. proximal pole)
Lowermost figure: equatorial view (expanded grain, surface and section)

Figure 283. DICOLPORATE. Tiliaceae, *Entelea aborescens,* 26μ. From Cranwell, 1942, pl. 54, Fig. 9.

Figure 284–285. DIORATE. Didymelaceae, *Didymeles madagascariensis,* 23μ. From Erdtman, 1952, fig. 82.

284. a, polar view; b, equatorial view (sexine not shown); f, aperture, 2000x (meridional section showing operculum and two ora); c, equatorial view
285. Detail of fig. 284c, 2000x.

Figure 286–287. EXINE RUGA. Polar view of two pollen. From Potonié, 1934, figs. 37, 40.

Figure 288. POLLEN WITH MERIDIONAL AREAS. From Potonié, 1934, fig. 14.

Figure 289. SULCUS. Equatorial view and polar view. Erdtman, 1946, fig. 1.

Figure 290. DIMENSIONS OF PORES. P-M, diameter of the pore; P-Lg, length of the pore; P-Lt, width of the pore; anl-M, annulus width. From Iversen and Troels-Smith, 1950, pl. 7.

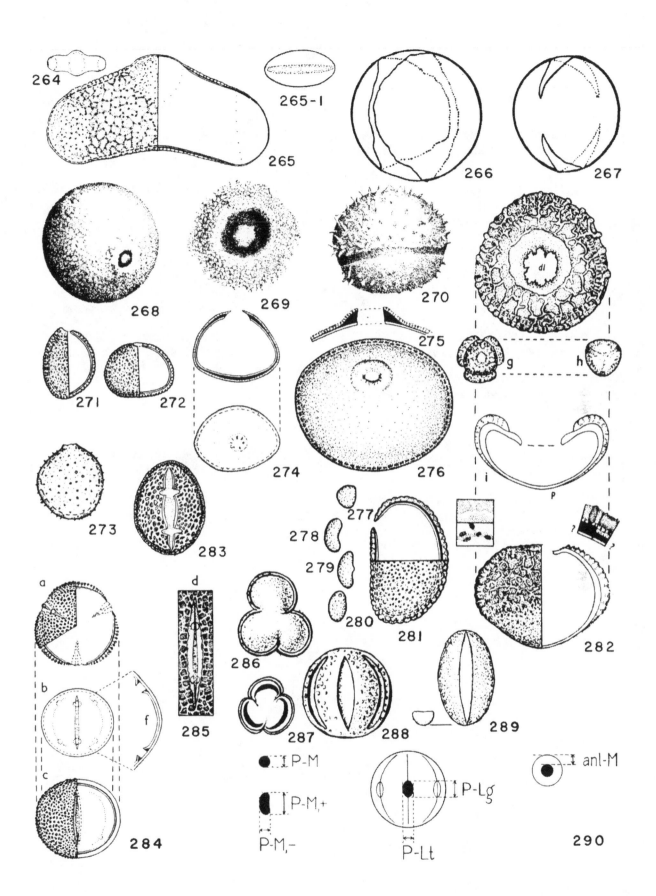

Plate 16

Figure 291–293. SULCULUS. Amaryllidaceae, *Crinum americanum,* 75μ.

291. Cf. polar view (equator and amb coincident)
292. Transverse cf. equatorial view (poles at top and bottom)
293. Longitudinal cf. equatorial view (longest axis = equator; poles at left and right)
From Erdtman, 1952, fig. 12a-c.

Figure 294. PSEUDOCOLPUS. Hydrophyllaceae, *Phacelia ciliata,* 24μ. From Erdtman, 1952, fig. 124.

Figure 295–297. SULCULUS. Monimiaceae, *Altherosperma moschata,* 41μ. From Erdtman, 1952, fig. 157E.

295. Polar view
296. Equatorial-transverse view
297. Equatorial-longitudinal view

Figure 298–300. SULCULUS, POLYPLICATE. Araceae, *Spathiphyllum patinii,* 33μ. From Erdtman, 1952, fig. 19G.

298. Polar view
299. Equatorial view. The exine probably cracks longitudinally (Erdtman, p. 56)
300. LO-pattern of sulculi

Figure 301–305. COLPUS. Major morphological types characterizing *Ephedra* pollen grains. From Steeves and Barghoorn, 1959, pl. 1.

301. *Ephedra aspera,* Type C, 1200x
302. *E. sinica,* Type A, 1200x
303. *E. trifurca,* Type D, 1200x
304. *E. californica,* Type B, 1200x
305. *E. distachya,* Type A, 1200x

Figure 305–1. PLICATE. From Norem, 1958, text-fig. 7.

Figure 306–311. COLPOROIDATE. From Erdtman, 1952, figs. 18B, 237C, and 18C.

306–307. Aquifoliaceae, *Phelline lucida,* ca. 21μ
306. Polar view
307. Equatorial view
308–309. Salicaceae, *Salix myrsinites,* 36μ
308. Polar view
309. Equatorial view
310–311. Aquifoliaceae, *Ilex canariensis,* 49μ
310. Equatorial view
311. Polar view

Figure 312–313. DIORATE. Euphorbiaceae, *Breynia fruticosa,* 25μ. From Erdtman, 1952, fig. 97C.

312. Polar view
313. Equatorial view (optical section)
314. Detail of fig. 313, 3500x

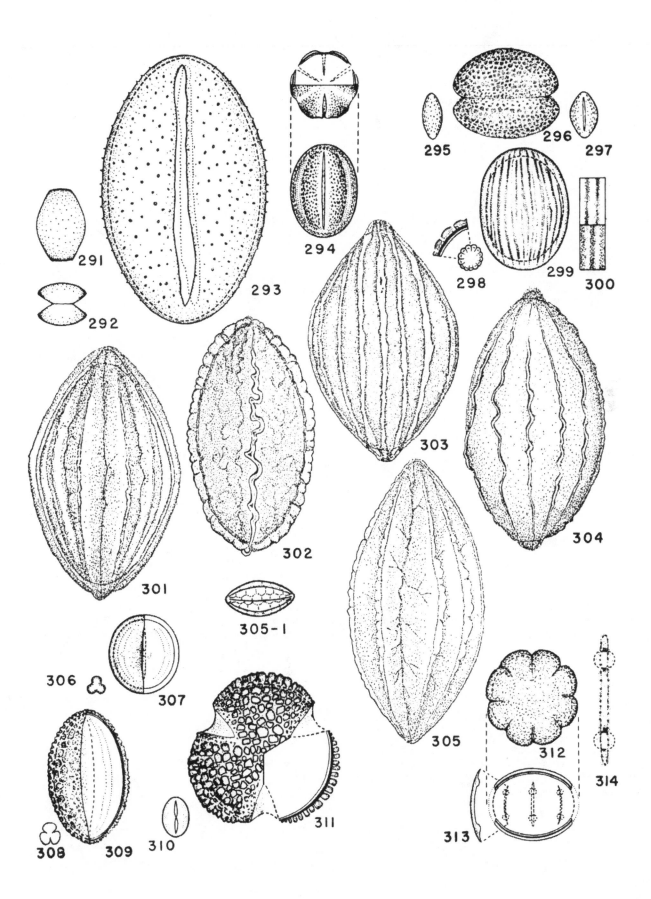

Plate 17

Figure 315–317. APERTURE. Sulcus, rimula, and vestibulum. Plan view and cross section of germinal of *Tilia*. From Potonié, 1934, figs. 30, 33, 38.

Figure 318–320. APERTURE. Exitus digitatus. From Potonié, 1934, pl. 2, fig. 35; pl. 3, fig. 26; and pl. 3, fig. 35.
318. *Pollenites exactus* Potonié, germinal apparatus
319. *P. pompeckji* Potonié, side view of grain, 28μ
320. *P. pompeckji* Potonié, germinal apparatus, oblique-vertical view

Figure 321. APERTURE. Vestibulum, rimula. Germinal of *Fagus* sp., plan view and cross section. From Potonié, 1934, fig. 35.

Figure 322–324. APERTURE. Exitus digitatus. From Potonié, 1934, pl. 4, fig. 33, 34; pl. 6, fig. 22.
322. *Pollenites megadolium digitatus* Potonié, germinal apparatus, somewhat lateral view
323. *P. digitatus* Potonié, germinal apparatus, vertical view
324. *P. exactus* Potonié, 16.5μ, equatorial view of pollen grain

Figure 325–328. APERTURE. Tenuitas. Germinal apparatus with equatorial ruga (e.g. *Sapota*). From Potonié, 1934, figs. 43, 44.
325. Meridional section
326–327. Tangential section
328. Cross section

Figure 329. APERTURE. Pollen with intexine ruga. From Potonié, 1934, fig. 42.

Figure 330. APERTURE. Tenuitas. *Pollenites megadolium* Potonié, vertical view of the germinal apparatus. From Potonié, 1934, pl. 4, fig. 32.

Figure 331. APERTURE. Sulcus, tenuitas. *Pollenites megadolium* Potonié, 42μ, equatorial view. From Potonié, 1934, pl. 6, fig. 30.

Figure 332. APERTURE. Meridional ruga, sulcus. *Pollenites megadolium digitatus* Potonié, 40μ. Equatorial view. From Potonié, 1934, pl. 6, fig. 31.

Figure 333. APERTURE. Altitude. *Pollenites pseudocruciatus* Potonié, 29μ. Equatorial view. From Potonié, 1934, pl. 6, fig. 13.

Figure 334–335. APERTURE. Tenuitas. *Pollenites megadolium sinuatus* Potonié, ca. 68μ. Equatorial view. From Potonié, 1934, pl. 5, figs. 4, 5.

Figure 336. GERM PORE. Polygonaceae, *Eriogonum gracile*. Side view, 32μ. From Wodehouse, 1935, pl. VIII, fig. 6.

Figure 337–339. COLPORATE. From Erdtman, 1952, figs. 6D, 10A.
337–338. Acanthaceae, *Mendoncia aspera*, 51μ
337. Polar view
338. Equatorial view
339. Aperture details in *Alangium platanifolium* var. *trilobatum*, Alangiaceae

Figure 340. COLPOIDORATE. Aperture details in *Alangium villosum* f. *vitiense*. From Erdtman, 1954, fig. 10B.

Figure 341–344. GENICULUS. From Stanley and Kremp, 1959, text-fig. 1.
341. Schematic drawing of a polar view of the pollen of *Quercus prinoides*
342. Section A, taken along the line A-A, showing the geniculus
343. Section B, taken along the line B-B, showing three deep colpi
344. Section C, taken along the line C-C, showing three shallow colpi

Figure 345–346. COLPORATE. Leguminosae, *Lathyrus vernus*, 41μ. From Erdtman, 1952, fig. 135D.
345. Polar view
346. Equatorial view

Figure 347–348. COLPORATE. LALONGATE ORA. Solanaceae, *Hyoscyamus albas*, 52μ. From Erdtman, 1952, fig. 235B.
347. Polar view
348. Equatorial view

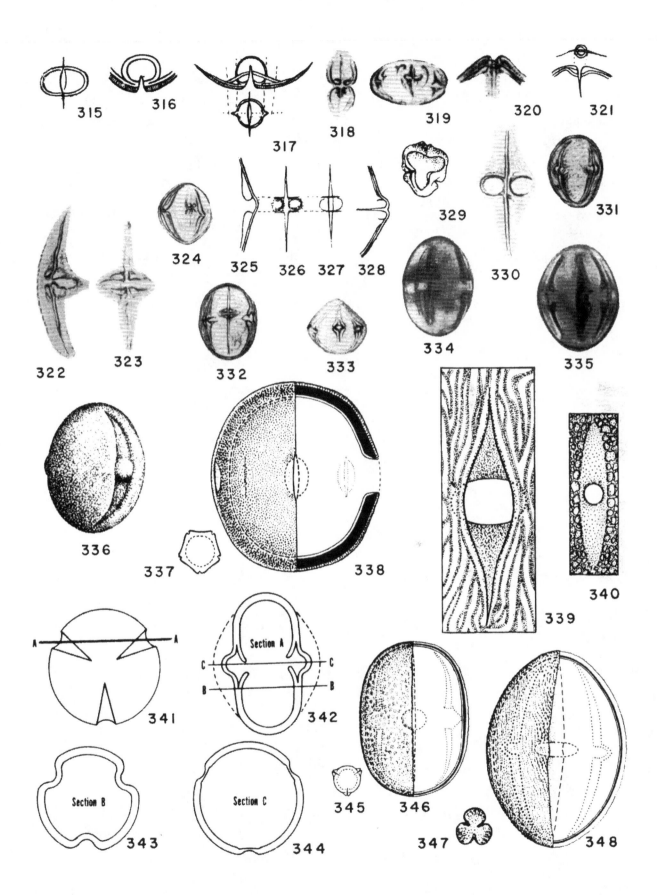

315

316

317

318

319

320

321

322

323

324

325 326 327 328

329

330

331

332

333

334

335

336

337

338

339

340

341

A——A

Section A

C——C

B——B

342

Section B

343

Section C

344

345

346

347

348

221

Plate 18

Figure 349–364. Terms used in describing various features of the apertures of pollen. [*See* also under COLPUS]

aeq, aequator, aequatorialis
anl, annulus, annulatus
ap, apex
C, colpus
cent, centrum
cost, costae
ekt, ektexina
ex, exine
inter C, intercolpium
inter P, interporium
l, limes
Lg, longitudo

Lt, longitudo transversa
M, mensura
max, maximum
mb, membrana
med, medianum
mg, margo, margines
O, absent
op, operculum
P, porus
polar, area polaris
tr, transversalis
vest, vestibulum

From Iversen and Troels-Smith, 1950, pl. I, VI, VII, VIII

349. COLPUS
350. PORE
351. DIMENSIONS (Interporium, intercolpium)
352. DIMENSIONS (Colpus)
353. DIMENSIONS (Pore)
354. DIMENSIONS (Colpus)
355. DIMENSIONS (Pore, annulus)
356. COLPUS
357. PORE
358. DIMENSIONS (Pore)
359. COLPUS (Colpus transversalis)
360. COLPUS (Costae transversalis)
361. DIMENSIONS (Apex colpi)
362–363. DIMENSIONS (Pollen grain)
364. DIMENSIONS (Polar area)

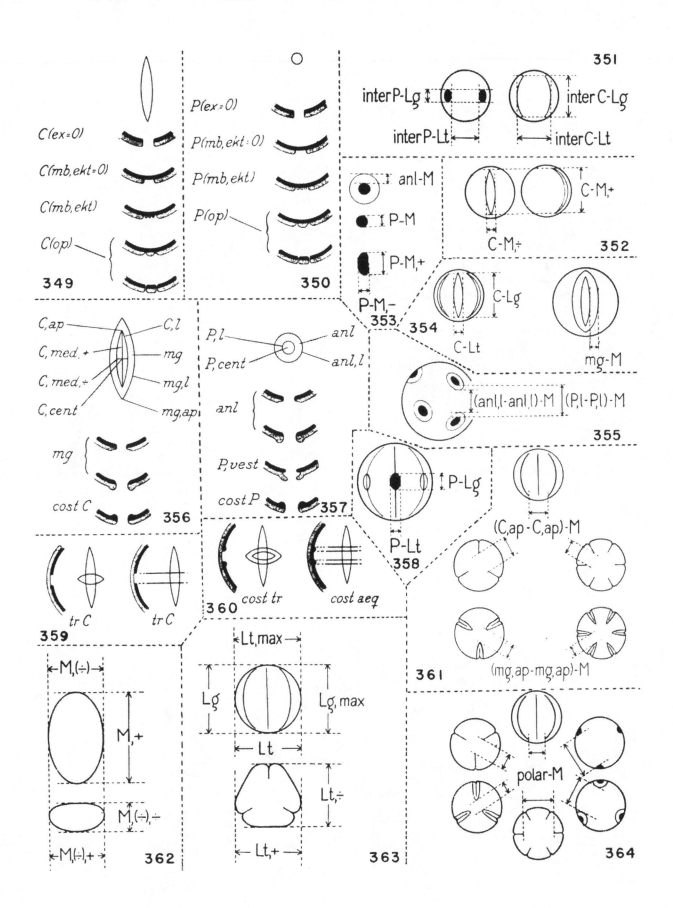

Plate 19

Figure 365–367. ASPIDOTE, BREVISSIMI-RUPATE. Haloragaceae, *Myriophyllum spicatum,* 28μ. From Erdtman, 1952, fig. 118A.

365. Polar view
366. Side view
367. Grain unrolled

Figure 368–371. APERTURE (Exitus). h, meridional diameter; b, equatorial diameter. From Thomson and Pflug, 1953, p. 39, and fig. 15.

368. Pore, circular
369. Pore, elliptical, meriodionally elongated
370. Pore, elliptical, equatorially elongated
371. Ruga

Figure 372–375. CAVERNA. From Thomson and Pflug, 1953, fig. 16.

372. Caverna circumscribing the porus; equatorial view, view toward the centrum; b: breadth of the caverna
373. Equatorial view, tangential view; t: depth of the caverna
374. Porus wider than the caverna; view toward the centrum
375. Polar view, section through the equatorial plane. The porus herein is not considered

Figure 376–388. APERTURE. Germinal types of brevaxonate pollen. From Thomson and Pflug, 1953, pp. 34–35 and fig. 8–10.

376–378. Ektexine development of brevaxonate pollen with annulus (fig. 9a-c).

376. Globe-like cross section
377. Drop-shaped, symmetrical cross section
378. Club-shaped cross section, centripetally developed

379–382. Germinal types of brevaxonate pollen, primary forms (Schema, fig. 8).

379. Germinal without splitting and solution (*Corylus* type)
380. Atrium: solution in the endexine. Endopore more than three times as large as exopore (*Myrica* type)
381. Vestibulum: splitting between ect- and endexine (*Betula* type)
382. Postvestibulum: splitting in the endexine (*Tilia* type)

383–385. Germinal types of brevaxonate pollen, some composite forms (Schema, fig. 10).

383. Combination of atrium and labrum (*Triatriopollenites bituitus* type)
384. Combination of atrium and annulus (*Triatriopollenites excelsus* type)
385. Combination of annulus, vestibulum, and postatrium (*Extratriporopollenites obexemplum* type)

386–388. Ektexine development of brevaxonate pollen (fig. 9d-f).

386. Club-shaped cross section, centripetally developed; with annulus
387. Labrum
388. Tumescence

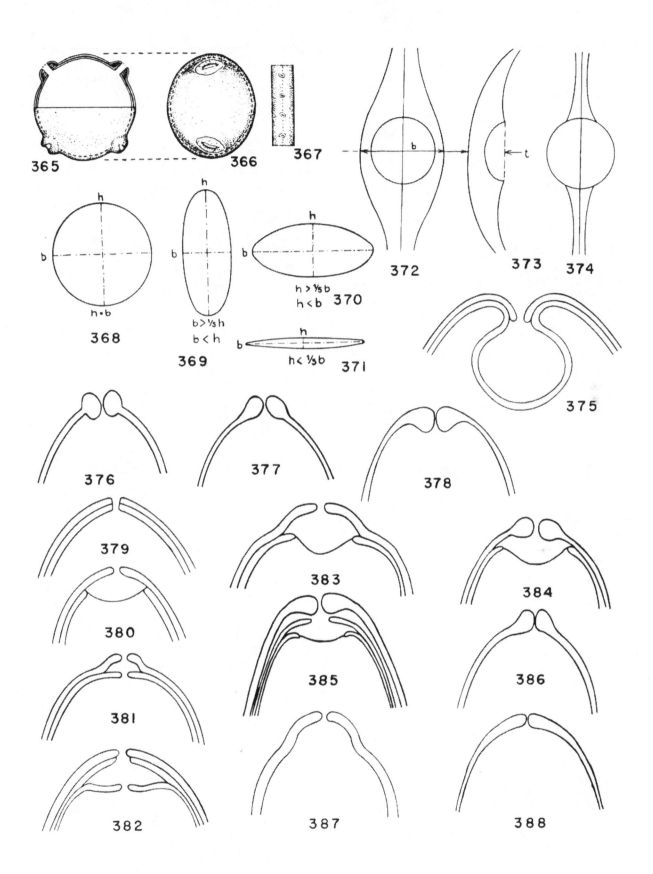

365

366

367

h•b
368

b>⅓h
b<h
369

h>⅓b
h<b   370

h<⅓b   371

372

373

374

375

376

377

378

379

380

381

382

383

384

385

386

387

388

Plate 20

Figure 389–399. APERTURE. From Potonié, 1934, fig. 19, 20, 22–29, 34.

389. Fovea. Pollen with equatorial exitus (p. 16)
390. Pollen with equatorial exitus (p. 16)
391. Porus vestibuli with lips (p. 19)
392. Cross section of porus vestibuli, e.g. *Corylus,* (p. 19)
393. As 392
394. Plan view of vestibulum-germinal of *Corylus* (p. 19)
395. *Alnus* pollen with arci (p. 20)
396. *Pollenites plicatus* Pot. with arci (p. 20)
397. Arci of *Pollenites microexcelsus* Pot. (*Betula*) (p. 20)
398. Plan view of germinal of *Alnus* (p. 20)
399. As 398

Figure 400–401. ASPIDOPORATE. From Erdtman, 1943, fig. 225, 224.

400. Juglandaceae, *Carya cordiformis.* Three-pored grain, polar view, abporal hemisphere, 50μ
401. Juglandaceae, *Carya cordiformis.* Polar view, poral hemisphere, 51μ

Figure 402–405. ARCUS. Rhoipteleaceae, *Rhoiptelea chiliantha,* 28μ. From Erdtman, 1952, fig. 222A.

402. Aperture, surface view, 1000x
403. Equatorial view
404. Polar view
405. Aperture, optical section, 2000x

Figure 406. ASPIDOPORATE. Betulaceae, *Betula nana.* Polar view 20μ. From Erdtman, 1943, pl. V, fig. 63.

Figure 407. ARCUS. Construction of pore and arci in *Alnus* pollen (schematic). eq, equator; M-M, meridian; EQ, equatorial; M, meridional section of pore. Upper left corner, construction of an arcus. From Erdtman, 1943, fig. 7.

Figure 408–409. ARCI. From Thomson and Pflug, 1953, fig. 12.

408. Polar view of the exine (*Alnus* type). a, arcus; v, vestibulum
409. Meridional section through the vestibulum, view toward the center: a₁-a₄, arci; e, endosporus; ekt, ectexine; end, endexine; v, vestibulum

Figure 410–412. Developmental stages of the SOLUTION MERIDIUM. From Thomson and Pflug, 1953, p. 32, fig. 7.

410. Solution notch (incidence)
411. Solution wedge (cuneus)
412. Solution channel (platea)

Figure 413–414. EXPANSION FOLD. The arrows mark the stronger swelling capacity of the pore regions as opposed to the centers of the sides. From Thomson & Pflug, 1953, fig. 11.

413. Plicae. p, expansion folds of the entire exine (*Triatriopollenites plicatus* type)
414. Endoplicae. e, expansion folds of the endexine (*Triatriopollenites excelsus* type); a, atrium

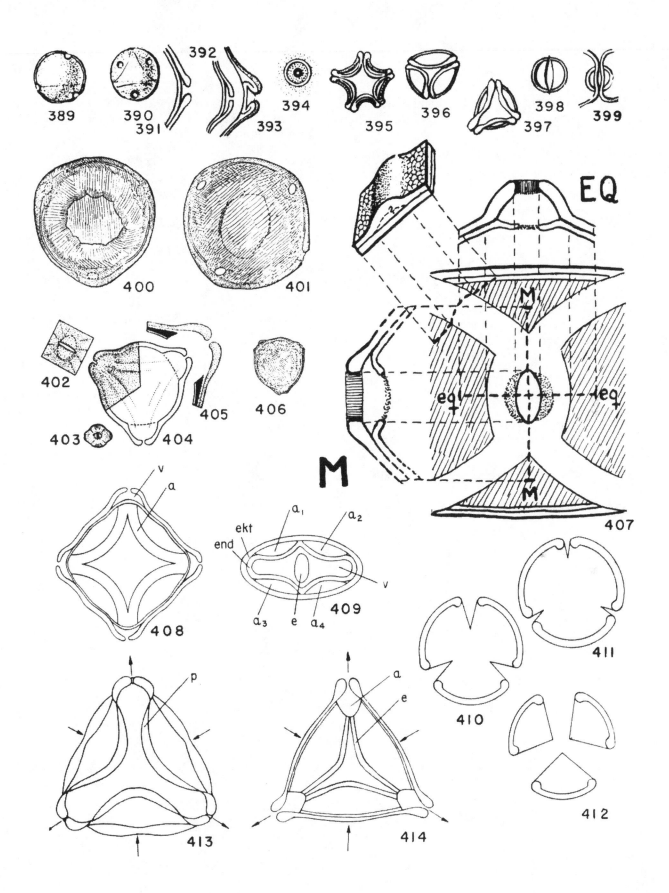

Plate 21

Figure 415. LALONGATE. Euphorbiaceae, *Leptonema venosum*. Equatorial view. From Erdtman, 1952, fig. 97B.

Figure 416–418. PONTOPERCULATE. Rosaceae, *Dendriopoterium menendezii*, 36μ. From Erdtman, 1952, fig. 223C.

416. Equatorial view
417. Polar view
418. Plan view of colpus

Figure 419–421. OPERCULUM. From Erdtman, 1952, fig. 75B and 130A.

419. Cucurbitaceae, *Cayaponia racemosa*. d, part of exine; e, LO-pattern of same at high adjustment; f, at medium adjustment; g, at low adjustment
420–421. Labiatae, *Teucrium capitatum* var. *Valentinum*. Polar and equatorial view.

Figure 421–1 and 2. TRILATIPORATE. Equatorial and polar view. From Norem, 1958, text-fig. 16.

Figure 422–423. LOLONGATE. Boraginaceae, *Borago officinalis*. Polar and equatorial view. From Erdtman, 1952, fig. 35.

Figure 424–425. LALONGATE. Euphorbiaceae, *Reverchonia arenaria*, 38μ. Polar and equatorial view. From Erdtman, 1952, fig. 97D.

Figure 426. TRANSCOLPATE. Aquifoliaceae, *Sphenostemon balansae*, 16μ. Erdtman, 1952, fig. 18A.

Figure 427–428. CONSTRICTICOLPATE. Bruniaceae, *Brunia laevis*. Polar and equatorial view. From Erdtman, 1952, fig. 38A.

Figure 429–430. BREVICOLPATE. Connaraceae, *Jollydora duparquetiana*, 51μ. Equatorial and polar view. From Erdtman, 1952, fig. 67A.

Figure 431–432. CRUSTATE. Didiereaceae, *Alluaudia humberti*, 65μ. Equatorial and polar view. From Erdtman, 1952, fig. 81.

Figure 433–435. BREVISSIMICOLPATE. From Erdtman, 1952, fig. 226Ac and 239E.

433. Rutaceae, *Erythrochiton brasiliensis*. Equatorial view, 76μ; colpi 18 × 4μ
434–435. Sterculiaceae, *Ayenia magna*, 32μ. Equatorial and polar view

Figure 436–438. RUGA. Caryophyllaceae, *Drymaria cordata*. Grains 12-rugate. From Erdtman, 1952, fig. 50D.

Figure 439–442. SPIRAPERTURATE. From Erdtman, 1952, fig. 94A.

439–440. Eriocaulaceae, *Eriocaulon septangulare*, 34μ
441–442. Berberidaceae, *Berberis dictyophylla*, 46μ

Figure 442–1. ZONACOLPATE. From Norem, 1958, text-fig. 22b.

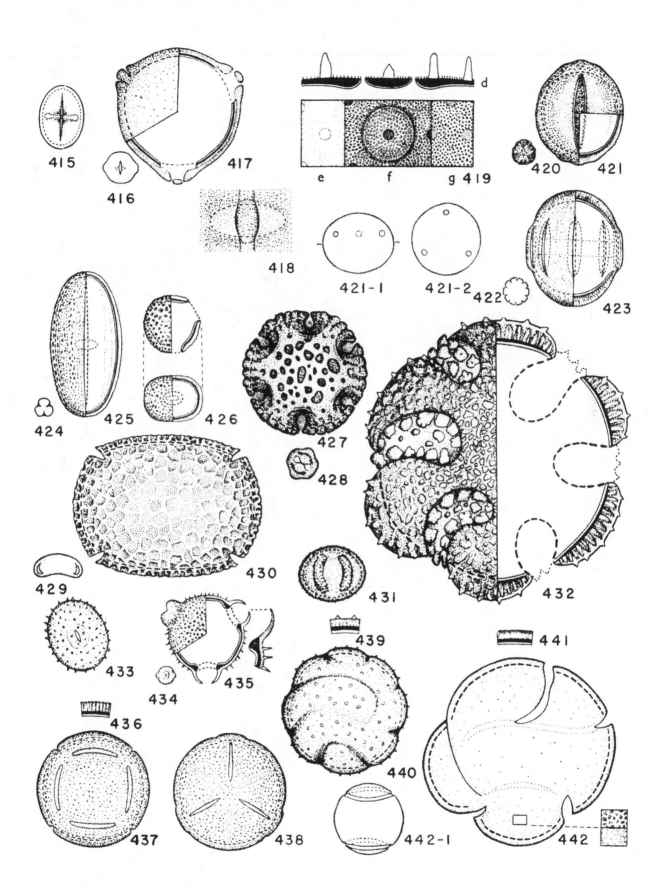

415

416

417

d

e   f   g 419

420   421

418

421-1   421-2   422   423

424   425   426   427   428   432

429   430   431

433   434   435   439   441

436   440

437   438   442-1   442

229

Plate 22

Figure 443–444. RUGA. Cactaceae, *Echinocactus tabularis,* 15-rugate. View of pollen grain and LO-pattern. From Erdtman, 1952, fig. 41.

Figure 445–451. SYNCOLPATE. From Erdtman, 1952, fig. 109C, 120, 164A.

445–446. Gentianaceae, *Nymphoides indica,* parasyncolpate, 33μ. LO-pattern, polar and side view (fig. 109C).

447–448. Heteropyxidaceae, *Heteropyxis natalensis,* 16μ, parasyncolpate. LO-pattern, polar and side view (fig. 120).

449–451. Myrtaceae, *Metrosideros linearis,* 20μ. (fig. 164A).

449. Grain subisopolar, 4-ruporate (parasynrupate) with ± rectangular aporupia at right angles to each other

450. Equatorial view of 451

451. Grain isopolar, 4-colporate (parasyncolpate), with square apocolpia

Figure 452. PARASYNCOLPATE. Sapindaceae, *Matayba apetala,* 3-colporate (angulaperturate), parasyncolpate, peroblate, 21μ. Polar and side view. From Erdtman, 1952, fig. 229A

Figure 453–454. APOCOLPIUM. Caryocaraceae, *Caryocar brasiliense,* 3-colporate, subprolate, some-

times parasyncolpate, apocolpium small, 60μ. Equatorial and polar view. From Erdtman, 1952, fig. 49.

Figure 455–456. SYNCOLPATE. Myrtaceae, *Eucalyptus ficifolia,* 35μ. Equatorial and polar view. From Erdtman, 1952, fig. 164B.

Figure 457–460. APOCOLPIUM. From Erdtman, 1952, fig. 47B, 234F.

457–458. Caprifoliaceae, *Triosteum pinnatifidum,* ora slightly lolongate, apocolpium large, 53μ. Equatorial and polar view (fig. 47B).

459–460. Pedaliaceae, *Sesamum capense,* 6-colpate, oblate spheroidal, apocolpium hexangular, 59μ. Polar and equatorial view (fig. 234F).

Figure 461–462. SYNCOLPATE. Simarubaceae, *Kirkia acuminata,* 3-colporate (syncolpate), ora circular. Polar and side view. From Erdtman, 1952, fig. 234F.

Figure 463–464. APOCOLPIUM. Acanthaceae, *Blechum laxiflorum,* 3-colpate (syncolpate), apocolpium lacking, oblate spheroidal, 48μ. Side and polar view. From Erdtman, 1952, fig. 6C.

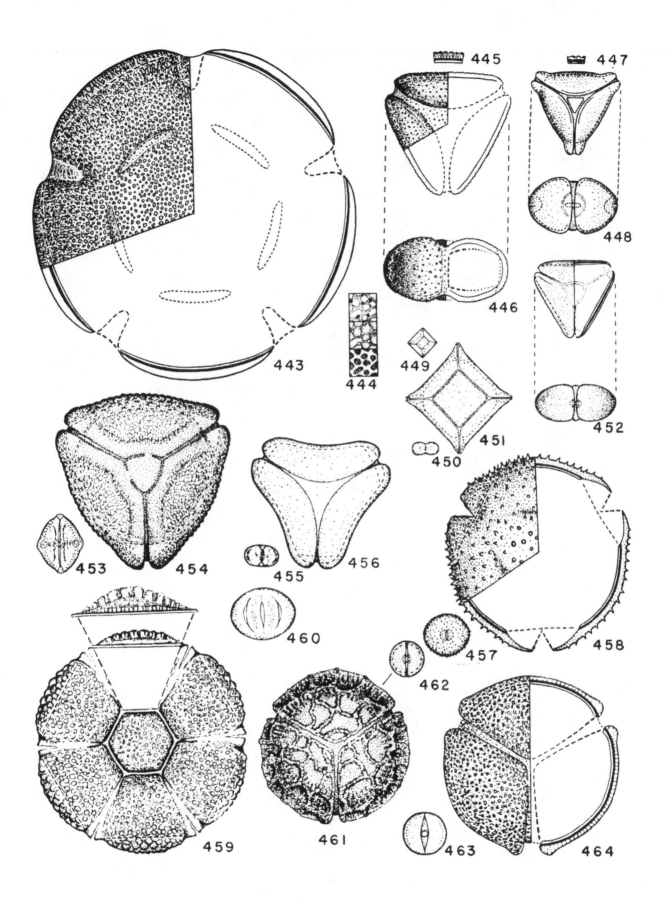

443

444

445

446

447

448

449

450

451

452

453 454

455 456

457 458

459

460

461

462

463 464

Plate 23

Figure 465–466. DIPLODEMICOLPATE. Loranthaceae, *Macrosolen bibracteolatus,* 3-colpate, 72μ. From Erdtman, 1952, fig. 146A.

Figure 467–469. SULCULI. Liliaceae, *Tofieldia calyculata,* 2-sulculate, 30μ. From Erdtman, 1952, fig. 140B.

467. Equatorial transverse view
468. Polar view
469. Equatorial longitudinal view

Figure 470–472. SYNDEMICOLPATE. Rosaceae, *Grielum tenuifolium.* "Tentative interpretation: oblate, . . . 44μ . . . (sexine not included); saccate [perisaccate: saccus provided with 2 polar three-armed openings (6 demicolpi?, anastomosing 3 and 3 at the poles)]. Nexine provided with 6 rounded openings [individual openings (ora) correspond to the branches of the 3-slit openings in the sexine]." From Erdtman, 1952, Fig. 223A and p. 381.
Fig. 470. Individual opening (os).

471. Polar view
472. Equatorial view (without sculptine)

Figure 473–474. DIPLODEMICOLPATE. Loranthaceae, *Amylotheca djamuensis,* 3-diplo-demicolpate. From Erdtman, 1952, fig. 146B.

473. Polar view
474. Vertical section through the dotted line in Fig. 473

Figure 475–479. SULCULI. From Erdtman, 1952, fig. 218.

475–476. Rapateaceae, *Caphalostemon angustatus,* zonisulculate, 85μ
475. Exine stratification
476. Equatorial longitudinal view
477–479. Rapateaceae, *Rapatea spectabilis,* zonisulculate, 40μ
477. Polar view
478. Equatorial transverse view
479. Equatorial longitudinal view

Figure 480–486. SULCULI. From Erdtman, 1952, fig. 101, 19A.

480–484. Eupomatiaceae, *Eupomatia laurina,* 35μ, ". . . slightly subisopolar, probably 2-sulculate, . . . or 3-sulculate, . . . sometimes ± zonisulcate. . . . sulculus-zone is sometimes ruptured and the halves of the pollen grain thus separated can easily be mistaken for 1-sulcate or trichotoma-sulcate grains. (p. 175)
483. One half of a 3-sulculate grain seen from the interior
484. One half of a 2-sulculate grain seen from the interior
485–486. Araceae, *Afroraphidophora africana,* 27μ, 3-sulculate. Equatorial and polar view

Figure 487–488. ZONORATE. Caprifoliaceae, *Abelia triflora,* ± brevicolpate, spheroidal, 63μ. From Erdtman, 1952, fig. 48.

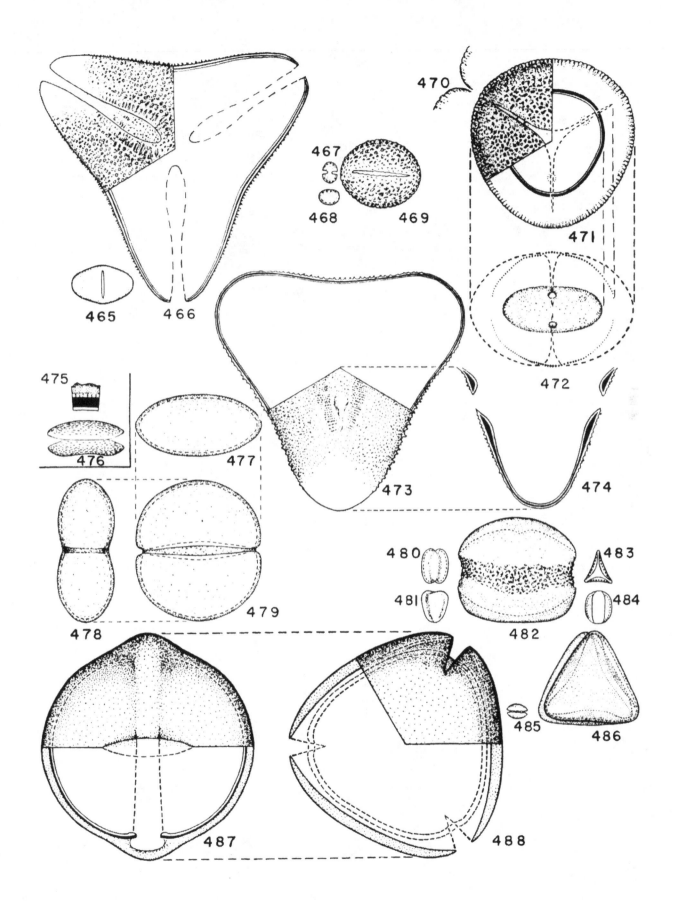

465

466

467

468 469

470

471

472

473

474

475

476

477

478

479

480

481

482

483

484

485

486

487

488

233

Plate 24

Figure 489. FORAMEN. Convolvulaceae, *Calystegia sepium*, 80μ. From Erdtman, 1952, fig. 68C.

Figure 490–492. FOVEA. From Potonié, 1934, fig. 21, 31, 32.
490. Pollen with many foveae, e.g. *Sagittaria*
491. Fovea, cross section
492. Derivation of the porus vestibulum from a fovea

Figure 493. FORATE. Chenopodiaceae, *Salsola tragus*, 27μ. The details in the figure are shown as they appear at low focus. From Erdtman, 1954, fig. 54B.

Figure 494. CRIBELLATE. Chenopodiaceae, *Salsola pestifer*, 27μ. From Wodehouse, 1935, pl. VIII, fig. 4.

Figure 495. FORAMEN. LO-pattern of Fig. 489. From Erdtman, 1952, fig. 68C.

Figure 496–501. AMB. Amb-types in tricolporate pollen grains. From Erdtman, 1952, fig. 2.
496–497. Angulaperturate
499. Planaperturate
500. Sinuaperturate
501. Fossaperturate

Figure 501–1. POLYPERIPORATE. From Norem, 1958, text-fig. 15.

Figure 501–2. POLYPERICOLPATE. From Norem, 1958, text-fig. 18.

Figure 502. FORAMEN. Convolvulaceae, *Ipomea purpurea*. Part of exine surface with six apertures. From Erdtman, 1952, fig. 68B.

Figure 503–504. AMB. Olacaceae, *Harmandia mekongensis*, angulaperturate, ca. 28μ. Equatorial and polar view. From Erdtman, 1952, fig. 172F.

Figure 505–506. AMB. Bombacaceae, *Bombax album*, planaperturate, 25μ. From Erdtman, 1952, fig. 32B.

Figure 507–508. AMB. Araliaceae, *Anomopanax schlechteri*, fossaperturate, 25μ. From Erdtman, 1952, fig. 20B.

Figure 509–515. POLE CAP. From Thomson and Pflug, 1953, fig. 13, 14.
509–512. Contours of the polar cap of prolate pollen.
509. Flattened
510. Subhemispherical
511. Hemispherical
512. Pointed
513–515. Sidelines of longaxonate pollen.
513. Straight
514. Slightly convex
515. Strongly convex

Figure 516. DIMENSIONS. Bipolar pollen grains.
Lt+, largest transverse dimension;
Lt−, smallest transverse dimension;
Lg, longitudo;
Lg max, maximal length. From Iversen and Troels-Smith, 1950, pl. VI.

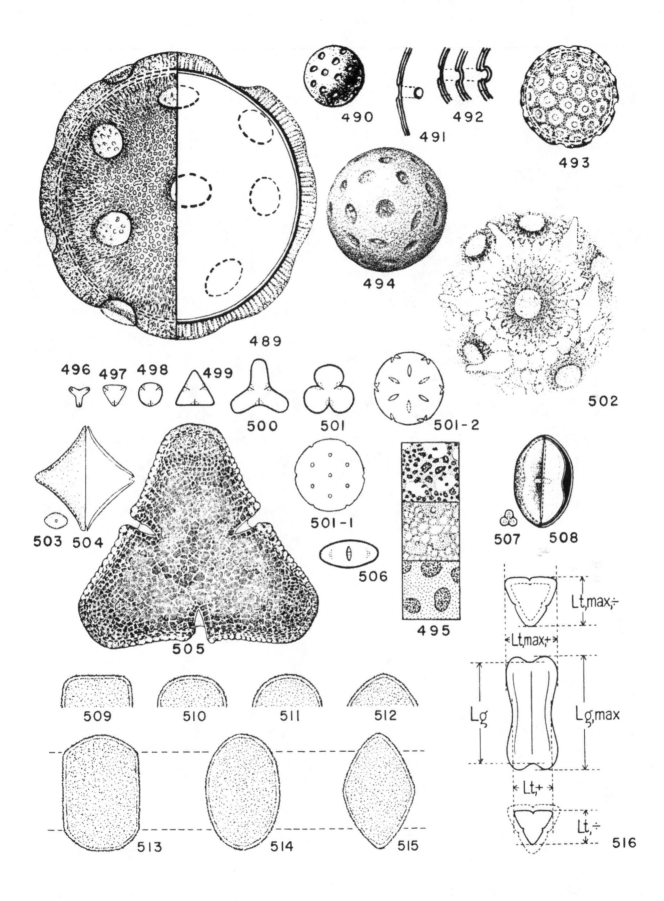

489

490 491 492 493

494

502

496 497 498 499 500 501 501-2

503 504 505 501-1 506 495 507 508

509 510 511 512

513 514 515

$L_{t,max,\div}$

$L_{t,max,+}$

L_{g} $L_{g,max}$

$L_{t,+}$

$L_{t,\div}$

516

Plate 25

Figure 517–518. FORM. From Potonié, 1934, fig. 1 and 2.

517. Basic type of dicotyledonous pollen
518. Derivation from basic type

Figure 519–520. OBLATOID. Proteaceae, *Spatalopsis confusa*, 23µ. Equatorial and polar view. From Erdtman, 1952, fig. 205D.

Figure 521. SHAPE CLASSES. Shapes of radiosymmetrical pollen grains. "With the polar axis of the grains (vertical in the diagram) comprising the *major* axis of the ellipse: 1 by 1, spherical; between 1 and 2, prolate spheroidal; between 2 and 3, subprolate; between 3 and 4, prolate; 4 by 4, perprolate. With the polar axis of the grains (horizonal in the diagram) comprising the *minor* axis of the ellipse: 1 by 1, spherical; between 1 and 2, oblate spheroidal; between 2 and 3, suboblate; between 3 and 4, oblate; 4 by 4, peroblate." From Erdtman, 1943, text-fig. 2.

Figure 522–523. OBLATOID. Proteaceae, *Synaphea polymorpha*, 73µ. Equatorial and polar view. From Erdtman, 1952, fig. 205A.

Figure 524–526. SUBPROLATE. From Erdtman, 1952, fig. 70C and 47C.

524. Coriariaceae, *Mastixia clarkeana*, 39µ. Polar and equatorial view

525–526. Caprifoliaceae, *Sambucus canadensis*, 50µ. Polar and equatorial view.

Figure 527–530. OBLATOID. From Erdtman, 1952, fig. 205C,B.

527–528. Proteaceae, *Spatalla currifolia*, 22µ. Equatorial and polar view
529–530. Proteaceae, *Stirlingia latifolia*, 60µ. Equatorial and polar view

Figure 531–532. BILATERAL. Myrtaceae, *Pileanthus peduncularis*, 42µ. Polar and equatorial view. From Erdtman, 1952, fig. 164E.

Figure 533. APOLAR. Aristolochiaceae, *Asarum arifolium*, 53µ. From Erdtman, 1952, fig. 21C.

Figure 534–537. SUBISOPOLAR. From Erdtman, 1952, fig. 185B, 154.

534–535. Peripterygiaceae, *Pteripterygium moluccanum*, 15µ. Polar and equatorial view
536–537. Medusagynaceae, *Medusagyne oppositifolia*, 19µ. Polar and equatorial view

Figure 538–539. APOLAR. From Erdtman, 1952, fig. 21B,A.

538. Aristolochiaceae, *Apama corymbosa*, 27µ
539. Aristolochiaceae, *Aristolochia grandiflora*, 73µ

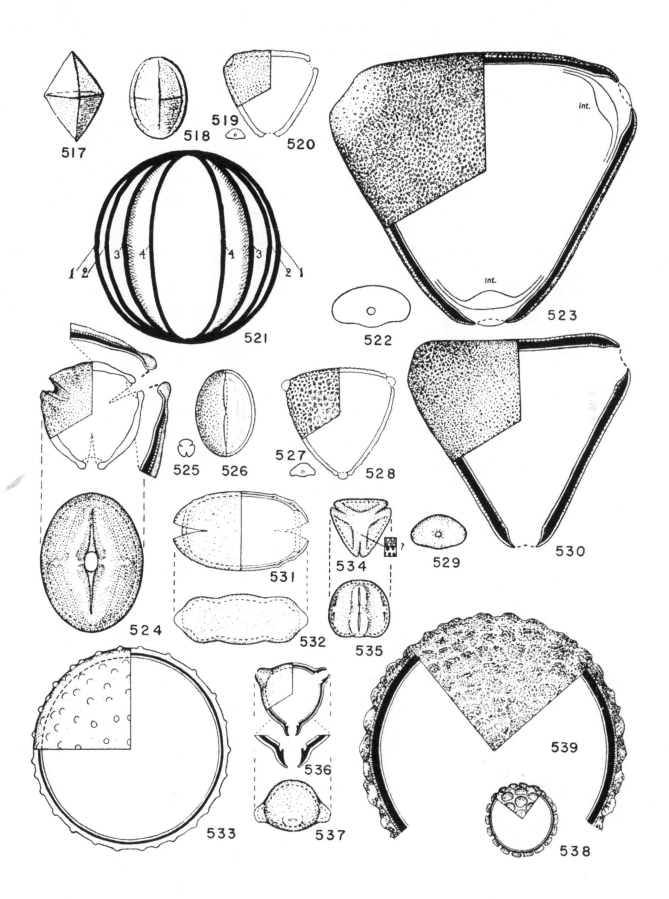

517

518

519

520

521

522

523

524

525

526

527

528

529

530

531

532

534

535

533

536

537

538

539

int.

int.

Plate 26

Figure 540. SPORODERMIS. Schematic presentation of a part of the pollen grain of *Tilia cordata* L. and names of the membrane layers given by different palynologists.
Redrawn from Kupriyanova, 1956, pl. 2, fig. 3.

Figure 541. SPORODERMIS. From Erdtman, 1948, fig. 144. For detail of terminology see text.

Figure 542. MATRIX. (gran, granula) From Iversen and Troels-Smith, 1950, pl. II.

Figure 543–544. LAMELLA. From Thomson and Pflug, 1953, fig. 6.

543. Orientation of brevaxonate pollen, polar view; ekt, ektexine; end, endexine; p, pore canal; pl, pole; s, center of the side; d, equatorial diameter

544. Structure elements of an extratriporate pollen grain with eight laminae conspicuae. a, anulus; ea, end-anulus; enda, endb, endc, endolamina a, b, c; ep, endoporus; exa, exb, exc, exd, exe, exolamina a, b, c, d, e; int, interloculum; oc, oculus; pk, pore canal; pr, praevestibulum; s, solution meridium; v, vestibulum.

Figure 545. SPORODERMIS. Exine stratification and fine structure based on electron microscopic studies. From Larson, Skvarla, and Lewis, 1962, fig. 1.

Figure 546. SPORODERMIS. Terminology. From Van Campo, 1954, p. 252.

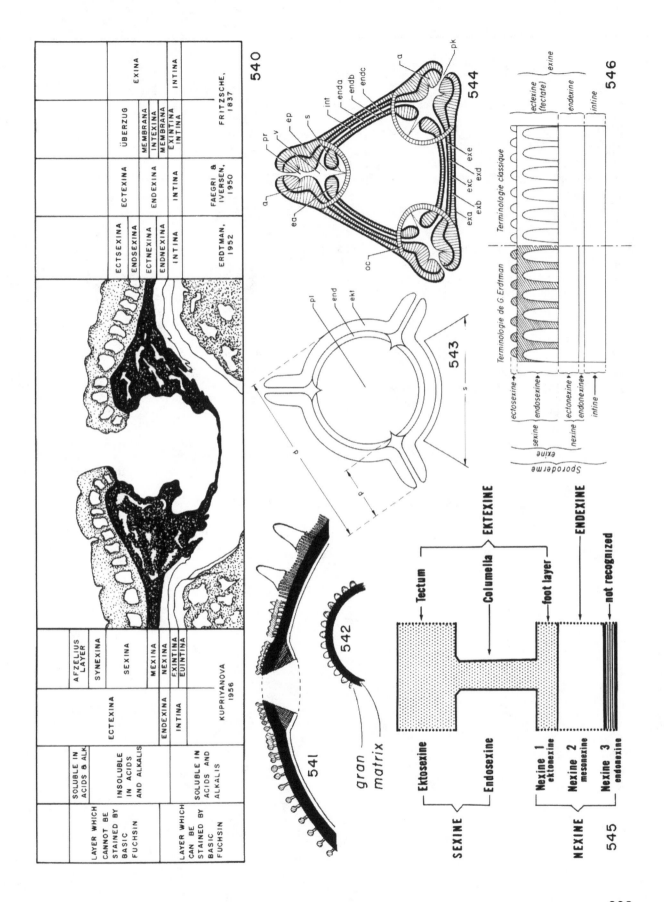

	LAYER WHICH CANNOT BE STAINED BY BASIC FUCHSIN	SOLUBLE IN ACIDS & ALK.					
		AFZELIUS LAYER	SYNEXINA			ECTSEXINA	ECTEXINA
			ECTEXINA	SEXINA		ENDSEXINA	
		INSOLUBLE IN ACIDS AND ALKALIS			MEXINA	ECTNEXINA	ENDEXINA
	LAYER WHICH CAN BE STAINED BY BASIC FUCHSIN	SOLUBLE IN ACIDS AND ALKALIS	ENDEXINA	NEXINA	NEXINA	ENDNEXINA	
				EXINTINA	EXINTINA	INTINA	INTINA
			INTINA	EUINTINA			
			KUPRIYANOVA 1956			ERDTMAN, 1952	FAEGRI & IVERSEN, 1950

EXINA

	ÜBERZUG	EXINA
	MEMBRANA INTEXINA	
	MEMBRANA EXINTINA	
	INTINA	INTINA

FRITZSCHE, 1837

540

544

546

543

542

541

gran
matrix

545

Tectum — EKTEXINE
Columella
foot layer

Ektosexine
Endosexine
SEXINE

Nexine 1 ektnexine
Nexine 2 mesoexine
Nexine 3 endoexine
NEXINE

ENDEXINE
not recognized

239

Plate 27

Figure 547. EXOLAMELLA. Separation of *Abies* air sacs. From Potonié, 1934, fig. 10.

Figure 548–550. PERISPORIUM. Spores of *Aspidium trifoliatum*. From Bower, 1923, fig. 257.

548–549. Ripe spore with prickly exospore (ex) and transparent perispore (psp), appearing like a loose sac (500x)

550. Part of the exospore and perispore more highly magnified

Figure 551–559. PERISPORIUM. Series of photomicrographs to show the presence of an outer layer presumably of similar origin in each case but hitherto regarded as a perispore only when present as an obviously independent layer, as in the first two examples. From Harris, 1955, fig. 2.

551. *Blechnum procerum*, 500x
552. *B. membranaceum*, 500x
553. *B. banksii*, 500x
554. Same slide and negative as 553 but showing the presence of a thin outer layer where it is fragmented, 500x
555–556. *Doodia media*, 500x
557. Same as 555–556 but with outer layer partly shed, 500x
558. *Cyathea medullaris* with outer layer complete, 1000x
559. Same with outer layer incomplete

Figure 560–562. SUBSACCATE. Proteaceae, *Grevillea bipinnatifida*, 74μ. From Erdtman, 1952, fig. 208.

560. Equatorial view
561. Polar view
562. LO-pattern

Figure 563. PUSTULATE. From Krutzsch, 1959, text-fig. 16.

Figure 564. VESICULATE. Cyatheaceae, *Pyrrosia serpens*, ca. 80μ. From Harris, 1955, pl. 5, fig. 8.

Figure 565–566. MESOSPORE. Diagrammatic reconstruction showing the main characters of *Duosporites congoensis* Heg, Bose, and Manum, ca. 600μ. From Heg, Bose, and Manum, 1955, fig. 1.

565. Axial section cut through one branch of the triradiate mark; corresponding to the horizontal line in Fig. 566. Exosporium dotted.
566. View of the proximal hemisphere, partly dissected.
   a. Exosporium cut along the equator and removed so that the proximal surface of the mesosporium is visible. The projections on the inner side of the mesosporium are shown as darker spots
   b. Equatorial section of exosporium and mesosporium
   c. Surface view of the proximal hemisphere. In the flattened fossil spores the mesosporium appears smaller in relation to the exosporium, due to folding

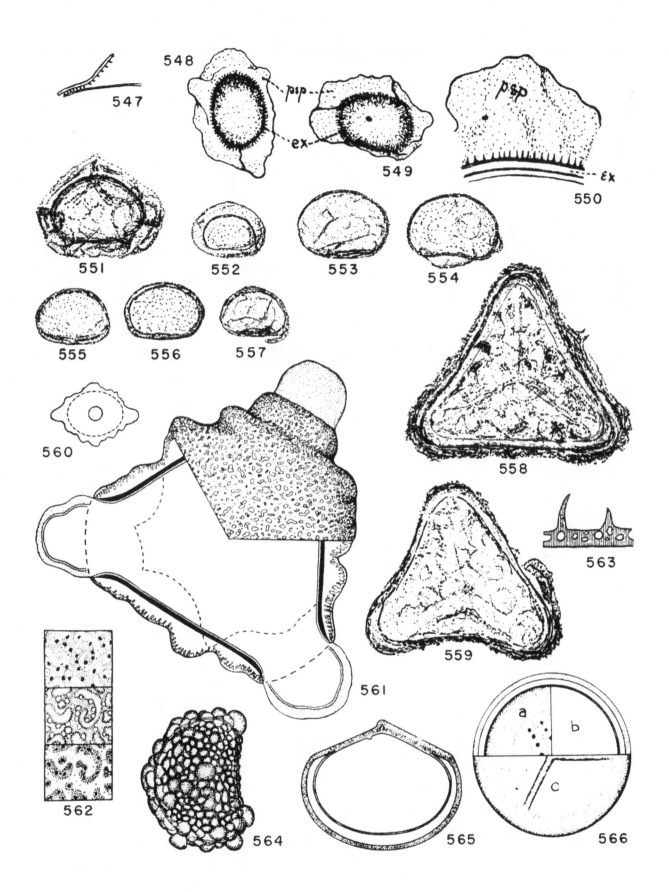

241

Plate 28

Figure 567–569. PERISPORIUM. *Triletes brasserti* Stach and Zerndt, 50x. After Dijkstra and van Viersen Trip, 1946, pl. 5, fig. 51–53.

567. Detached zona of the spore body
568. Central body without zona
569. Spore with zona and central body

Figure 570–581. SPORODERMIS; MEDINE. Diagrammatic representation of wall differentiation in various types of pollen grains:

in, intine
m me, modified medine
me, medine
se, sexine
ne, nexine
ex, exine
ap. m, aperture membrane. From Saad, 1963, pl. VI.
570–571. Non-aperturate pollen grains
572–573. Spirotreme pollen grains
574–575. Colpate pollen grains
576–577. Colporate pollen grains

578–579. Few-porate pollen grains
580–581. Pantoporate pollen grains

Figure 582–588. ONCUS. All grains drawn in optical section to show exine (in outline only) and intine (shaded). From Hyde, 1954, fig. 1.

582–583. *Corylus avellana,* polar and lateral view, 1000x
584–585. *Fraxinus ornus,* polar and lateral view, 1000x
586–587. *Ulmus glabra,* polar and lateral view, 1000x
588. *Frangula alnulus,* polar view, 1000x

Figure 588a. MESINE. The exinous sculptural elements which occur, with variation, in the Commelinaceae are shown diagrammatically. The terms adopted by Erdtman (1952) for sporoderm stratification (right) are compared with the terms used in . . . [Rowley's] study (left). From Rowley, 1959, fig. 1.

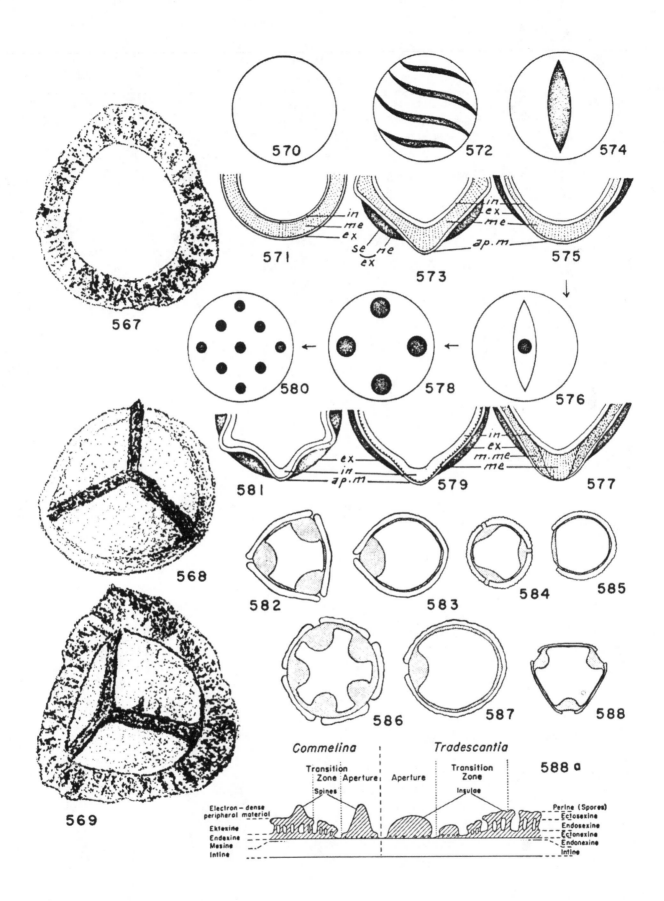

567

570

572

574

571

in
me
ex

573

se me
ex

in
ex
me

ap.m

575

568

580

578

576

581

ex
in
ap.m

579

577

in
ex
m.me
me

569

582

583

584

585

586

587

588

Commelina   *Tradescantia*

588 a

Transition Zone   Transition Zone
Aperture   Aperture
Spines   Insulae

Electron-dense peripheral material   Perine (Spores)
                                            Ectosexine
Ektexine   Endosexine
Endexine   Ectonexine
Mesine   Endonexine
Intine   Intine

Plate 29

Figure 589–590. CRASS-EXINOUS. Gyrostemonaceae, *Gyrostemon racemigerus,* 34μ. Polar view, equatorial view, and exine stratification. From Erdtman, 1952, fig. 117.

Figure 591–592. CRASSISEXINOUS. From Erdtman, 1952, fig. 130A,B.

591. Labiatae, *Teucrium pseudochamaepitys,* exine section
592. *T. capitatum* var. *valentinum,* equatorial view

Figure 593–596. CRASSINEXINOUS. From Erdtman, 1952, fig. 150A,B.

593–594. Malpighiaceae, *Aspicarpa rosei,* 53μ. 593, exine section. 594, view of grain
595–596. Malpighiaceae, *Bunchosia media,* 50μ. 595, exine section. 596, view of grain

Figure 597. SECONDARY FOLD. *Calamospora hartungiana* Schopf, ca. 88μ. From Potonié and Kremp, 1955, fig. 6.

Figure 598–599. TENUI-EXINOUS. Iridaceae, *Syringodea montana,* ca. 62μ. From Erdtman, 1952, fig. 127.

598. Exine section. 599. View of grain.

Figure 600–601. TENUISEXINOUS. Proteaceae, *Conospermum longifolium,* 63μ. From Erdtman, 1952, fig. 201A.

Figure 602–603. CRASSITEGILLATE. Elaeagnaceae, *Elaeagnus argentea,* 44μ. From Erdtman, 1952, fig. 89B.

Figure 604–607. TENUISEXINOUS. Proteaceae, *Conospermum densiflorum,* 75μ. From Erdtman, 1952, fig. 201B.

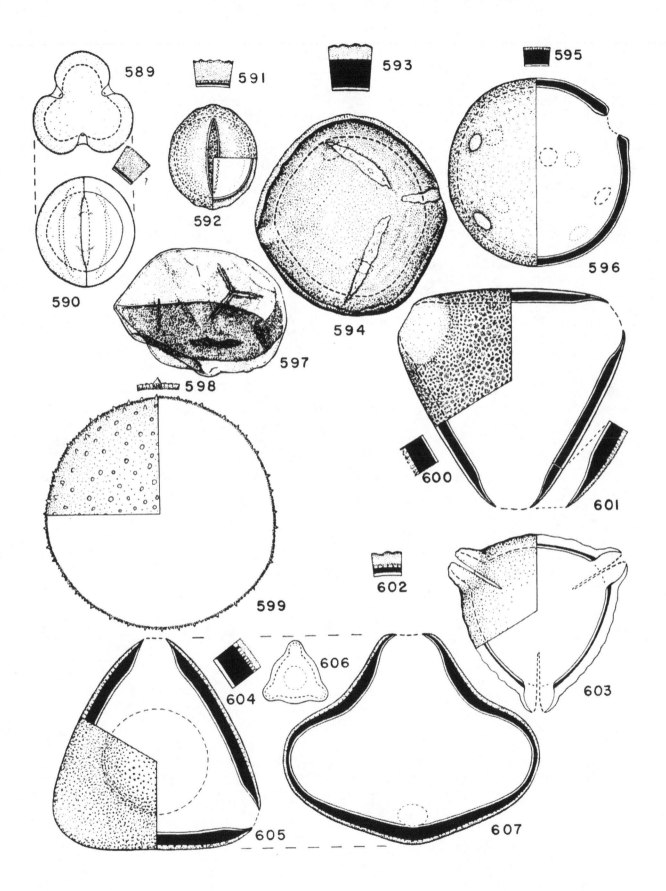

589

591

593

595

592

594

596

590

597

598

600

601

599

602

603

604 606

605

607

Plate 30

Figure 608. INTERCOLPAR THICKENING. Polygonaceae, *Chorizanthe pungens,* optical section, 25μ. From Wodehouse, 1935, p. 405, pl. VIII, fig. 3.

Figure 609–619. CRASSIMARGINATE. Pollen of the Loganiaceae. 609–612, *Anthocleista parvifolia;* 613, 615, *Labordia helleri;* 614, *L. tinifolia;* 616–618, *L. hedysomifolia.* From Erdtman, 1952, fig. 145A, B.

609. Equatorial view of *A. parvifolia,* 35μ
610. Polar view of 609
611. Exine section of 609
612. LO-pattern of 609
613. Exine section of *L. helleri,* 1000x
614. LO-pattern of *L. tinifolia,* 1000x
615. LO-pattern of *L. helleri,* 1000x
616. Equatorial view of *L. hedysomifolia,* 36μ
617. Polar view of 616
618. Exine section of 616
619. LO-pattern of 616

Figure 620–621. ORNAMENTATION. From Iversen and Troels-Smith, 1950, pl. III.

620. Inordinate
621. Ordinate

Figure 622. DIMENSIONS OF EXINE

ekt M+ (ekt M−), ektexina mensura, maximum (. . . minimum)
end M, endexina mensura
ex M, exina mensura
tec M, tectum mensura
From Iversen and Troels-Smith, 1950, pl. IX.

Figure 623–626. ORNAMENTATION. Pollen sculpture types.

bac: bacula, baculatus
cla: clavae, clavatus
col (conj): columellae conjunctae
col (dig): columellae digitatae
col (simple): columellae simplices
ech: echini, echinatus
gem: gemmae, gemmatus
intec: intectatus
psi: psilatus
tec: tectatus
tec (perf): tectatum
ver: verrucae, verrucatus
From Iversen and Troels-Smith, 1950, pl. II, III and Faegri and Iversen, 1950, pl. I.

Figure 627. DIMENSIONS OF SCULPTURE ELEMENTS. cla, clava; D, diameter, foc. 5, focus 5 [see Fig. 636]; H, altitude; t, maximum; −, minimum. From Iversen and Troels-Smith, 1950, pl. IX.

Figure 628–629. LO-ANALYSIS.

628. Exine with punctitegillate, spiniferous sexine (optical section)
629. LO-analysis of the exine in 628. From Erdtman, 1952, fig. 5A.

Figure 630–635. LO-ANALYSIS. Various sexine patterns at high and low focus (upper and lower row, respectively).
630. Sexine pilate
631. Retipilate
632. Simplibaculate
633. Duplibaculate
634. Scrobiculate
635. Areolate
From Erdtman, 1952, p. 22, fig. 5B.

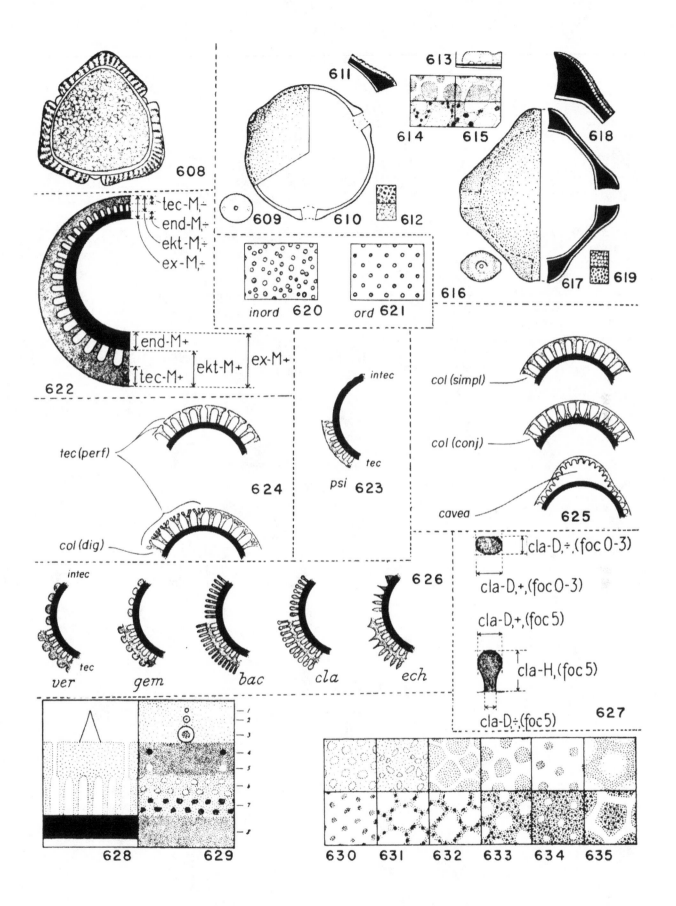

608

611

613

tec-M,÷
end-M,÷
ekt-M,÷
ex-M,÷

609   610   612

614   615   618

617   619

616

end-M+
ekt-M+   ex-M+
tec-M+

622

inord   620   ord   621

col (simpl)

col (conj)

intec

cavea   625

tec (perf)   psi   623

624

col (dig)

626

cla-D,÷,(foc 0-3)

cla-D,+,(foc 0-3)

cla-D,+,(foc 5)

cla-H,(foc 5)

intec

tec

ver   gem   bac   cla   ech

cla-D,÷,(foc 5)   627

628   629

630   631   632   633   634   635

247

Plate 31

Figure 636–637. LIMES EXTERIOR. Statement of focusing (foc) in relation to upper or lower boundary level of pollen grain. From Iversen and Troels-Smith, 1950, pl. X.

Figure 638. LO-PATTERN. *Microreticulatisporites* sp., ca. 450x. From Potonié and Kremp, 1955, fig. 24.

Figure 639. ORNAMENTATION. Types of exine ornamentation.
1, ornamentation in optical section (lateral view); 2, in surface view at high adjustment; 3, in surface view at low adjustment; A, echinate exine (with spines); B, granular exine (with grana: there is a gradual transition from grana to the warts of verrucate exines); C, piliferous exine (with pila); D, pitted, or scrobiculate, exine; E, reticulate exine; F, exine with "negative reticulum" (exine isles separated from each other by a network of grooves).
From Erdtman, 1943, text-fig. 5.

Figure 640. ORNAMENTATION. Diagrammatic drawing of various types of spore coat ornamentation: A, levigate; B, granulose; C, papillate; D, punctate; E, punctatereticulate; F, reticulate; G, vermiculate; H, obervermiculate; I, verrucose; J, rugose; K, lobate; L, striate; M, spinose; N, setaceous; O, processes-projections. From Kosanke, 1950, fig. 2.

Figure 641–647. ORNAMENTATION. Sculptures and structures in side view. All elements mentioned have a more or less rounded plan-view. Cavities are stippled. From Thomson and Pflug, 1953, fig. 2.

641. Intrabaculate
642. Intragranulate
643. Baculate
644. Verrucate
645. Gemmate
646. Clavate
647. Echinate

Figure 648. SPORODERMIS. Sexine patterns (A-D) and exine stratification (E,F). Exine crassinexinous (E:b), crassisexinous (E:a), tenuinexinous (E:a), tenuisexinous (E:b). Sexine pilate (A, E:b; ± sympilate in E:a), reticulate (B), striate (C), ornate (D), tegillate-baculate (E:c, F:d-g), tegillate-ramibaculate (F:f,g), ± psilate (E:c), subsaccate (F:d), spinulose (F:e), verrucose (F:f), spinose (F:h,i), crass-ectosexinous (tenui-endosexinous) in E:A. Sexine stratification ± obscure in F:h,i. From Erdtman, 1952 fig. 4.

Figure 649–657. ORNAMENTATION. Sculpture elements of the exine. From Potonié, 1934, fig. 3–9, 41, 39, and p. 11.

649. Grana
650. Bastionate
651. Bacula
652. Spinae
653. Pila
654. Pila
655. Verrucae
656. Spinae. Composite pollen with obvious sculpture
657. Reticulum cristatum. Comb-like exine sculpture

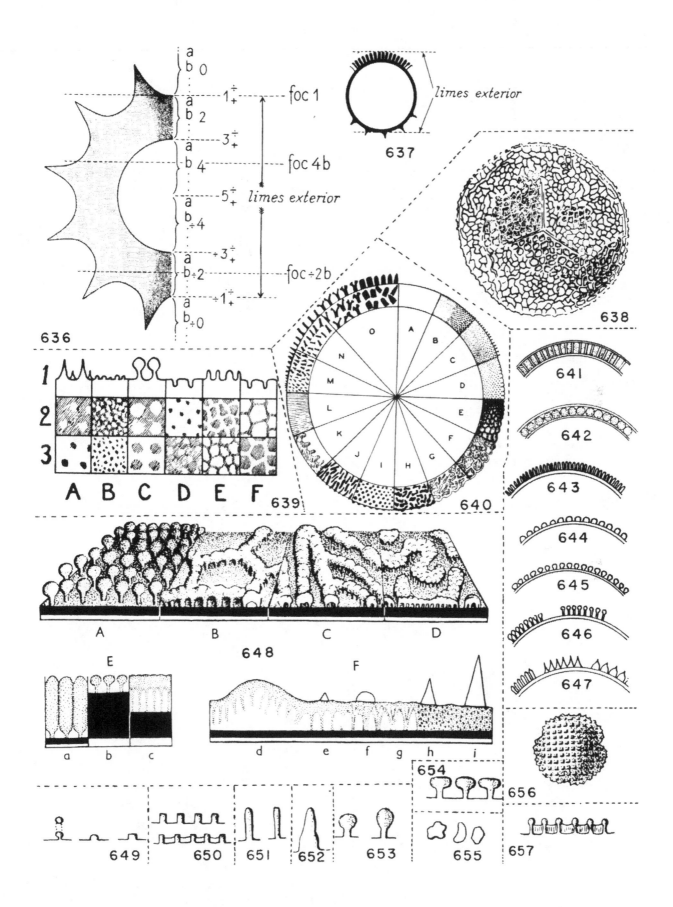

636

a
b 0

1 ÷
+   foc 1
a
b 2

3 ÷
+
a
b 4   foc 4b

5 ÷
+   *limes exterior*
a
b ÷4

÷3 ÷
+
a
b ÷2   foc ÷2b

÷1 ÷
+
a
b ÷0

637

limes exterior

638

639

1

2

3

A B C D E F

640

O A
N B
C
M D
L E
K F
J G
I H

641

642

643

644

645

646

647

648

A B C D

E

a b c

F

d e f g h i

656

654

649   **650**   **651**   **652**   **653**   **655**   **657**

Plate 32

Figure 658–669. ORNAMENTATION. From Ku-priyanova, 1948, pl. V.

658. Thorn-shaped (spinosa, echinata)
659. Tuberculous (tuberculata)
660. Wart-like (verrucosa)
661. Club-like (piliformis clavata)
662. Negatively net-like (areolata reticulum fossulare)
663. Granular (granulata)
664. Rough shaped (textura scabra)
665. Pit-like (foveolata)
666. Simple net sculpture having unbroken walls (reticula sepibus densis)
667. Simple net-like sculpture having rod-like walls (reticulata sepibus evallis)
668. Complex net-like sculpture having rod-like walls (reticula composita sepibus evallis)
669. Wrinkle-like sculpture (corrugata)

Figure 670–677. ORNAMENTATION. The elevated elements are light, the depressed areas and the cavities are drawn in dark. From Thomson and Pflug, 1953, fig. 3.

670. Rugulate and intrarugulate
   a. rugulate, side view
   b. rugulate or intrarugulate, plan view
   c. intrarugulate, side view
671. Striate and intrastriate
   a. striate, side view
   b. striate or intrastriate, plan view
   c. intrastriate, side view
672. Cicatricose

   a. side view
   b. plan view
673. Reticulate and intrareticulate
   a. reticulate, side view
   b. reticulate or intrareticulate, plan view
   c. intrareticulate, side view
674. Foveolate
   a. side view
   b. plan view
675. Fossulate
   a. side view
   b. plan view
676. Canaliculate
   a. plan view
   b. side view
677. Corrugate
   a. side view
   b. plan view

Figure 678–687. ORNAMENTATION. Sculpture elements and patterns of spores. From Couper and Grebe, 1961, fig. 22–31.

678. Verrucae, verrucate
679. Pila, pilate
680. Bacula, baculate
681. Spinae, spinose
682. Coni, conate
683. Capilli, capillate
684. Cristae, cristate
685. Rugulae, rugulate
686. Vermiculi, vermiculate
687. Foveolae, foveolate

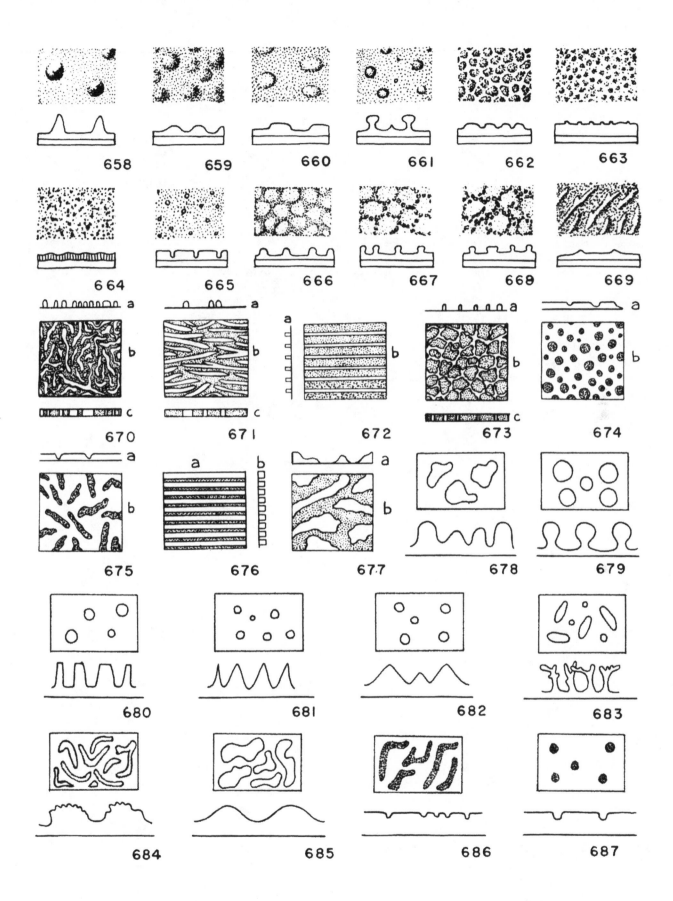

658 659 660 661 662 663

664 665 666 667 668 669

670 671 672 673 674

675 676 677 678 679

680 681 682 683

684 685 686 687

251

Plate 33

Figure 688–709. ORNAMENTATION. Sculpture of the exine and pollen sculpture types. From Iversen and Troels-Smith, 1950, pl. III-V, and Faegri and Iversen, 1950, pl. II, III.

688. Ridges and lumen. Lum, lumen; pl, platea; val, vallae [valli]. Sculptural elements dotted.

689. Reticulum. Lum, lumen; mur, murus.

690. Structural types. Intec, intectate; tec, tectate.

691. Foveolate, e.g. *Lycopodium selago*

692. Fossulate, e.g. *Ledum palustre*

693–697. Exine seen in varying focus. Tectate-perforate pollen type, branched columellae (*Stellaria longipes*)

693. External spines bright

694. External spines dark, diffuse. "Perforations" dark, distinct

695. "Perforations" bright

696. Branches of columellae dark, the basal trunks rather bright. Space between columellae bright

697. Columellae (basal trunks) dark

698–700. Exine seen in varying focus. Tectate-reticulate pollen type (*Galeopsis tetrahit*)

698. Protruding reticulum bright

699. Reticulum dark, columellae bright

700. Reticulum indistinct, columellae dark

701–709. Sculpturing types, radial projections elongated. Sculptural elements dotted

701. Rugulate, e.g. *Sedum rosea*

702. Rugulate, e.g. *Nymphoides peltata*

703. Rugulate, e.g. *Prunus insititia*

704. Striate, e.g. *Menyanthes trifoliata*

705. Striate, e.g. *Acer pseudoplatanus*

706. Striate, e.g. *Saxifraga oppositifolia*

707. Reticulate, e.g. *Iris pseudoacorus*

708. Reticulate, e.g. *Lysimachia thyrsiflora*

709. Reticulate, e.g. *Thesium ebracteatum*

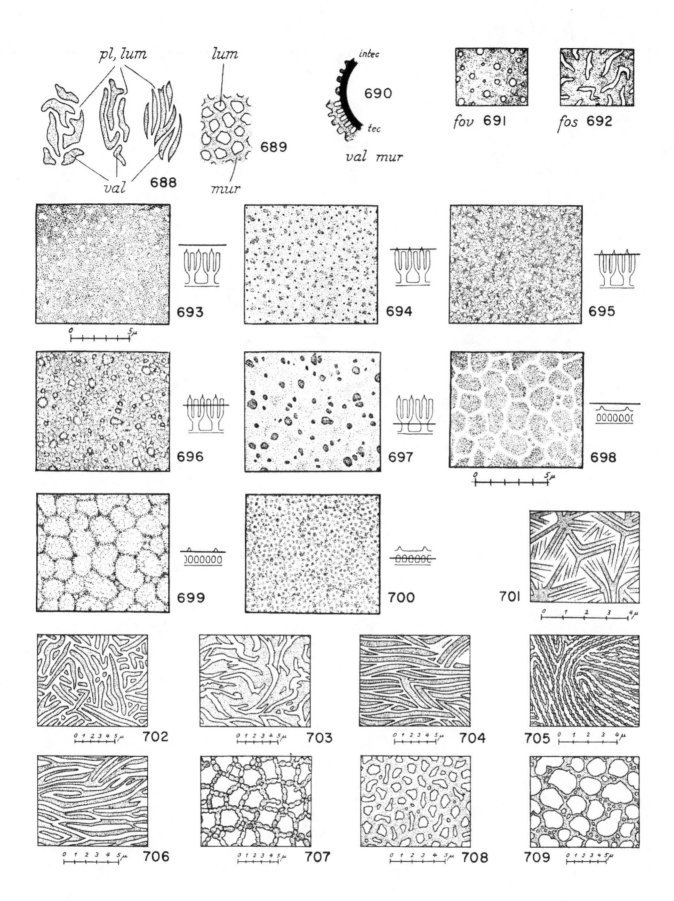

pl, lum

lum

688

val

mur

689

intec

690

tec

val mur

fov 691

fos 692

693

0   5µ

694

695

696

697

698

0   5µ

699

700

701

0   1   2   3   4µ

702   0 1 2 3 4 5µ

703   0 1 2 3 4 5µ

704   0 1 2 3 4 5µ

705   0 1 2 3 4µ

706   0 1 2 3 4 5µ

707   0 1 2 3 4 5µ

708   0 1 2 3 4 5µ

709   0 1 2 3 4 5µ

253

Plate 34

Figure 710. PSILATE. Polygonaceae, *Rumex acetosella,* polar view, 23μ. From Wodehouse, 1935, pl. VIII, fig. 1.

Figure 711. VESTIGIAL SPINE. Ambrosieae, *Xanthium pennsylvanicum,* side view 23μ. From Wodehouse, 1935, pl. XIV, fig. 7.

Figure 712–714. SPINA. Malvaceae, *Kydia calcina,* spines ca. 17μ. From Erdtman, 1952, fig. 151A.
712. Exine stratification, electron microphotograph, 2000x
713. Plan view
714. LO-pattern

Figure 715. PSILOTEGILLATE; ASPIDOTE. Strasburgeriaceae, *Strasburgeria calliantha,* 60μ, polar and equatorial view. From Erdtman, 1952, fig. 240.

Figure 716. PUNCTITEGILLATE. Caryophyllaceae, *Agrostemma githago.*

a. Foramen, section view, ca. 1200x
b. Foramen, surface view, ca. 1200x
c-e. LO-analysis (spinules bright in c, very dark in d; puncta dark in c, bright with dark contour in d) From Erdtman, 1952, fig. 50C

Figure 717–720. DUPLIBACULATE. Erythropalaceae, *Erythroxylon emarginatum,* 44μ. From Erdtman, 1952, fig. 94C.
717. Exine stratification
718. Polar view
719. Equatorial view
720. LO-pattern

Figure 721–722. SPINA. Malvaceae, *Althaea rosea.* Grains polyforate (132μ), provided with diamorphic spines: long, pointed (13μ), and short, blunt (5μ). From Erdtman, 1952, fig. 151C.
721. Exine stratification, electron microphotograph, 2000x
722. Plan view

Figure 723. PSILOTEGILLATE, ASPIDOTE, ASPIS. Betulaceae, A, *Betula nana,* 2100x; B, *B. tortuosa,* 2100x; C, *B. pubescens,* 2100x. From Erdtman, 1952, fig. 30.

Figure 724–727. SIMPLIBACULATE. Resedaceae, *Reseda suffruticosa,* 25μ. From Erdtman, 1952, fig. 219.
724. LO-pattern
725. Exine stratification
726. Polar view
727. Equatorial view

Figure 728. RAMIBACULATE. Compositae, *Echinops sphaerocephalus.* The ramifications of the bacula are probably extended ± to the outer surface of the exine. From Erdtman, 1952, fig. 64.

Figure 729. SPINA. Malvaceae, *Cenocentrum tonkinense.* Exine stratification, electron microphotograph, 2000x. Length of spine 19μ, basal diameter 9.5μ. From Erdtman, 1952, fig. 151B.

Figure 730. RAMIBACULATE. Goodeniaceae, *Camphusia glabra* var. *kauaiensis,* 80μ. From Erdtman, 1952, fig. 113.

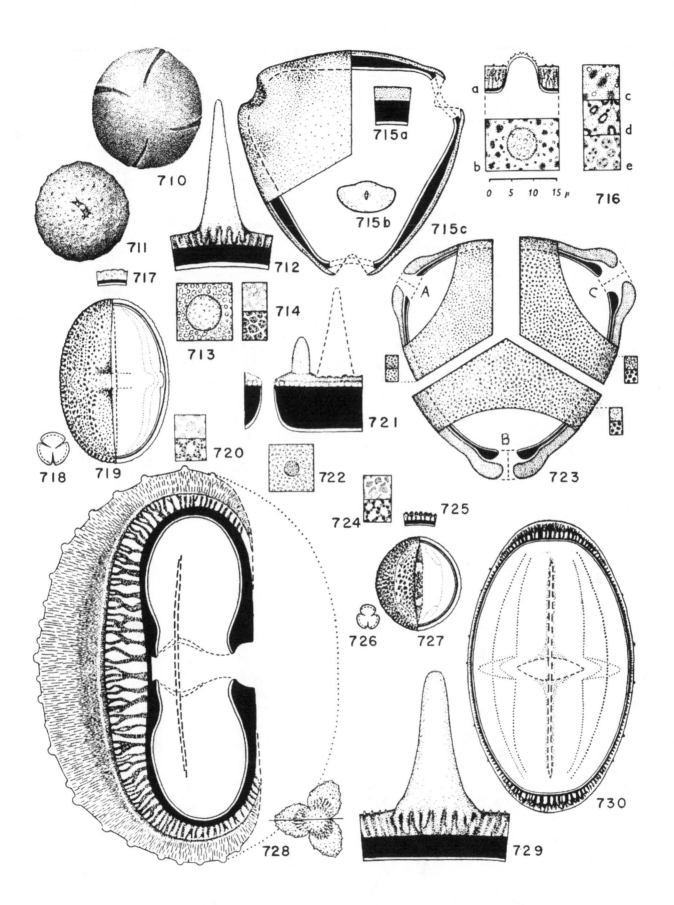

715a

715b

715c

716

a

b

c

d

e

0   5   10   15 μ

710

711

717

712

713

714

718

719

720

721

722

723

724

725

726

727

728

729

730

A

B

C

Plate 35

Figure 731. ECHINATE. Ambrosieae, *Oxytenia acerosa,* 18μ, side view. From Wodehouse, 1935, pl. XIV, fig. 1.

Figure 732. GRANULA. *Cyclogranisporites* sp., ca. 620x. From Potonié and Kremp, 1955, fig. 9.

Figure 733–734. GRANULA. From Erdtman, 1952, fig. 227A,B.

733. Salicaceae, *Populus tremula,* 30μ.
734. Salicaceae, *P. tremula* f. *gigas,* 40μ.

Figure 735. HAMULATE. From Krutzsch, 1959, text-fig. 12.

Figure 736. SUBECHINATE. Ambrosieae, *Ambrosia elatior,* 18μ. From Wodehouse, 1935, pl. XIII, fig. 7.

Figure 737. ECHINATE. Veronieae, *Solidago speciosa,* 23μ. From Wodehouse, 1935, pl. XII, fig. 4.

Figure 738–746. ORNAMENTATION. From Potoníe and Kremp, 1955, fig. 8, 10, 11, 18–20, 27, 37.

738. Granula. *Granulatisporites* sp., ca. 760x (fig. 8)

739. Negative reticulum. *Verrucosisporites* sp., ca. 500x (fig. 11)
740. Spinae. *Acanthotriletes* sp., ca. 1000x (fig. 18)
741. Verrucae. *Converrucosisporites* sp., ca. 1500x (fig. 10)
742. Cristae, *Cristatisporites* sp., ca. 800x (fig. 27)
743. Coni. *Lophotriletes* sp., ca. 700x (fig. 14)
744. Fimbriae. *Setosisporites* sp., ca. 55x (fig. 37)
745. Bacula, *Raistrickia* sp., ca. 700x (fig. 19)
746. Bacula, *Raistrickia grovensis* Schopf, ca. 54μ (fig. 20)

Figure 747. SUBLOPHATE. Cichorieae, *Catananche caerulea,* side view, 27μ. From Wodehouse, 1935, pl. XI, fig. 3.

Figure 748–749. POLAR LACUNA. From Wodehouse, 1935, pl. XII, fig. 7, 6.

748. Cichorieae, *Barnadesia venosa,* polar view, 67μ
749. Cichorieae, *B. trianae,* polar view, 50μ

Figure 750. LOPHATE. Cichorieae, *Pacourina edulis,* 50μ. From Wodehouse, 1935, pl. XII, fig. 1.

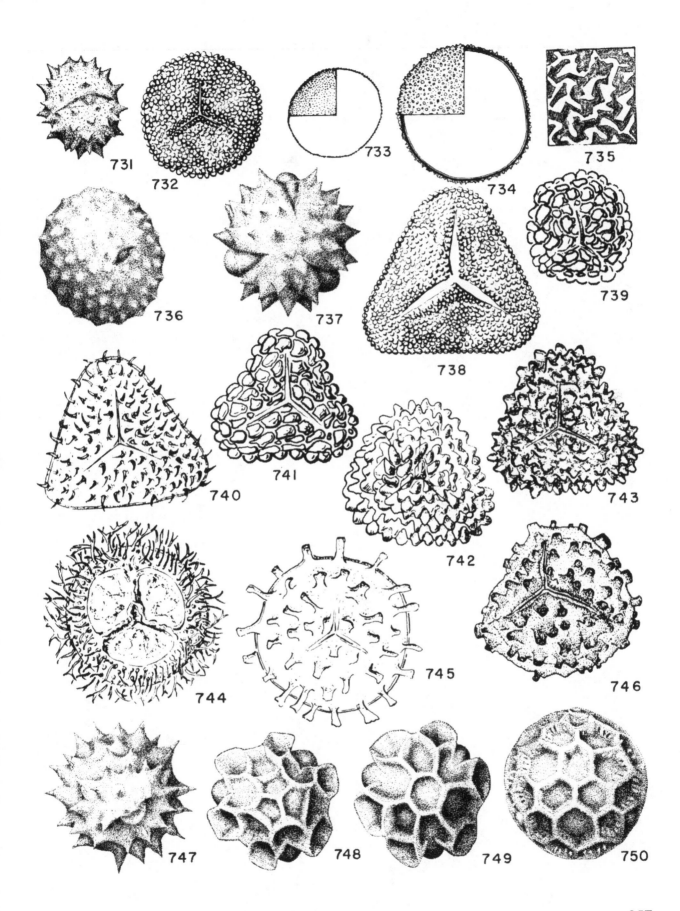

Plate 36

Figure 751–753. RETICULUM, INTECTATE-RE-
TICULATE: HOMOBROCHATE. Geraniaceae,
Viviania parvifolia, 29μ. From Erdtman, 1952, fig.
110C.

751. Exine stratification
753. LO-pattern

Figure 754–756. SUPRARETICULATE. Labiatae,
Catopheria chiapensis, ca. 124μ. From Erdtman,
1952, fig. 130C.

754. Exine stratification
755. LO-pattern
756. Polar view

Figure 757. BROCHAL, POLYFORATE Zygophyl-
laceae, *Kallstroemia maxima,* 65μ. From Erdtman,
1952, p. 457, fig. 261A.

Figure 758–760. RETICULUM, NONAPERTU-
RATE. Geraniaceae, *Balbisia meyeniana,* 32μ. From
Erdtman, 1952, fig. 110D.

758. Exine stratification
760. LO-pattern

Figure 761. SUBECHINOLOPHATE. *Stokesia laevis,* polar view, 55μ. From Wodehouse, 1935, pl. XII,
fig. 2.

Figure 762–765. LACUNA. Pollen grains of Cichorieae. From Wodehouse, 1935, pl. XI, fig. 4, 5, 1, 2.

762. Equatorial lacuna and paraporal ridge. *Tragopogon pratensis,* polar view, 32μ
763. Equatorial lacuna. *Scorzonera hispanica,* polar
view, 37μ
764. Abporal and poral lacunae. *Scolymus hispani-
cus,* polar view, 35μ
765. Interlacunar ridge; Poral, paraporal and abporal lacunae. *Taraxacum officinale,* polar view,
26μ

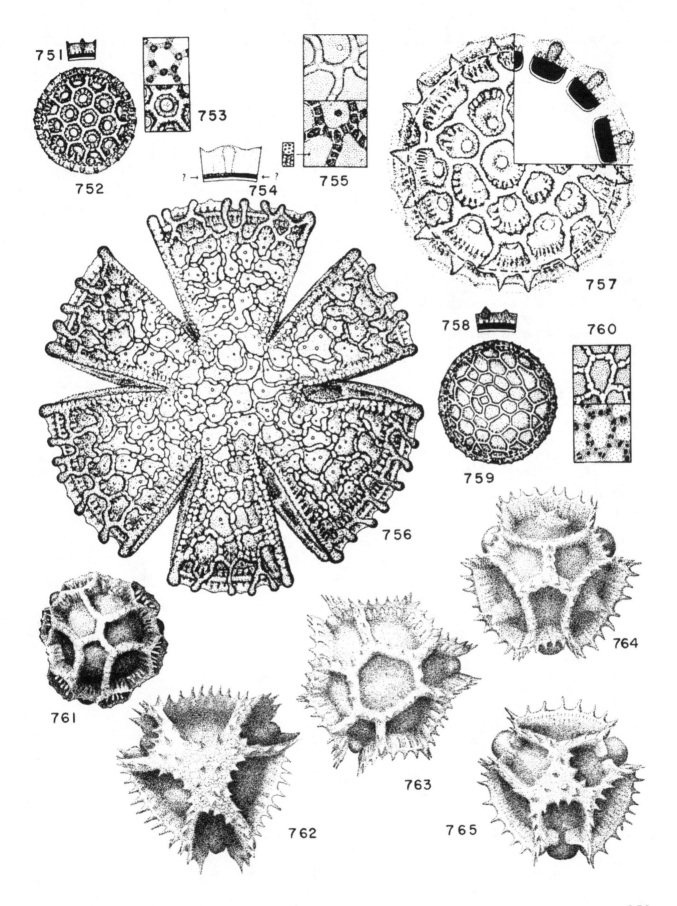

751

752

753

754

? → ← ?

755

757

758

759

760

756

761

762

763

764

765

Plate 37

Figure 766–768. ORNAMENTATION. From Couper and Grebe, 1961, fig. 32.

766. Reticulate
767. Striate
768. Muri in optical section. Tops of elevations rounded, pointed, flat, or serrated

Figure 769–772. ORNAMENTATION. From Krutzsch, 1959, fig. 13–15, 17.

769. Spongy
770. Acusporide
771. Foveate
772. Foraminate

Figure 773–774. ORNAMENTATION. Reticulum of smooth, low stripes. *Dictyotriletes* sp., ca. 900x. From Potonié and Kremp, 1955, fig. 28, 29.

773. Distal view
774. Meridial view

Figure 775–777. AREOLATE. Bignoniaceae, *Catalpa speciosa.* From Erdtman, 1952, fig. 31B.

775. Exine stratification
776. LO-pattern

Figure 778. RUGULATE. Liliaceae, *Chrysobactron Hookeri,* EM (unexpanded), 550x, sect.; exine in low focus below surface also. From Cranwell, 1953, pl. 5, fig. 8.

Figure 779–781. FOVEOLA. Rubiaceae, *Tocoyena formosa,* 53μ. From Erdtman, 1952, fig. 224B.

779. Exine stratification
780. Equatorial view
781. Polar view

Figure 782. RETICULATE. Oleaceae, *Ligustrum ibota,* polar view, 30μ. From Wodehouse, 1935, pl. X, fig. 8.

Figure 783–785. AREOLATE. Bignoniaceae, *Sparattosperma* vernicosum. From Erdtman, 1952, fig. 31A.

783. Exine stratification
784. LO-pattern

Figure 786–787. AREOLATE. Aristolochiaceae, *Apama corymbosa,* 27μ. From Erdtman, 1952, fig. 21B.

786. Exine stratification
787. Side view of the non-aperture grain

Figure 788–789. SCROBICULUS. Restionaceae, *Hypolaena lateriflora,* ca. 55μ. From Erdtman, 1952, fig. 220.

788. Exine stratification
789. Equatorial view of the 1-aperturate grain

Figure 790–792. ORNATE. Bombacaceae, *Ceiba aesulifolia,* 81μ. From Erdtman, 1953, fig. 33.

790. Equatorial view, of 3-colporate grain
791. Polar view
792. LO-pattern

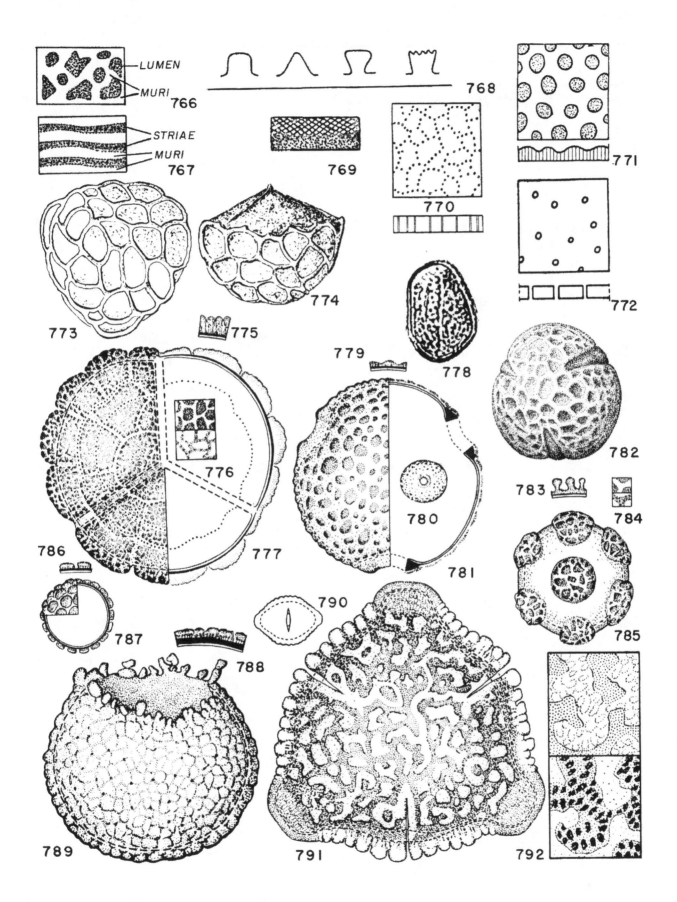

LUMEN
MURI
766

STRIAE
MURI
767

768

769

770

771

772

773

774

775

776

777

778

779

780

781

782

783

784

785

786

787

788

789

790

791

792

Plate 38

Figure 793–794. CARINIMURATE. Magnoliaceae, *Illicium floridanum,* 33μ. From Erdtman, 1952, fig. 148C.

792. Polar view
793. Equatorial view (section)
794. Exine patterns at different adjustments of the microscope from high (left) to low (right)

Figure 795–797. SPINULA. Liliaceae, *Aphyllanthes monspeliensis,* 46μ. From Erdtman, 1952, fig. 140C.

795. Exine stratification
796. Polar view
797. LO-pattern

Figure 798–801. BACULARIUM. Leguminosae, *Aprevalia floribunda,* 61μ. From Erdtman, 1952, fig. 135A.

798. Polar view of 3-colpate (brevicolpate) grain
799. Exine stratification
800. LO-pattern
801. Equatorial view

Figure 802. ORNAMENTATION. Reticulum with high ridges (muri). *Reticulatisporites* sp., ca. 500x. From Potonié and Kremp, 1955, fig. 30.

Figure 803–806. SPINULA. Proteaceae, *Xylomelum salicinum,* 37μ. From Erdtman, 1952, fig. 214.

803. Equatorial view
804. Exine stratification
805. LO-pattern
806. Polar view

Figure 807–810. HETEROBROCHATE. Euphorbiaceae, *Chrozophora rottleri,* 70μ. From Erdtman, 1952, fig. 98D.

807. Exine stratification
808. Equatorial view
809. LO-pattern
810. Polar view

Figure 811–814. PILARIUM. Oleaceae, *Jasminum humile,* 57μ. From Erdtman, 1952, fig. 173.

811. Equatorial view
812. Exine stratification
813. LO-pattern
814. Polar view

Figure 815–818. POLYBROCHATE, Proteaceae, *Buckinghamia celsissima.* From Erdtman, 1952, fig. 213B.

815. Exine stratification
816. Equatorial view
817. Polar view
818. LO-pattern

Figure 819–822. POLYBROCHATE. Rubiaceae, *Carlemannia griffithii,* 45μ. From Erdtman, 1952, fig. 224G.

819. Equatorial view
820. Exine stratification
821. LO-pattern
822. Polar view

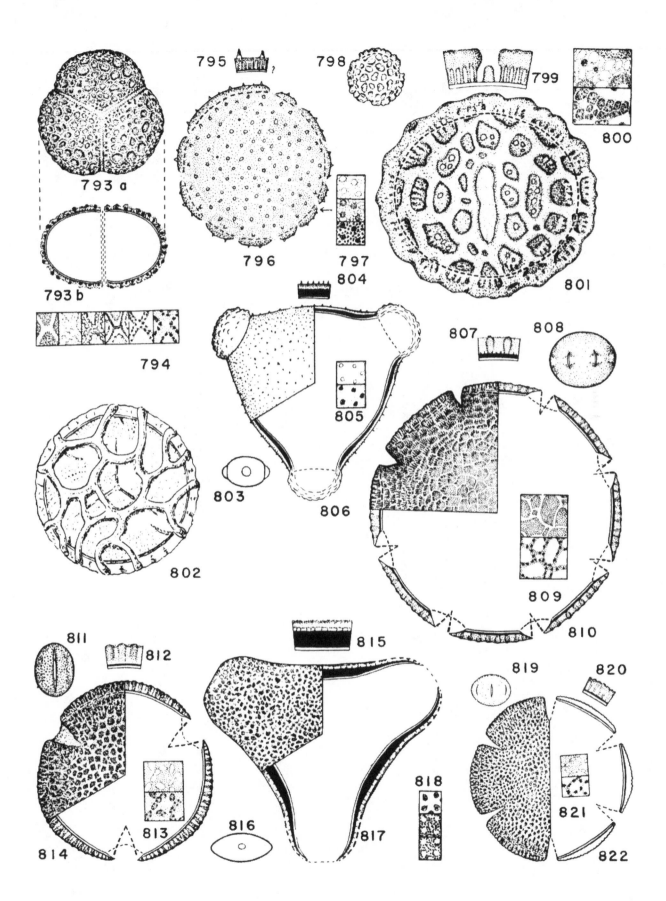

793 a

793 b

794

795

796

797

798

799

800

801

802

803

804

805

806

807

808

809

810

811

812

813

814

815

816

817

818

819

820

821

822